**Magnetic Field Measurement with Applications to Modern Power Grids**

# Magnetic Field Measurement with Applications to Modern Power Grids

*Qi Huang*

School of Mechanical and Electrical Engineering, University of Electronic Science and
Technology of China (UESTC), P.R. China

*Arsalan Habib Khawaja*

U.S-Pakistan Center for Advanced Studies in Energy (USPCASE), National University of
Sciences and Technology (NUST), Pakistan

*Yafeng Chen*

School of Mechanical and Electrical Engineering, University of Electronic Science and
Technology of China (UESTC), P.R. China

*Jian Li*

School of Mechanical and Electrical Engineering, University of Electronic Science and
Technology of China (UESTC), P.R. China

The right of Qi Huang, Arsalan Habib Khawaja, Yafeng Chen and Jian Li to be identified as the authors of this work has been asserted in accordance with law.

*Registered Offices*
John Wiley & Sons, Inc., 111 River Street, Hoboken, NJ 07030, USA
John Wiley & Sons Ltd, The Atrium, Southern Gate, Chichester, West Sussex, PO19 8SQ, UK

*Editorial Office*
The Atrium, Southern Gate, Chichester, West Sussex, PO19 8SQ, UK

For details of our global editorial offices, customer services, and more information about Wiley products visit us at www.wiley.com.

Wiley also publishes its books in a variety of electronic formats and by print-on-demand. Some content that appears in standard print versions of this book may not be available in other formats.

*Library of Congress Cataloging-in-Publication Data*
Names: Huang, Qi, 1976- author.
Title: Magnetic field measurement with applications to modern power grids /
   Qi Huang (University of Electronic Science and Technology of China) [and
   three others].
Description: Hoboken, NJ : Wiley-IEEE Press, 2020. | Copyrighted by John
   Wiley & Sons, Ltd. | Includes bibliographical references and index. |
   Identifiers: LCCN 2019015135 (print) | LCCN 2019017754 (ebook) | ISBN
   9781119494461 (Adobe PDF) | ISBN 9781119494508 (ePub) | ISBN 9781119494515
   (hardcover)
Subjects: LCSH: Electromagnetic fields. | Electromagnetic fields–Industrial
   applications. | Electric fields–Measurement. | Magnetic
   fields–Measurement. | Electric power distribution.
Classification: LCC QC665.E4 (ebook) | LCC QC665.E4 M335 2019 (print) |
   DDC 621.31028/7–dc23
LC record available at https://lccn.loc.gov/2019015135

Cover Design: Wiley
Cover Images: © alengo/Getty Images, © Yelantsevv/Getty Images

Set in 10/12pt Warnock by SPi Global, Pondicherry, India
Printed and bound in Singapore by Markono Print Media Pte Ltd

10  9  8  7  6  5  4  3  2  1

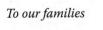

*To our families*

# Contents

# Foreword

Most contemporary electrical engineering at a high level can be modelled by the Maxwell equations:

$$\nabla \cdot B = 0$$

$$\nabla \cdot D = \rho$$

$$\nabla \times E = -\frac{\partial B}{\partial t}$$

$$\nabla \times H = \frac{\partial D}{\partial t} + J$$

where $B$ and $D$ are the magnetic and displacement fields, respectively, $E$ and $H$ are the electric and magnetizing fields, and $J$ and $\rho$ are the current and electric charge densities. In the compact vector notation shown, the dot and cross notations refer to divergence and curl, respectively. A significant part of the physics of these fields is the *magnetic field, B*, which plays an especially important role in electric power engineering both contemporaneously and historically. Reference [1] gives a concise discussion of magnetic fields as described by Maxwell's equations and how these mathematical models have evolved. Perhaps the close nexus of magnetic fields and power engineering is due to the history of the development of motors and generators. Traditionally, these devices rely on the conversion of mechanical, electrical, and magnetic energy. These have been the 'work horses' of world industry, but the nexus goes beyond motors and generators: sensors, measurement instruments, controls, electromagnets of a wide variety, and energy converters of all types entail applications of the phenomenon of magnetism. As such, power engineers need to appreciate not only the theory of magnetism, but also the practicalities, i.e. how applications are realized in a modern contemporary setting. These practicalities include *measurement*. This book focuses on *sensing* and *measurement* of magnetic fields as commonly encountered in electric power engineering. However, the present book does not lose sight of the underpinnings of electromagnetic field theory, for example the use of the Poynting vector $S$,

$$S = E \times H,$$

to describe the interrelationship of $E$ and $H$. How $S$ describes the spatial motion of power is explained in some detail in Chapter 1.

The present book covers a range of topics on magnetic field measurement and applications in power engineering, from the history of this part of physics to actual measurement methods and commercial applications in instrumentation. Practical measurement is discussed in detail. Monitoring the utilization of magnetics and human health (e.g. electromagnetic field exposure) are also discussed because the standardization of limits of field strengths has taken on important international acceptance. That is, if a device cannot operate in the presence of a human being, it should not be considered operable at all. These standards are a necessary consequence of international commercialization. New topical areas have implications for the integration of renewable generation resources. This is a consequence of the importance of the sustainability of electric energy resources. Because power engineering is undergoing a transformation from a traditional science to a consumer oriented, efficient, highly controlled, commercial engineering venture (i.e. the 'smart grid'), most topics in the book have a specialized focus on smart grid topics such as 'big data', renewable generation integration, smart sensory applications (especially at the secondary distribution consumer level, i.e. up to 1 kW), and enhancement of efficiency. Large-scale system applications, especially system-wide control, are included. The book has an emphasis on measurement techniques, and these might be classified as various forms of magnetoresistive techniques (e.g. anisotropic magnetoresistance (AMR), giant magnetoresistance (GMR), tunnel magnetoresistance (TMR), colossal magnetoresistance (CMR), and extraordinary magnetoresistance (EMR), methods suitable for transmission line current measurement, electro-optical phenomena (e.g. the Faraday effect), and how instrumentation and measurement relate to the objectives of the smart grid. In many of these cases, shielding and noise have a significant impact and need specialized methods to render measurements accurate and usable. This is the case since magnetic field strengths less than 1 mTesla are usual in typical measurement applications. It is evident that measurements made some distance from high power equipment require significant sensitivity. This is especially true when a high level of resolution is needed so that low field strengths at high spectral frequencies can be detected. In electric power engineering, in view of the large number of solid state power converters in commercial use, the subject of harmonic voltages and currents has captured significant attention (e.g. 'power quality engineering'). Some of these 'harmonics' are asynchronous to the 'power frequency' and this may make magnetic field and electric current measurements especially difficult. For example, the popularity of pulse width modulated (PWM) converters (even at levels above 1 MW) for energy flow control results in frequency spectra that contain linear combinations and multiples of both the power frequency and the PWM carrier switching frequency. These PWM spectral components may go well above 20 kHz. As a guide, one may expect that unipolar PWM converters cause frequency spectral components approximately of the order of $kf_s$, where $f_s$ is the PWM carrier frequency and $k = 2, 4, \ldots$ .Recent advances in power electronic switches have resulted in applications above $f_s = 10$ kHz. Reference [2] describes some of the mathematics involved. Power quality applications of magnetic field measurements are discussed in Chapter 5 of this book.

In this book, magnetic field measurements in substations, transmission systems, and distribution systems are considered in some detail separately. In this way, the specialized applications of magnetic field measurements in these venues can be discussed, including the detail of typical sensors used, the field strength levels and spectral frequencies involved, and the mathematics that are needed to process data measurements.

The book concludes with a very interesting outline of challenges, trends, and needs for future magnetic measurement systems. These are categorized again mainly by transmission, substation, and distribution applications. Included are remarks on the required levels of standardization, smart grid applications, and innovative sensors.

The general topic of electromagnetics often presents challenges to electrical engineering students because of the complexity of vector mathematics and the inability to easily 'visualize' electromagnetic fields. This book should be a significant help in education relating to magnetic fields, especially as a complement to a university course in power engineering. The technologies of measurement are a bridge between mathematical models and application-oriented practice. The book should be a guide to that bridge.

**1** J. C. Rautio, "The long road to Maxwell's equations," *IEEE Spectrum*, December 2014, pp. 38–56.

**2** G. R. Ainslie-Malik, Mathematical analysis of PWM processes, Doctoral Thesis, University of Nottingham, Nottingham, 2013.

Arizona State University                                             *G. Thomas Heydt*
Tempe, Arizona, USA

# Preface

Magnetism, an interaction among moving charges, is one of the oldest branches of science and is still under constant active study with great implications for modern industrial applications, especially energy and environment. Electric and magnetic phenomena have been recognized since ancient times, but the means to measure, generate, control, and use the phenomena to develop practical devices became adequately understood only in the past 200 years. In recent years, with progress in the material, electronic packaging, and other related circuit technologies, the applications of magnetics have evolved and are promising. Due to their non-contact character, magnetics-related applications have great potential in detection in some atrocious environments or where human access is difficult. Non-contact and non-intrusive detection technologies are finding more and more applications in many fields. Another potential application is wireless power transmission with magnetic effects, which will potentially revolutionize power utilization, for example wireless charging for electric vehicles and cell phones is already available.

Voltage and current are the fundamental parameters in an electric power system. It is known that the magnetic field is always associated with current, therefore magnetics-related phenomena play an important role in power systems.

This book presents the most updated technological developments in magnetic-field-based measurement solutions in modern power systems under the smart grid environment. Smart grids are revolutionarily transforming the modern power grid. This initiative is one of most prevalent trends in power systems all over the world, becoming the national level strategic development plan in many countries. The aims of smart grids include improving efficiencies, increasing grid flexibility, and reducing power outages, as well as other market, consumer, and societal needs, such as integration of diverse generation and storage options, integration of electric vehicles, and competitive electricity pricing, etc. It is known that information and communication/control technology is the enabler for smart grids, therefore non-contact magnetic-field-based measurement will provide many solutions for smart grids.

Although there is a wide range of applications for non-contact measurement, this book mainly focuses on the research conducted in the authors' research laboratory. The concepts of magnetism and magnetic fields are reviewed and their potential applications in modern industry, especially in power systems, are discussed. Advanced magnetic sensor technology, principally magneto-resistive sensors, is reviewed. In Chapters 3 to 6 the applications of magnetic-field-based measurement solutions are presented according to the electricity production chain, i.e. generation, transformation, transmission, and distribution. Magnetism is extremely useful for converting energy from one form to another. About 99% of the power generated from fossil fuels, nuclear and hydro-electric energy, and wind comes from systems that use magnetism in the conversion process, therefore magnetic-field-based solutions can find applications in any aspect of modern power grids. According to the development of modern electric power systems, potential applications for renewable energy generation are discussed. In addition, as electricity utilization increases in every aspect of daily life, people are becoming aware of the harmful effect of electromagnetic exposure, and this is discussed in the chapter on distribution. Finally, future visions are presented in Chapter 7.

The authors welcome all readers to discuss the book with us and contribute to this research field.

# Acknowledgments

The authors are grateful to the many people who made this book possible.

Special thanks go to Mr Louis Manohar from Wiley. Without his encouragement and help, it would have been impossible to finish the writing of the book. The authors also appreciate the efforts of all professors and graduate students whoever worked in the authors' research lab. Most of the contents are from the research projects in our research lab. These personnel include Mr Youliang Lu, Mr Fuchao Li, Mr Xiaohua Wang, and many more who cannot be mentioned here for reasons of space.

Special thanks go to Mr Wei Zhen and the Sichuan Electric Power Research Institute, Sichuan Power Company, State Grid of China. Without their help and continuous funding support, we would not have been able to persist in the research of magnetic field measurement. There are many other research scientists and engineers who provide assistance during the 10 years of research. Dr Guiyun Tian's research group is collaborating with the authors to extend electromagnetic sensors to the field of electromagnetic measurement. Mr Yang Yang from No. 9 Research Institute of China Electronics Technology Group Corporation, also known as the Southwest Applied Magnetics Research Institute, helped on sensor manufacturing and advocating the concepts proposed in this book to industrial applications. A group of research engineers at Cheng Dian Da Wei Ltd are trying to transit the prototypes into market products. The authors would also like to acknowledge the collaboration of Professor Philip Pong, from the University of Hong Kong.

Thanks also go to the publishers for granting the permission for reprinting some of the authors' publications in this book.

Finally, it is our great honor that Dr G. T. Heydt, of the US National Engineering Academy, has written a high-impact foreword for us, which will definitely contribute to the success of the book.

Qi Huang
Arsalan Habib Khawaja
Yafeng Chen
Jian Li

# 1

# Introduction

Electricity is an electromagnetic phenomenon in nature. Electrical power engineering is one of most successful utilizations of this natural phenomenon by humankind. Currents and voltages, as well as transport of signals or power, cannot exist without magnetic and electric fields. The fields of electricity and magnetics are more fundamental than currents and voltages, and fields or waves exist in the space far away from currents in conductors. Right after the electromagnetic phenomenon was discovered and mastered by engineers, various applications (including electrical power engineering itself) were created. It is important to understand the electromagnetic phenomenon in modern electric power systems and use it to improve the performance of power systems by taking advantage of modern technological advances, such as sensor and data communication and signal processing.

## 1.1 Magnetism and Magnetic Fields: A Historical View

Magnetism is the branch of physics that deals with the forces of attraction and repulsion produced by specific materials known as magnets. Magnetism is also a branch of electromagnetism, which is also a branch of physics involving the study of the electromagnetic force, a type of physical interaction that occurs between electrically charged particles.

### 1.1.1 A Historical View of Magnetism

Magnetism has been known for many centuries and was initially used mainly for navigation purposes. The compass was the first application of a magnet, used first by traders and later by sailors. The scientific community agrees that the magnetic phenomenon dates back to the creation of the universe and our solar system. However, in recorded times Thales of Miletus, in about 585 CE, stated that the lodestone or magnetite attracts iron. In ancient China, the earliest literary reference to magnetism lies in a 4th-century CE book named after its author, *The Sage of Ghost Valley*. After the discovery of lodestone, in the 12th century, a compass was invented that could be used for navigation, by sculpting a directional spoon from lodestone in such a way that the handle of the spoon always pointed south. In modern times, a comprehensive analysis of magnetism

*Magnetic Field Measurement with Applications to Modern Power Grids*, First Edition.
Qi Huang, Arsalan Habib Khawaja, Yafeng Chen and Jian Li.
© 2020 John Wiley & Sons Ltd. Published 2020 by John Wiley & Sons Ltd.

**Figure 1.1** Experimental scientists or physicists contributed to the development history of magnetism.

was performed by William Gilbert. He revealed that our planet Earth behaves as a huge magnet. Magnetism has an influential impact on every walk of life, from consumer electronics, industrial applications, navigation, and electric current measurement to radio communication for military and high-speed applications.

The science of electromagnetism developed rapidly in the quest for the fundamental forces, i.e. atomic force. It is now well understood that electromagnetism is one of four fundamental forces that do not appear to be reducible to more basic interactions, i.e. gravitational and electromagnetic interactions, which produce significant long-range forces whose effects can be seen directly in everyday life, and strong and weak interactions, which produce forces at minuscule, subatomic distances and govern nuclear interactions.

Many experimental scientists or physicists have contributed to the development of magnetism (Figure 1.1). In 1820, Andre Marie Ampere discovered that the magnetic field circulating in a closed path was related to the current flowing through the perimeter of the path, and Jean-Baptiste Biot and Flix Savart came up with the Biot–Savart law giving an equation for the magnetic field from a current-carrying wire. Michael Faraday found that a time-varying magnetic flux through a loop of wire induced a voltage, known as Faraday's law of induction, in 1831. In 1834, Emil Lenz observed that the induced electromotive force and rate of change in magnetic flux have opposite directions and can be represented by different signs. In 1895, Hendrik Lorentz derived the equation for a charged particle moving through a magnetic field, knows as the Lorenz force. James

Clerk Maxwell synthesized and expanded these insights into Maxwell's equations, unifying electricity, magnetism, and optics into the field of electromagnetism.

Electromagnetism has continued to develop into the 21st century, being incorporated into the more fundamental theories of gauge theory, quantum electrodynamics, electroweak theory, and finally the standard model. The electromagnetic force plays a major role in determining the internal properties of most objects encountered in daily life. Ordinary matter takes its form as a result of intermolecular forces between individual atoms and molecules in matter, and is a manifestation of the electromagnetic force. Electrons are bound by the electromagnetic force to atomic nuclei, and their orbital shapes and their influence on nearby atoms with their electrons are described by quantum mechanics. The electromagnetic force governs all chemical processes, which arise from interactions between the electrons of neighboring atoms.

To understand the phenomenon of magnetism, it is necessary to first understand the internal structure of atoms. The drastically different modern understanding of the structure of atoms was achieved in the course of the revolutionary decade stretching from 1895 to 1905. Recalling the atomic structure of Niels Bohr's model, an atom is made up of the nucleus, consisting of neutrons and protons, with electrons moving around it. These electrons are not only revolving around the nucleus but also spinning around their axes. Later on, quantum theory played an important role in describing the discrete magnetic nature of the elements by proposing that two electrons with opposite spins pair up around the nucleus, cancelling each other out. An element with an unpaired electron exerts a force on another unpaired electron of the element depending upon the direction of the spin, which determines whether they repel or attract each other. Elements with unpaired electrons in their outermost shell therefore have are magnetic as they arrange themselves according to the Earth's spinning. This is why, when one attaches a magnet to a string it aligns itself according to the Earth's magnetic field. The strength of the magnetism possessed by any element depends on the electron moving in its outermost shell.

Later on, Pierre Curie demonstrated the effect of heat on magnetism, which decreases below a certain temperature, known as the Curie temperature, and Wilhelm Weber invented a method of detecting and measuring magnetism. In the 20th century, Paul Langevin further explained Curie's work through the theory of how heat affects the magnetism. French physicist Pierre Weiss proposed the domains theory of magnetism in which magnetron particles are present in materials and cause their magnetic properties. This was explored further Samuel Abraham Goudsmit, who showed how the magnetic properties of materials result from the spinning motion of the electrons inside them.

Magnetism arises from two sources: electric current and the spin magnetic moments of elementary particles. The magnetic properties of materials are mainly due to the magnetic moments of their atoms' orbital electrons. Sometimes, either spontaneously or owing to an applied external magnetic field, each of the electron magnetic moments will be, on average, lined up. A suitable material can then produce a strong net magnetic field.

Depending upon the strength of the magnetism, elements are divided into three categories: paramagnetic, ferromagnetic, and diamagnetic. Paramagnetic elements have moderate magnetism and are magnetized whenever they are placed in the magnetic field and demagnetized as they are moved away from the field. Ferromagnetic elements have a strong magnetism that means they remain magnetized even if they are removed from the magnetic field. Diamagnetism refers to elements that are not affected by the magnetic field around them. These elements have paired electrons in their outermost shells.

The most familiar effects occur in ferromagnetic materials, which are strongly attracted by magnetic fields and can be magnetized to become permanent magnets, producing magnetic fields themselves. This can be well explained and observed experimentally by the theory of magnetic domain.

The most common observed magnetism phenomenon is seen in a magnet bar. Magnets are the natural occurring elements found in the form of ore that are mainly magnetite and lodestone or can be created by different types of materials such as iron, nickel, cobalt etc. Magnets consist of two ends, known as the north and south poles, that are created by the Earth's magnetic field, aligning the north pole of the magnet towards the Earth's North Pole and similarly for the south pole. This same phenomena has been used in compasses for navigation for decades [1].

### 1.1.2 Magnetic Field

Although magnets and magnetism had been studied much earlier, research into magnetic fields really began in 1269 when French scholar Petrus Peregrinus de Maricourt mapped out the magnetic field on the surface of a spherical magnet using iron needles. What convinced physicists that they needed this new concept of a field of force? A question now naturally arises as to whether there is any time delay in this kind of communication via magnetic (and electric) forces [2].

The traditional Newtonian concept of matter interacting (theory) via instantaneous forces at a distance states that the interaction energy arises from the relative positions of objects that are interacting via forces. With this assumption, Isaac Newton established classical mechanics under absolute space and time. Newton thought that there is no time delay in electric or magnetic interaction, since he conceived of physics in terms of instantaneous action at a distance. During the 1820s, when explaining magnetism, Michael Faraday inferred a field filling space and transmitting that force. If it takes some time for forces to be transmitted through space, then apparently there is some thing that travels through space. Nowadays, it is common sense that a magnetic field is a vector field that describes the magnetic influence of electrical currents and magnetized materials. Magnetic fields are widely used throughout modern technology, particularly in electrical engineering and electromechanics.

Applied magnetism can be broadly divided into two branches, one of which caters for magnetic interactions at higher frequencies where electric and magnetic fields are coupled to behave as electromagnetic waves. At lower frequencies electric and magnetic fields remain uncoupled, and can be treated and interpreted separately. Magnetic field measurement has a proven role in magnetic rotating disks, electromechanical drives, and relays, to name just a few applications. Low-frequency magnetic field measurement has emerged as a potential candidate for a role in power systems. Power systems are rich in electric current-carrying conductors right from generation down to consumption. The flow of electric current in generation units, transmission circuits, and distribution circuits as well as in loads produces a magnetic field signature distinct to each of these devices. Furthermore, these signatures also adopt a distinguishable pattern for the state and health of these devices.

### 1.1.3 The Mathematics of Magnetism

Originally, electricity and magnetism were considered to be two separate forces. This view changed, however, with the publication of James Clerk Maxwell's *A Treatise on Electricity and Magnetism* in 1873, in which the interactions of positive and negative charges were shown to be mediated by one force.

Magnetic fields are typically conceptualized with so-called "flux lines" or "lines of force". The basic quantities of a magnetic field are two vector fields, magnetic field intensity $\vec{H}$ and intensity of magnetization or flux density $\vec{B}$, based on the effects it has on its environment, with the relationship,

$$\vec{B} = \mu\vec{H} \tag{1.1}$$

where $\mu$ is the permeability of the material or substance where $\vec{H}$ and $\vec{B}$ coexist.

With these quantities, together with the quantities in the electric field, the Maxwell equations in integral and differential form, respectively, are as follows:

$$\begin{cases} \oiint_{\partial\Omega} E \cdot dS = \frac{1}{\varepsilon_0} \iiint_\Omega \rho dV \\ \oiint_{\partial\Omega} B \cdot dS = 0 \\ \oint_{\partial\Sigma} E \cdot dl = -\frac{d}{dt} \iint_\Sigma B \cdot dS \\ \oint_{\partial\Sigma} B \cdot dl = \mu_0 \left( \iint_\Sigma J \cdot dS + \varepsilon_0 \frac{d}{dt} \iint_\Sigma E \cdot dS \right) \end{cases} \tag{1.2}$$

$$\begin{cases} \nabla \cdot E = \frac{\rho}{\varepsilon_0} \\ \nabla \cdot B = 0 \\ \nabla \times E = -\frac{\partial B}{\partial t} \\ \nabla \times B = \mu_0 J + \mu_0 \varepsilon_0 \frac{\partial E}{\partial t} \end{cases} \tag{1.3}$$

This is a comprehensive model for study of electromagnetic fields, especially electromagnetic waves. Yet magnetic and electric fields can be studied separately at low frequency. Coulomb force is used to describe in an electric field the electrostatic interaction between two point charges, attraction or repulsion. Magnetic forces are interactions between moving charges, occurring in addition to the electric forces. The mathematics of magnetism are significantly more complex than the Coulomb force law for electricity.

As magnetism is usually caused by the motion of charge, so the electricity, i.e. the flow of electrons, flowing through a conductor can also produce a magnetic field around it. The magnetic field produced by the current flowing in a simple straight wire [3] (theoretically infinitely long) is shown in Figure 1.2.

A magnetic field is also produced in the surrounding space when current charges interact with a fixed magnet and it exerts magnetomotive force on the things that come into its magnetic field.

This gives rise to another term in the field of magnetism, magnetic flux, which is the number of lines passing through a specific area. As discussed above, the orientation of magnetic lines is very important in measuring the strength of the magnetic field as the spaces between the magnetic lines show the intensity of the magnetic field in that area, i.e. an area with closely packed magnetic lines represents a strong magnetic field, e.g. at the ends of a magnet, whereas an area where the magnetic lines are far apart represents

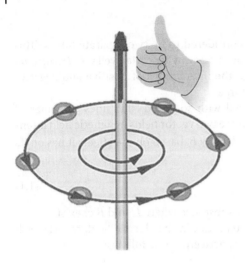

**Figure 1.2** Magnetic field around a conductor wire.

a weak magnetic field, e.g. in the middle of a magnet. Magnetic flux is also calculated by the following equation:

$$\phi = \vec{B} \cdot \vec{A} = BA(cos\theta) \tag{1.4}$$

where $\phi$ is the magnetic flux, $B$ is the magnetic field, and $\theta$ is the angle between the surface and magnetic lines. Equation (1.4) shows that magnetic flux is not only dependent upon the magnetic lines passing through an area but also on the angle between the magnetic lines passing through a specific area. If a flat surface is placed perpendicular to the magnetic lines then the magnetic flux is only equal to $BA$.

Magnetic flux is measured in Webers whereas the SI unit of the magnetic field is the Tesla, after the famous Serbian inventor and engineer Nikola Tesla. Magnetic flux is a measure of the force exerted on the moving charges present in a magnetic field [4].

Another branch of magnetism was discovered independently and named electromagnetic induction. This has played an important role in the power sector around the world. In the 19th century, a British scientist observed an amazing phenomena: any change in the magnetic environment of a coil induces voltage in the coil where that change can be produced by changing the strength of the magnetic field by moving the magnet position or orientation which result in change of the flux:

$$E = N\frac{d\phi}{dt} \tag{1.5}$$

Later, the Russian physicist Heinrich Friedrich Emil Lenz formulated Lenz's law, which advances Faraday's law and states that the magnetic field of the induced current opposes the magnetic field of the source that has produced it.

These two laws have been become the basic working principle of most power instruments, such as the motor, whose armature is placed between the magnets as the current flows through it. The armature experiences some force that rotates it and it converts this electrical energy into mechanical energy, which is further connected to a rotating load such as a fan. The same phenomenon is used in generators but in the opposite direction, which is an example of Faraday's electromagnetic induction that whenever a conductor moves in a magnetic field it induces current in a conductor, which generates electricity [5].

Furthermore, mutual induction is the basic principle of a transformer: whenever alternating voltage is applied to the primary coil, current starts to flow, resulting in a magnetic field around it that induces voltage in a secondary coil according to Faraday's law:

$$E = \frac{d\phi}{dt} \tag{1.6}$$

If there is more than one turn of the coil then

$$E = N\frac{d\phi}{dt} \tag{1.7}$$

The induced electromagnetic force or voltage is directly dependent upon the magnetic flux, i.e. the sum of magnetic field $B$ over the surface area $A$:

$$\phi = B/A \tag{1.8}$$

By putting this into (1.7), we get:

$$E = N\frac{d(BA)}{dt} \tag{1.9}$$

It is clear that the current produced in the secondary coil depends on the magnetic field that is produced by the current in the primary coil, as stated by Ampere's law:

$$\oint \vec{B}d\vec{l} = \mu_0 I_{enc} \tag{1.10}$$

For any closed-loop path, the sum of the length element times the magnetic field in the direction of the length element is equal to the permeability times the electric current enclosed in the loop. So if we have current flowing from the straight wire, the path of its magnetic field will be the circle and sum of all the elements along the path, resulting in the circumference of the circle. Then (1.10) becomes

$$B(2\pi r) = \mu_0 I_{enc} \tag{1.11}$$

$$B = \frac{\mu_0 I_{enc}}{2\pi r} \tag{1.12}$$

### 1.1.4 Magnetism in Daily Life

All the study and research into magnetics has made our lives very easy, with the invention of many appliances that we use daily. Some of these appliances (heating, wireless communication, and wireless energy transfer) and the magnetic phenomena they use are discussed below.

#### 1.1.4.1 Induction Heating and Microwave Ovens

Widely used across Europe and North America, with the rest of the world rapidly catching up with the convenience that is offered by them, electric induction hotplates are a practical example of magnetism being used in our daily lives. Induction coils are used to heat up a metal pot of food placed on the hotplate via heat generated from the electromagnetic effect through the induction in the coils. This heat cooks the food. Induction coil electric stoves are efficient and can boil water in as little as 5 minutes compared to typical gas stoves which take approximately 10 minutes. Induction cooking involves the electrical heating of a cooking vessel by magnetic induction, instead of by radiation or thermal conduction from an electrical heating element or a flame. Because inductive heating directly heats the vessel, very rapid increases in temperature can be achieved and changes in heat settings are instantaneous.

Microwave ovens heat and cook food by exposing it to electromagnetic radiation in the microwave frequency range.

These applications of electromagnetism are widely available and are very efficient, being ahead of the competition in terms of overall performance and convenience.

### 1.1.4.2 Cell Phones and Wireless Communications

Cell phones (or mobile phones) have become a must-have for most people in the modern world, and are a major force in changing people's modern lives, especially with the development of smart phones. Another type of wireless communication, short-range Wi-Fi, is also used in daily life by smartphones as it offloads traffic from cell networks onto local area networks. The fundamental technology of wireless communication is electromagnetic, or radio, waves, which are used to transfer information or power between two or more points that are not connected by an electrical conductor. Depending on the frequency band or protocol adopted, the distance and speed of communication will be different, and can be categorized into many types, such as Wi-Fi, Bluetooth etc. In terms of the technology involved in the communication part of cell phones, not accounting for the various applications, developments from first to fifth generation have been observed in the last 20 years. Other examples of applications of radio wireless technology include GPS units, garage door remote controls, wireless computer mice, keyboards, and headsets, headphones, radio receivers, satellite television, broadcast television, and cordless telephones.

### 1.1.4.3 Loudspeakers and Headphones

Headphones and loudspeakers use electromagnets to translate or convert an electrical signal into vibration, which is the sound that we hear through the speaker. The audio input from the source is an electrical signal or amplifier that travels to the loudspeaker or headphone, which receives the electrical signal at an electromagnet. The current flows through the electromagnet, which in turn creates a vibrating signal with a fixed magnet based on the varying magnetic field coming off the amplifier to recreate the sound from the input that was transported via the cord to the speaker or headphone. This application of magnetism is very common and we encounter it many times nearly every day.

### 1.1.4.4 Maglev Train and Wireless Power Transfer

Levitation has always been a very fascinating concept and we have now reached a point where it has been made possible by magnetism. Magnetic levitation trains use magnetic repulsion, the mechanism through which the same type of poles repel each other, to levitate the trains. The idea of a hyperloop, which some refer to as the fifth mode of transportation, is dependent on magnetic levitation to achieve the ultra-high speeds that make it the fastest terrestrial transportation in existence. Hyperloops are currently being built in the United Arab Emirates and the United States. Other than hyperloops, in developed countries like Germany, China, Japan, Spain, and France Maglev trains are used for the inter-city travel and are very promising and safe. Although this application of magnetism is not widespread, its reliability and potential for the future are remarkable, and this should ensure its widespread use in the future. Magnetic field based power transfer has developed rapidly in recent years and wireless charging (for cell phones and electric vehicles) can be found in most modern cities.

### 1.1.5  Magnetic Fields in Industry

Magnetism not only plays a role in the domestic household but is also important in bulk industries and acts as the backbone of many industrial machines. These machines can be just a part of an industry, like generators, or can form the basis of the industry.

#### 1.1.5.1  Magnetic Resonance Imaging

Originally named nuclear magnetic resonance imaging (NMRI) and now known as magnetic resonance imaging, this is a very common medical diagnostic procedure. It applies magnetic field and radio waves to the human body to change the spin of the protons in the body under the influence of a magnetic field or a varying magnetic field via radio waves. This produces an image of the body tissues and enables them to be compared and examined to give an insight into different body tissues and parts, thus improving diagnostics. This application of magnetism in medical diagnostics is widely used across the world and its discoverers were awarded the Nobel Prize in Physics. There are different techniques that can be used in MRI but the principal of using the variation in the quantum behaviour of body tissues under the influence of varying magnetic field is always the same.

#### 1.1.5.2  Relays

One of the most extensive industrial applications of magnetism is in relays. Relays have an electromagnetic core that is driven by a small current to control the flow of current to a larger circuitry or system. Automated industrial control units and manual control rooms all, directly or indirectly, employ relays to trigger or halt various operations, circuits, valves or switches etc. Mechanization and automation have made relays, and thus the use of magnetism, vital and they are extensively used in industrial systems.

#### 1.1.5.3  Electricity Meters

Electricity meters employ electromagnetism in the form of electromagnetic induction to read the revolutions of a metallic element that rotates with the consumption of power. They are extensively used in domestic as well as industrial settings across the globe. This application of magnetism is widely used by power distribution and generation companies to monitor domestic power consumption and impose tariffs on consumers, and is used in industrial systems to monitor the power used for different purposes depending upon the nature of the industry.

#### 1.1.5.4  Giant Cranes

Giant cranes, used at ports and on construction sites to carry containers or objects from one place to another, have a very powerful electromagnet driven by an electric current applied by the operator as needed. When the electromagnet is close to the object to be lifted it is magnetized by the current applied by the operator and is able to lift ton weights with ease. This application of magnetism is very widely used and we all indirectly consume things that we receive through shipping via ports where magnetic cranes transport containers from the port to shipping or other vessels for further transportation.

### 1.1.5.5 Microphones

Another widespread application of magnetism is in microphones, which are the opposite of the loudspeakers and headphones that we discussed earlier. Microphones register the a diaphragm vibrational motion as an electrical signal via an electromagnetic coil, then transport the electrical signal to a speaker, pre-amp, or recorder to be amplified and heard, enhanced, or stored, respectively.

## 1.2 Magnetic Fields in Modern Power Systems

Power systems are made up of generation, transmission, and distribution systems. The generation system can be powered either by a traditional method using fossil fuels that is available 24 hours and can be used for the production of electricity, or by renewable sources, also known as the modern power system, which the power sector of every country is shifting towards. The major concern with these modern power systems is that most of the sources they rely on are not available all the time so they cannot be used for the base load [6].

### 1.2.1 Components of Modern Power Systems

Electric power systems are essential in the modern world and dominant every system in the world, therefore avoiding any interruption in this system is of utmost concern to every country. Electric power systems deal with the generation, transmission, and utilization of electrical power, which was discovered by the great scientist Michael Faraday in the 18th century. They are composed of a bulk generator that generates electrical power from different sources, a transmission system through which electrical energy is transmitted from one end to another, and a distribution system that distributes electrical energy to end users with different load requirements. A typical electric power system is shown in Figure 1.3.

#### 1.2.1.1 Electric Generation System

The electric generation system is the preliminary system of an electric power system that generates electrical energy from different renewable and non-renewable sources. There are various methods of electricity generation dependent on the type of energy required. However, the basic principle of a generation system is the utilization of electromagnetic theory, as shown in Figure 1.4. The primary energy from natural sources such as fossil fuels and renewable sources is used to run a turbine that converts the potential energy of the source into mechanical energy. The generator is then used to convert the mechanical energy into electrical energy.

The fundamental principles of electricity generation were discovered in the 1820s and early 1830s by Faraday and his method is still used today. The most important piece of equipment is the generator. The generator is the machine that uses the magnetic principle to generate electricity and it is the backbone of the electric generation system. The basic function of the generator is to convert mechanical energy into electrical energy.

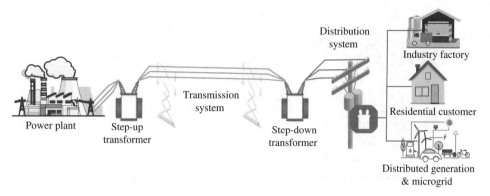

**Figure 1.3** A typical electric power system.

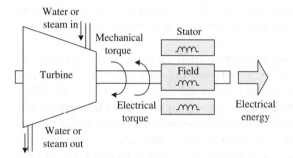

**Figure 1.4** A typical electric generation system.

The turbine is the part of electric generation system that produces mechanical energy for the generators. This mechanical energy is generated by two types of energy: electro-chemical energy produced from heat energy in the form of steam or gas produced by the combustion of fossil fuels, geothermal energy or nuclear fission reaction, or kinetic energy from water or wind. The rotating turbine is connected to the armature in which either the magnetic field is varied or the conductor rotates, depending on the nature of the generator. This satisfies the law of electromagnetic induction and produces electricity, which is further drawn by the load connected to it. Figure 1.5 shows the coils and associated magnetic field in a simple generator. The main electrical and mechanical components of generators are the magnetic field, armature, prime mover, rotor, stator, slip rings, shaft, and bearings. The field in an AC generator consists of coils of conductors within the generator that receive a voltage from a source (called excitation) and produce a magnetic flux. A strong magnetic field is produced by a current flowing though the field coils of the rotor due to the rotation. The magnetic flux in the field cuts the armature to produce a voltage. This voltage is ultimately the output voltage of the generator. The armature is the part of a generator in which voltage is produced. This component consists of many coils of wire that are large enough to carry the full-load current of the generator.

The rotor's magnetic field may be produced by induction (as in a "brushless" alternator), by permanent magnets (as in very small machines), or by a rotor coil energized

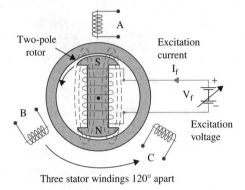

Two-pole rotor

Excitation current $I_f$

Excitation voltage $V_f$

B

A

C

Three stator windings 120° apart

**Figure 1.5** Magnetic field and coils in a simple generator.

with direct current through slip rings and brushes, the usual solution for industrial generators. The rotor's magnetic field may even be provided by a stationary field coil, with moving poles in the rotor.

There are two types of the generators depending upon the nature of the current produced, i.e. DC and AC generators. In AC generators, the rotor produces the rotating magnetic field and is placed in the stator on which conductor coils are wrapped. When the rotor rotates due to the turbine it creates a rotating magnetic field that exerts a force on the electrons of the conductor present on the stator and cause them to flow, which is further connected to the load. Due to the rotating magnetic field the north and south poles vary, which changes the direction of the current and produces an AC current. However, in a DC generator the magnetic field is fixed while the coils rotate in the field, producing a current in one direction, i.e. DC current.

### 1.2.1.2 Transmission Systems

Extensive energy transport systems are scattered across the globe movings energy from where it occurs naturally to where it can be put to good use. Electrical power transmission is the most flexible and efficient way to transport energy. The transmission system is used to transmit the generated electrical energy from the source to the load in a reliable and economical manner over a long distance. There are two types of transmission system used worldwide: overhead transmission systems, which transmit electrical energy through power cables that are suspended between the poles, and underground transmission lines, in which power cables are underground, not exposed to the environment [7]. Overhead high-voltage transmission lines (HVTLs) serve as the backbone of the existing power grid framework. An important part of this process includes transformers, which are used to increase voltage levels to make long distance transmission feasible, i.e. they transmit power at high voltage and with low loss. Since power plants are most often located outside densely populated areas, the transmission system must be fairly large.

A power transmission system includes following main components of HVTLs and associated facilities: transmission towers (for overhead HVTLs), conductors (power lines), substations (usually where the transformer is located), transformer, and transmission corridor (right of way).

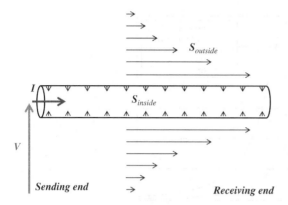

**Figure 1.6** The Poynting vector in a power transmission system.

Since voltage and current are present in the power transmission line, the electromagnetic field is always associated with the transmission system, but in fact it is the electromagnetic field that makes the energy transmission possible. According to the Poynting's theorem, the energy flux, described by the Poynting vector, is the cross-product of the magnetic field and the electric field:

$$\vec{S} = \vec{E} \times \vec{B} \tag{1.13}$$

where $\vec{S}$ is the Poynting vector and $\vec{E}$ is the electric field.

As any type of energy has a direction of movement in space, as well as a density, it is customary to use the Poynting vector to represent rates of flow of energy and momentum in electromagnetic waves. The Poynting vector represents the particular case of an energy flux vector for electromagnetic energy. It shows that power flows in the space surrounding a conductor and not our preconception that it should be inside the conductor. This is widely used for educational purposes and urges us to understand that the magnetic field is very important in the transmission system. Other than this, Poynting's theorem may provide another explanation of electrical energy transmission: the current is set up by the the energy transmitted through the medium around it, in addition to our common sense that the current is set up by the magnetic induction. The distribution of the Poynting vector around a transmission line is shown in Figure 1.6. Inside the conductor the vector points inward because of the resistive loss. Outside the conductor the vector points along the direction of the conductor and decays as the distance increases.

Overhead HVTLs are deployed in diverse geographic regions and their safety is continuously challenged by the varying outdoor environment. Hence, it is desirable to monitor the operating conditions of overhead power lines in a fully integrated smart grid. Non-contact sensors adapted for transmission lines are used to monitor electrical and spatial parameters using various sensing technologies such as magnetic field sensors, distance measuring lasers, and camera-based sensors. With rapid development in microelectromechanical systems (MEMSs) and material technology, magnetic field sensors based on the magnetoresistance (MR) effect have found potential applications in power systems and can be used to carry out point measurements of magnetic fields with high accuracy. It is also important to note that for any power line the magnetic field for a current-carrying conductor is dependent on the distance between the conductor

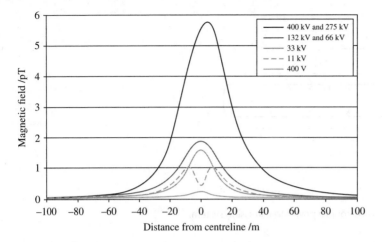

**Figure 1.7** Magnetic field strength of different overhead HVTLs.

and the sensing point. Magnetic field strength in close vicinity to different voltage carrying overhead HVTLs is demonstrated in Figure 1.7. It can be inferred that the field strength can be interpreted for electric and spatial parameter monitoring purposes.

### 1.2.1.3 Distribution Systems

The electric transmission system is used in combination with power plants, distribution systems, and substations to form the electrical grid. The electric distribution system is the final system of the electric power system and distributes the electric energy from the transmission system to different types of end user. It takes electricity from the highly meshed transmission system and steps down the voltages in distribution substations according to the nature of the end users, which residential, industrial, and commercial consumers [8].

### 1.2.1.4 Transformers

The transformer is one of the most important parts of the electric power system. It steps up and steps down the voltage of the electricity according to its placement. Voltage is stepped up from the generating station to the transmission line to avoid $I^2R$ losses in the complex networked grid of power cables. In contrast, voltages is stepped down to the required voltage of end users from transmission lines to the distribution system [9].

A transformer is an electrical device that transforms voltages from one value to another by operating on the principle of electromagnetic induction. This principle states that when a current flows in a coil it magnetically induces voltages in another coil. A transformer usually consists of two coils: a primary one that takes power and a secondary one that delivers power. These two coils are not electrically in contact with each other, but are wrapped on the stack of the core, which reduces its core losses [10].

Transformers can be divided into three different criteria: construction, services, and power utility. There are two types of construction transformer: core and shell. Core transformers have a coil wrapped around the side limbs and a core surrounding the coil, whereas shell transformers have both the coils wrapped on a central limb and a coil surrounding the core in this case. However, on the basis of power utility, transformers can

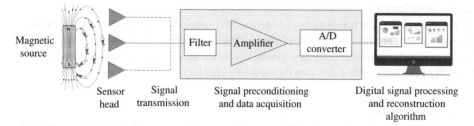

**Figure 1.8** Typical magnetic field detection system.

be either single phase or three phase depending upon the nature of the load connected at the end user. On the basis of services they can be categorized as power transformers or distribution transformers. Power transformers are usually used in high-voltage transmission networks to step up and step down the voltage for transmission. Distribution transformers, also called utility transformers, are connected directly to the end user side that steps down the voltage coming from the substation of 11 or 10 kV to 400 or 230 V depending on the nature of the load connected at the secondary side [11].

## 1.2.2 Magnetic Field Detection and Interpretation

To avoid any kind of disturbance in an electric power system, protective devices which either detect the fault or remove the fault, depending upon the demand of the system, are commonly used. Magnetic field measurement, inspired by the development of magnetic field sensor technology, is developing quickly and is widely used in various applications.

A typical magnetic field measurement system consists of the sensor, signal conditioning system, and processing algorithm, as shown in Figure 1.8. The sensor may be deployed within the field to form an array for special detection purposes. The sensed signal is transmitted, processed, and sampled. The sampled signal is in digital form and can be easily processed. Generally, in an electric power system, the measured magnetic field signal will be related to the operation state of the power system components, e.g. the current source reconstruction, sag (of conductor) estimation etc.

The sensor is the most important part in the measurement system. Many systems and sensors have been created in recent years for the detection of magnetic fields. Among theses is the MEMS magnetic field sensor, which because of its small size, broad bandwidth, high sensitivity, and fast response is widely used in smart grids. Some sensors are discussed below.

### 1.2.2.1 Hall Effect Sensor

Magnetic sensors are intended to react to an extensive range of positive and negative magnetic fields in many applications. One magnet sensor whose output is a function of the magnetic field density around it is known as the Hall effect sensor. It is the simplest and most basic method for magnetic field measurement and detection, and is made of the semiconductor known as Hall element. Lorentz forces act on the charges of the semiconductor when placed in a magnetic field, creating a potential difference at the ends of the Hall element that can be measured by a voltmeter [12].

### 1.2.2.2 AMR Sensor

The anisotropic magnetoresistance (AMR) sensor consists of a silicon chip on which an alloy of ferromagnetic material is placed whose resistance changes when it is placed in any external magnetic field depending upon its direction, according to which the strength of the magnetic fields can be measured [13].

### 1.2.2.3 GMR Sensor

The giant magnetoresistance (GMR) sensor is another type of MR sensor whose output is a function of resistance. It is made up of ferromagnetic layers that have a thin non-magnetic conductive layer sandwiched between them. The non-magnetic layer is made of copper, which is a very good conductor of the charges. Due to antiferromagnetic coupling, the magnetic moment in the alloy layer faces the opposite direction, which increase the resistance of the copper. When it is placed in the external magnetic field, the charges align themselves according to the direction and result in decreased resistance of the copper [14].

### 1.2.2.4 TMR Sensor

Tunnel magnetoresistance (TMR) sensors have the same construction as GMR sensors, except the middle layer is made up of insulating material. The working principle is almost similar, i.e. when the free layer and the pin layer have parallel magnetization the resistance is low whereas if they are in opposite directions the resistance is very high and only a small amount of current flows through it [15].

### 1.2.2.5 Magneto-optical Sensor

Magneto-optical sensors work on the principle of the Faraday effect, which is an interaction of the magnetic field and the beam of light that states that when the light propagates through a gyrotropic material parallel to the direction of the external magnetic field, the plane of polarization is rotated. The rotation of the plane depends on the nature of the gyro material and mainly on the strength of the magnetic field.

### 1.2.2.6 Lorentz Force-based MEMS Sensors

The MEMS sensor known as a microelectromechanical system used as an attractive field sensor is a small-scale microelectromechanical framework devices for recognizing and estimating magnetic fields (magnetometer). Most of these devices work by recognizing the impact of the Lorentz forces, that is, when the current-carrying conductor is placed in the magnetic field it experiences some kind of force on it. An adjustment in voltage or frequency might be estimated electronically or a mechanical displacement might be estimated optically [16].

### 1.2.2.7 Fluxgate Magnetometer

A fluxgate compass is an electromagnetic device consisting of small coils that are wrapped around highly ferromagnetic material. It is used for measuring and detecting the horizontal component of the Earth's magnetic field. A fluxgate compass differs from a normal compass in that its output is electronic and can be interpreted and transmitted easily.

### 1.2.2.8 Interpretation of Measured Magnetic Field: The Inverse Problem

After magnetic field data have been obtained, measurement technologies and analysis techniques should be applied for the quantitative characterization of the system under study. This is generally an ill-posed problem, and so is called an inverse problem, e.g. the location of magnetic sources by measurements of their magnetic fields is a typical ill-posed inverse problem. The objective of an inverse problem is to find the best model parameter $X$ such that (at least approximately)

$$Y = F(X) \tag{1.14}$$

where $F$ is an operator (generally nonlinear) describing the explicit relationship between the observed data, $Y$, and the model parameters. In various contexts, the operator $F$ is called the forward operator, the observation operator, or the observation function. In the most general context, $F$ represents the governing equations that relate the model parameters to the observed data (i.e. the governing physics).

An inverse problem in science is the process of calculating from a set of observations the causal factors that produced them, e.g. source reconstruction in acoustics or calculating the density of the Earth from measurements of its gravitational field. These are some of the most important mathematical problems in science because they tell us about parameters that we cannot directly observe.

The inverse problem can be conceptually formulated as from data to model parameters, as against the forward problem, which relates the model parameters to the data that we observe. The transformation from data to model parameters (or vice versa) is a result of the interaction of a physical system with the object that we wish to infer properties about. In other words, the transformation is the physics that relates the physical quantity (i.e. the model parameters) to the observed data. Inverse problem theory is used extensively in weather predictions, oceanography, hydrology, and petroleum engineering. Inverse problems are typically ill-posed, as opposed to the well-posed problems more typical when modeling physical situations where the model parameters or material properties are known.

The forward direction of magnetic field measurement is straightforward and is given by the physical laws governing the electromagnetic phenomena, as given in the mathematics section. However, for discrete magnetic fields, the laws, such as Biot–Savart law, are generally highly nonlinear, therefore it is very hard to solve the inverse problem. This problem is further complicated due to the fact the sensor cannot differentiate the magnetic field from various sources (a sensor would read all the magnetic field caused by the superposition of any sources presented in the field). Any pair of source readings has a nonlinear relationship. In a continuous field, we may need to describe the model by means of Laplace's equation, i.e.

$$\nabla^2 \Phi = 0 \tag{1.15}$$

where $\nabla^2$ is the Laplace operator and $\Phi$ is a scalar function, the potential of the magnetic field.

The general theory of solutions to Laplace's equation is known as potential theory. The solutions of Laplace's equation are the harmonic functions under the spherical coordinates

$$\Phi = \sum_{n=0}^{\infty} \sum_{m=-n}^{n} [k_{ni}^m r^n + \xi_{ni}^m r^{-(n+1)}][Y_n^m(\theta, \phi)] \tag{1.16}$$

where $k_{ni}^m$ and $\xi_{ni}^m$ are constants to be determined by boundary conditions. $Y_n^m(\theta, \phi)$ is the spherical harmonic functions defined by

$$Y_n^m(\theta, \phi) = \sqrt{\frac{2n+1}{4\pi} \frac{(n-m)!}{(n+m)!}} [P_n^m(\cos\theta)]e^{jm\phi} \tag{1.17}$$

where $P_n^m(\cos\theta)$ is associated Legendre functions, and $n$ and $m$ are integers with $-n \leq m \leq n$.

With this model, it is almost impossible to solve the inverse problem directly. Even for the forward problem, one needs to solve the field with the boundary conditions.

In the sense of functional analysis, the inverse problem is represented by a mapping between metric spaces. While inverse problems are often formulated in infinite dimensional spaces, the limitations of a finite number of measurements, and the practical consideration of recovering only a finite number of unknown parameters, may lead to the problems being recast in discrete form. In this case the inverse problem will typically be ill-conditioned. In these cases, regularization may be used to introduce mild assumptions to the solution and prevent over-fitting. Many instances of regularized inverse problems can be interpreted as special cases of Bayesian inference. In magnetic field measurement based reconstruction, many numerical solution approaches have been developed, and these are discussed in subsequent chapters.

### 1.2.2.9 Limitations of Magnetic Field Detection Based Applications

The novel approach based on non-contact measurement of a magnetic field has advantages over traditional approaches. Utilizing magnetic field sensing for estimation of the electric and spatial parameters of an electric power system is a diverse area. In particular, the robust monitoring of electric current estimation can be performed using non-contact magnetic field sensing. Some other extended applications use sensitive magnetic field sensors for transient and uniform current estimation for utility at smart substations. Apart from the diagnosis of power lines itself, a great deal of this sensing can be used to quantify the effect of these electromagnetic fields on human populations nearby. Moreover, MEMS sensors have been developed based on AMR effect materials and the software-supported geographical information system (GIS) interface makes the fault location intuitively clear and convenient. However, the existing solutions based on magnetic field sensing have limitations in terms of sensor technology and interpretation methods. These can be divided into two main problems, as discussed below [17].

### *Limitations of AMR-based Magnetic Field Based Sensor Technology*

The use of magnetic field measurements has huge potential in power system applications. Recently, researchers demonstrated the use of MR sensors for operation state monitoring and fault detection of overhead power transmission lines. Furthermore, applications of magnetic field measurements in smart grid and electronic substations have been studied extensively. A portable device for transient magnetic field measurements with an AMR sensor head has been developed. The physical characteristics of AMR sensors impose restrictions on their magnetic field detection range and disorientation effect in the presence of a strong magnetic field. The impact of a transient magnetic field (TMF) on sensor response time also need to be determined as its detection and characterization provide valuable information in a power system.

*Limitations of Existing Non-contact Magnetic Field Based Estimation Systems*
One currently popular method is interpretation of a magnetic field that is radiated by current-carrying elements. A transmission line monitoring system (TLMS) based on the non-contact principle is commercially available from Promethean Devices Ltd. It measures the magnetic field using sensing coils, then estimates the conductor to sensor clearance, sag, and conductor temperature. The device operates on the surface of the ground and requires placement of sensing units under each phase conductor. However, deviations in electric current and conductor clearance estimation for TLMS have been reported. Alternatively, magnetic field sensing can be realized with MR sensors favoured by a low power requirement in microwatts and high sensitivity ranging up to picoTeslas.

## 1.3 Magnetics in Smart Grids

'Grid' refers to the electric grid that generates electric power using traditional and non-traditional resources that is transmitted through a complex network of transmission lines and distributed to different kinds of end user, i.e. residential, commercial or industrial. Smart grids are modern and efficient grids that include sensing and monitoring devices for instrument conditioning and a communication network for connecting the utility and the customers. Smart grids move power systems into a new era of reliability, sustainability, flexibility, and efficiency that contributes to the efficient transmission of electricity, reducing the maintenance and operational costs. Smart grids involve magnetics in three categories. First, electromagnetic interference, is important as it causes electromagnetic disturbance and propagation, composed of the electrical and magnetic field in many of the electrical instruments that may occur deliberately or accidently. To remove the electromagnetic interference from the environment, the electromagnetic compatibility ensures that the instruments do not influence the electromagnetic environment to the extent that the function of the devices present in that environment is disturbed [18].

Second, many sensing and monitoring systems involve the phenomena of magnetics to measure the electrical and spatial parameters of the power system. The casual and direct relation between the current and the magnetic field gives a wide perspective to design the control or metering system. These sensors can detect or measure the different parameters of the instrument by the variation in the magnetic field in the environment. Many magnetic switches have been designed to protect the transmission system that work on magnetization. These switches are valuable because of their non-mechanical structure, which reduces the maintenance cost. The reed switch is a well-known magnetic switch that controls the flow of electricity. It consists of two electrical contacts made of ferromagnetic material that remain in open in the normal state and close when a magnet is positioned near to them [19].

Furthermore, for the generation and delivery of electricity magnetic storage devices have been shown to be a solution for electricity transmission and storage.

### 1.3.1 Magnetic Field in Lieu of Smart Grid Objectives

In past few decades, the use of magnetics has drastically increased and they are now being used in many applications. Those that are discussed below illustrate how magnetics are used in all branches of the sciences, for communication, health, and military purposes.

### 1.3.1.1 Monitoring in Smart Transmission Networks

Any kind of fault in a power system, whether minor or major, will result in power interruption or blackout, respectively. In areas that include industries, hospitals, and factories even a small fluctuation in electricity can cause catastrophic damage. The detection of a fault before any damage or permanent loss not only delivers continuous power but also increases the life and reduces the maintenance cost of the system. The only way to avoid failures and interruptions in a power system is to detect faults before the system collapses and make arrangement to avoid permanent damage. Monitoring and metering systems are designed to systematize and automate any system that uses sensors and switches to detect and measure faults, and then communicate and report to the control system to facilitate precautions and make alternative arrangements. Traditionally, current and potential transformers have been used to measure the current and voltage in high circuit voltages but due to their bulky size, incompatibility with DC, and narrow bandwidth they are inadequate in smart grids. Magnetic field based measuring devices and sensors that measure the current and other parameters without being in physical contact with instruments are easy to install and carry [20]. With the novelty of being contactless, these magnetic sensors have attracted the attention of many researchers and many sensors have been designed for monitoring and conditioning power systems. For current measurements in the generation and transmission of power systems, spintronic technology-based MR sensors are emerging, including AMR sensors and GMR sensors. For more sensitive cases, TMR sensors are highly promising for current sensing, especially for measurements in the picoTesla range [21].

### 1.3.1.2 Measurement of Transient Magnetic Fields

Transients are the surges produced in an electric circuit due to faults and interruption. These faults can occur due to the internal sources, such as switching, static discharge, and arcing, or external sources, such as lightning, poor connections, and normal utility operations. Transients occur when the normal sinusoidal wave of the voltage has abrupt spikes due to the large amount of current being drawn. As the transient is dependent on the increase and decrease of the current, which is directly related to the strength of the magnetic field, which transient it is can be detected by measuring the magnitude of the magnetic field using AMR sensors [22].

### 1.3.1.3 Permanent Magnet Generators and Motors

Permanent magnet (PM) generators and motors are famous for their excellent magnetic properties, such as maximum magnetic energy product, and can bear high temperatures, which is very useful for electromagnetic devices. PM generators, also known as alternators, use permanent magnets for coil excitation [23]. The magnet is attached to a rotor and generates a rotating magnetic field that produces AC electricity in the stationary coil wrapped on the stator, and vice versa for the PM motor. It is a very good example of magnetics playing an important role in a generation system due to high efficiency, high power factor, compact size, large rating, high controllability, and stable operation [24].

#### 1.3.1.4 Magnetic Fault Current Limiter

The short circuit current is the threshold current range of any device that it can easily bearable. Short circuit in a system can be due to a number of reasons and researchers focus on removing or controlling it. Fault current limiters (FCLs) are used to protect the flow of an excessive amount of current and hence contribute to the reliability, resilience, and responsiveness of smart grids. There are two types of the FCLs: magnetic and superconductive. In FCLs, electrical cables are wrapped around the magnetic core, which is saturated. When the DC current flow is more than the capacity the magnetic core tends to be saturated and greatly increases the impedance. In superconductive FCLs, the coils are made up of superconductor material through which the DC current flows. When the current is above its maximum rate, it quenches and the resistance increases, which reduces the fault current [25].

#### 1.3.1.5 Magnetic Energy Storage

Superconducting magnetic energy storage (SMES) systems store energy in DC electricity in the form of a magnetic field in a superconducting material below their cryogenic temperature. This is a superconducting critical temperature at which material has zero electrical resistance and ejection of the magnetic flux fields occurs in certain materials. These SMES systems have been a revelation in the field of magnetics in the power sector for the storage of electricity in mostly renewable power systems [26].

#### 1.3.1.6 Wireless Power Transmission

Wireless power transmission (WPT) via microwave (long distance), resonance (medium distance), and induction (short distance) is another area that holds great promise. An example of successful short distance WPT is wireless charging of electric vehicles. Magnetic induction conducts the transmission between the coil on the ground and the one in the electric vehicle. The addition of capacitance to the system can help extend it for transmission over medium distances, resulting in a system exhibiting a magnetic resonance frequency that is the product of capacitance and inductance of the system. Transfer of energy from the transmission coil to the receiving coil can take place only if the coils have the same resonant frequency and are placed only a few meters apart. Such an non-radiative energy transfer does not spread in all directions and is quite confined. Long distance WPT has been proposed that theoretically increases the possibility of transmission over longer distances spanning many miles. The microwave energy that the transmitter emits is collected by antennae at the receiving side, with rectification carried out by diodes that provide DC electricity.

#### 1.3.1.7 Smart Components by Soft Magnetic Material

The hysteresis loop of ferromagnetic materials is determined by the kind of microstructure and particularly grain size that affect the magnetic behavior of the material. Generally, larger grain size ($D > 100$ nm) gives softer magnetic properties whereas smaller grain sizes ($D < 20$ nm) in nanocrystalline alloys lead to lower coercivities. The permeability exhibits an inverse relationship to the coercivity, which shows $D^6$ dependence at smaller grain sizes. When the dimensions of a material approach the nanometer range, it exhibits exceptional magnetic properties. Hence, nanocrystalline microstructures are used to manufacture the highly permeable common mode chokes used in EMC filters and low coercivity transformer cores that exhibit soft magnetic properties. Such a soft magnetic nanomaterial is the Fe-Cu-Nb-Si-B alloy, which offers high permeability, low losses, high saturation magnetization (up to 1.3 T), and low magnetostriction. Saturation magnetization can be enhanced further due to the higher percentage of Fe in these nanocrystalline materials (up to 1.7 T). A few of the commonly available nanocrystalline alloys used for smart grid assemblies are METGLAS, NANOPERM, FINEMET, VITROPERM, and VITROVAC [27].

### 1.3.2 Magnetic Field Measurements for Innovative Applications

Conventional linear Hall sensors and switches only detect and measure the magnetic field that is perpendicular to the chip whereas GMR angle sensors only measure the planar-oriented field components. However, the TLE493D-W1B6 sensor is an innovative sensor that provides a three-dimensional image of the magnetic field by determining the three-coordinate plane and any variation and change in any one of the components is sensed by these sensors. Furthermore, the advancement of weak magnetic field estimation procedures in view of the superconducting gadgets known as SQUIDS has prompted great interest in research in both traditional applications and new areas of use. Specifically, the weak magnetic field related to the human body is in effect progressively studied.

Fibre Bragg grating (FBG) is used to sense the variation in external magnetic field by the prorogation of the optical waveguide that is sensitive to it. These sensors are easy to install and can provide high spatial resolutions maps of the magnetic field. This technology has a number of advantages and applications range from military security to navigation and geographical surveys.

## Bibliography

1 D. C. Giancoli, *Physics for Scientists and Engineers*. Upper Saddle River, NJ: Prentice Hall, 2008.

2 Wikipedia. (2018) Magnetic field. [Online]. Available: https://en.wikipedia.org/wiki/Magnetic_field

3 C. P. Steinmetz, "On the law of hysteresis," *Proceedings of the IEEE*, vol. 72, no. 2, pp. 197–221, 1984.

4 A. R. Hambley, *Electrical Engineering: Principles and Applications*. Upper Saddle River, NJ: Prentice Hall, 2011, vol. 2.

5 R. W. Erickson and D. Maksimović, "Basic magnetics theory," in *Fundamentals of Power Electronics*. Springer, 2001, pp. 491–537.

6 A. J. Khan, *The Pakistan Development Review*. Pakistan Institute of Development Economics, 2014, ch. Structure and regulation of the electricity networks in Pakistan, pp. 505–528.

7 L. L. Grigsby, *Electric Power Generation, Transmission, and Distribution*. Boca Raton, FL: CRC Press, 2016.

8 J. Aubin, R. Bergeron, and R. Morin, "Distribution transformer overloading capability under cold-load pickup conditions," *IEEE Transactions on Power Delivery*, vol. 5, no. 4, pp. 1883–1891, 1990.

9 L. H. Dixon, *Magnetic Field Evaluation in Transformers and Inductors*. Citeseer, 2004.

10  W. R. Taylor and M. A. Dubravec, "Evaluation of electrical energy consumption and reduction potential at the 7th Army Training Command (ATC), US Army, Europe," Contruction Engineering Research Lab, Champaign, IL, Tech. Rep., 1990.

11  P. S. Georgilakis, *Spotlight on Modern Transformer Design*. Springer, 2009, ch. Conventional Transformer Design, pp. 45–122.

12  E. Ramsden, *Hall-effect Sensors: Theory and Application*. Elsevier, 2011.

13  D. He, *AMR Sensor and its Application on Nondestructive Evaluation*. InTech, 2017.

14  J. Pelegri, J. Alberola, R. Lajara, and J. Santiso, "Vibration detector based on GMR sensors," in *Proceedings of the IEEE Instrumentation and Measurement Technology Conference*, 2007, pp. 1–5.

15  T. Lin, "Tunneling magnetoresistive (TMR) sensor with a Co-Fe-B free layer having a negative saturation magnetostriction," Patent, Nov. 9, 2010, US Patent 7,830,641.

16  M. Thompson, M. Li, and D. Horsley, "Low power 3-axis Lorentz force navigation magnetometer," in *IEEE 24th International Conference on Micro Electro Mechanical Systems (MEMS)*, 2011, pp. 593–596.

17  U. Topal, H. Can, O. M. Celik, A. Narman, M. Kamis, V. Citak, D. Cakrak, H. Sözeri, and P. Svec, "Design of fluxgate sensors for different applications from geology to medicine," *Journal of Superconductivity and Novel Magnetism*, vol. 1, pp. 1–6, 2018.

18  C. P. Gooneratne, B. Li, and T. E. Moellendick, "Downhole applications of magnetic sensors," *Sensors*, vol. 17, no. 10, p. 2384, 2017.

19  H. Akagi, Y. Kanazawa, and A. Nabae, "Instantaneous reactive power compensators comprising switching devices without energy storage components," *IEEE Transactions on Industry Applications*, no. 3, pp. 625–630, 1984.

20  A. Cataliotti, D. Di Cara, A. Emanuel, and S. Nuccio, "Characterization of current transformers in the presence of harmonic distortion," in *Proceedings of the IEEE Instrumentation and Measurement Technology Conference*, 2008, pp. 2074–2078.

21  Q. Huang, C. Zhang, Q. Liu, Y. Ning, and Y. Cao, "New type of fiber optic sensor network for smart grid interface of transmission system," in *IEEE Power and Energy Society General Meeting*, 2010, pp. 1–5.

22  J. Guttman, J. Niple, R. Kavet, and G. Johnson, "Measurement instrumentation for transient magnetic fields and currents," in *IEEE International Symposium on Electromagnetic Compatibility*, vol. 1, 2001, pp. 419–424.

23  A. J. Pawar and A. Patil, "Design and development of 48V PMBLDC motor for radiator fan application by using ANSYS Maxwell software," in *3rd International Conference on Sensing, Signal Processing and Security (ICSSS)*, 2017, pp. 247–252.

24  A. Rezig, M. R. Mekideche, and A. Djerdir, "Effect of rotor eccentricity faults on noise generation in permanent magnet synchronous motors," *Progress in Electromagnetics Research*, vol. 15, pp. 117–132, 2010.

25  Y. Zhang and R. A. Dougal, "State of the art of fault current limiters and their applications in smart grid," in *IEEE Power and Energy Society General Meeting*, 2012, pp. 1–6.

26  W. Yuan, W. Xian, M. Ainslie, Z. Hong, Y. Yan, R. Pei, Y. Jiang, and T. Coombs, "Design and test of a superconducting magnetic energy storage (SMES) coil," *IEEE Transactions on Applied Superconductivity*, vol. 20, no. 3, pp. 1379–1382, 2010.

27  K. Suzuki, N. Kataoka, A. Inoue, A. Makino, and T. Masumoto, "High saturation magnetization and soft magnetic properties of bcc Fe–Zr–B alloys with ultrafine grain structure," *Materials Transactions, JIM*, vol. 31, no. 8, pp. 743–746, 1990.

# 2

# State of the Art Magnetoresistance Based Magnetic Field Measurement Technologies

## 2.1 Introduction

Magnetic sensors have been utilized for different practical purposes over the last few centuries. Their diverse application scope ranges from suspended magnets used in navigation to smaller thin-film technologies. The techniques currently used for sensing magnetic fields have evolved from and are driven by the requirements to improve sensitivity, reduce size, reduce power consumption, and ensure compatibility with electronic equipment or devices. Magnetic field measurements are generally conducted to find other parameters of interest. For example, vehicle detection or wheel speed are measured indirectly from the analysis of the behavior of magnetic fields. Figure 2.1 shows the difference between conventional sensors and magnetic sensors [1]. In the first case, the output obtained is the desired parameter. On the other hand, magnetic field sensors measure the magnetic field to indirectly detect direction, rotation, angle, or electric currents. Once the sensor detects the magnetic field, the output signal then requires some signal processing steps to obtain the desired parameter value. Although this complicates the process, it provides accurate and reliable data. Figure 2.2 shows various sensing technologies and provides the magnetic field sensing ranges [2].

## 2.2 Progress in MR Sensing Technologies

Magnetoresistance (MR) is the change in resistance due to an externally applied magnetic field. The sensors developed on the principle of MR can be manufactured in small sizes using photolithography fabrication techniques. Such sensors have mid-range magnetic field detection and are made from thin-film material. Research and development in the field of MR sensors have played a crucial role in the last few decades. There are several types of effects related to MR. Some of them occur in bulk semiconductors and non-magnetic metals, such as geometrical MR. Other effects occur in magnetic metals, such as anisotropic magnetoresistance (AMR). In multi-layer systems, e.g. magnetic tunnel junctions (MTJs), giant magnetoresistance (GMR), tunnel magnetoresistance (TMR), colossal magnetoresistance (CMR), and extraordinary magnetoresistance (EMR) are observed.

*Magnetic Field Measurement with Applications to Modern Power Grids*, First Edition.
Qi Huang, Arsalan Habib Khawaja, Yafeng Chen and Jian Li.
© 2020 John Wiley & Sons Ltd. Published 2020 by John Wiley & Sons Ltd.

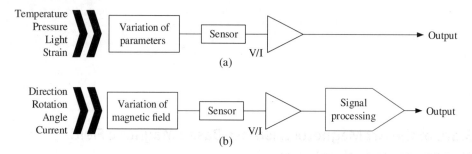

**Figure 2.1** Sensing methods: (a) conventional sensing and (b) magnetic sensing.

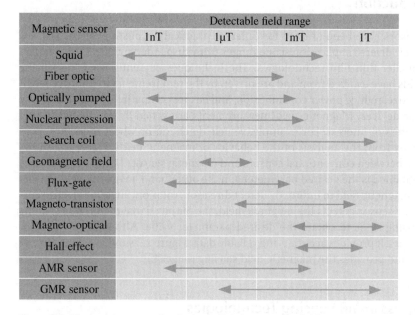

| Magnetic sensor | Detectable field range | | | |
|---|---|---|---|---|
| | 1nT | 1μT | 1mT | 1T |
| Squid | | | | |
| Fiber optic | | | | |
| Optically pumped | | | | |
| Nuclear precession | | | | |
| Search coil | | | | |
| Geomagnetic field | | | | |
| Flux-gate | | | | |
| Magneto-transistor | | | | |
| Magneto-optical | | | | |
| Hall effect | | | | |
| AMR sensor | | | | |
| GMR sensor | | | | |

**Figure 2.2** Magnetic sensor technologies and their respective field ranges.

## 2.2.1 AMR Sensors

### 2.2.1.1 AMR Effect

The MR effect that occurs in ferromagnetic materials such as transition metals is known as the AMR effect. In this effect the resistivity of a material is dependent on the orientation of magnetization with respect to the electric current flow. The common material used in the process is permalloy, which consists of nickel (80%) and iron (20%). The AMR effect in ferromagnetic metals was first discovered by William Thomson in 1857. In his paper entitled "Effects of magnetization on the electric conductivity of nickel and of iron", he discussed the anisotropic behavior that states that iron acquires an increase in resistance to the conduction of electricity along the lines of magnetization and a diminution of resistance to the conduction of electricity across the lines of magnetization.

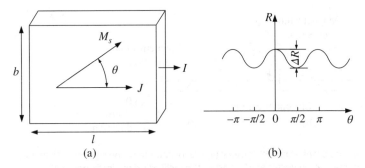

**Figure 2.3** An AMR element and its behavior. (a) Rectangular thin film representing current density and spontaneous magnetization. (b) Change of resistance due to the applied magnetic field.

#### 2.2.1.2 Resistance and Magnetization

The AMR effect arises from the action of magnetization and spin-orbit interaction occurring at the same time. The effect is based on the anisotropic scattering of conduction electrons, e.g. the three-dimensional orbit of transition metals (Fe, Co, and Ni) with uncompensated spins. The difference in energies of magnetic spin moment between two states is given by quantum-mechanical exchange energy. An important aspect of using MR sensors is the change in the resistance and its dependence on the externally applied magnetic field. The anisotropic thin film is uniaxial and its magnetization vector is directed along the anisotropy axis (also known as the easy axis of magnetization) when no external magnetic field is applied. Figure 2.3 [3] represents a two-dimensional rectangular thin-film ferromagnet (AMR element). In the presence of an external magnetic field, the thin film is magnetized and the magnetization vector is directed by the magnetic field component perpendicular to the anisotropy axis. An applied field $H_x$ rotates $M_s$ (spontaneous magnetization) along the axis direction shown in Figure 2.3a. The resistivity depends on the angle $\theta$ between $M_s$ and $J$. Therefore,

$$\rho(\theta) = \rho_o + (\rho_p - \rho_o)\cos^2\theta \tag{2.1}$$

$$\rho(\theta) = \rho_o + \Delta\rho\cos^2\theta \tag{2.2}$$

where $\rho = \rho_p$ for $M_s$ parallel to $J$ and $\rho = \rho_o$ for $M_s$ orthogonal to $J$. From Figure 2.3, we can write (2.2) in terms of resistance as follows:

$$R = \rho(\theta)\frac{l}{bd} = R + \Delta R\cos^2\theta \tag{2.3}$$

Figure 2.4 shows three different methods in which thin-film MR sensors are aligned to obtain linear characteristics [4]. The first one is the inclination of the MR path at 45° with respect to the anisotropy axis. In the second method, the path is along the anisotropy axis and the current flows inclined at an angle of 45° due to the appropriate shape of the electrodes. This is called a barber-pole structure. In the third method, instead of current direction the magnetization direction is initially inclined by 45° with respect to the anisotropy axis.

#### 2.2.1.3 Principle of AMR Sensors

The AMR sensor is composed of a silicon wafer, which is a thin film of permalloy fixed on the board. Permalloy consists of metals such as nickel (Ni) and iron (Fe) [5].

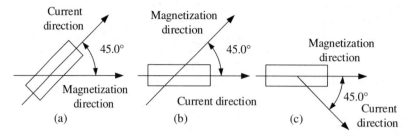

**Figure 2.4** Linear characteristics of thin-film MR sensors: (a) the inclination of the current path at 45°, (b) the inclination of the magnetization vector at 45°, and (c) the inclination of electrodes at 45°.

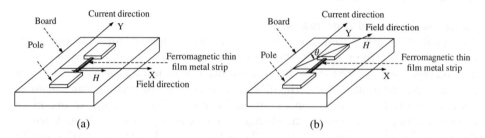

**Figure 2.5** Magnetic field direction and ferromagnetic thin-film metal element. (a) Field strength along the *x* axis. (b) Field direction makes an angle $\theta$ with current direction.

The resistance of the formed ferromagnetic thin-film metal changes according to the field strength of the externally applied magnetic field. Based on this phenomenon, the strength of the magnetic field can be measured [6]. The sensor designed to perform this effect is known as the AMR sensor. Figure 2.5 [7] shows the magnetic field applied along the *x* direction while the current in the ferromagnetic thin-film element flows along the perpendicular *y* direction. In such a case, the resistance of the AMR sensor is decreased according to the strength and direction of magnetic field.

In AMR sensors, the rate of change of resistance is within 3% corresponding to the applied magnetic field [8]. In Figure 2.6 [7] the relationship between the resistance and the applied magnetic field is shown. The outer region is defined to be the saturated sensitivity region and is given by (2.2). In the inner region, the resistance change rate remains constant and is within 3% when the field strength exceeds a certain threshold value. In another case, the magnetic field is applied at an angle $\theta$ to the current direction. The resistance change will move towards saturation in this condition.

$$\Delta R \propto H^2 \tag{2.4}$$

The resistance change ($\Delta R$) is a maximum when the current direction is vertical to the magnetic field direction ($\theta = 90°, 270°$) and is a minimum when the current direction is horizontal to the field direction ($\theta = 0°, 180°$). The resistance varies with respect to the angle between the current direction and the field direction, as shown previously (see Figure 2.3). In the saturated sensitivity region, $\Delta R$ remains constant and does not vary according to the magnetic field strength.

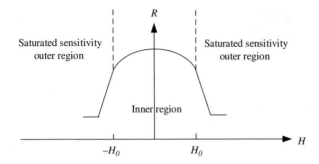

**Figure 2.6** The relationship of resistance and applied magnetic field.

#### 2.2.1.4 Applications of AMR Sensors

AMR sensors are used in a wide range of consumer applications, but mainly for open–close detection operation, position detection, and rotation detection. Consumer applications for open–close detection include smartphones, note/tablet PCs, hand-held game consoles, refrigerators, dishwashers, water heater panels etc. Rotation detection finds application in smart gas meters, smart water meters, smartphones, etc. Similarly, application examples for position detection include cylinder switches, smartphones, level sensors, electric toothbrushes, electric shavers, sliding doors, security buzzers etc.

### 2.2.2 GMR Sensors

#### 2.2.2.1 GMR Effect

The GMR effect is the change in specific resistance of multi-layers, consisting of alternate ferromagnetic and non-magnetic conductive layers, in the presence of an external magnetic field. The GMR effect is noted as a significant resistance change (typically 12–20%) and it is dependent on the magnetization of adjacent ferromagnetic layers in a parallel or antiparallel alignment (see Figure 2.7 [9]). This effect was discovered in the late 1980s by two scientists, Albert Fert and Peter Gruenberg. They observed very large changes in the resistance of materials composed of alternate thin layers of different metallic elements.

#### 2.2.2.2 Principle of GMR Sensors

Electric current ($J$) flowing in a magnetic multi-layer consisting of a sequence of thin layers separated by equally thin non-magnetic metallic layers is strongly influenced by the relative orientation of the magnetizations of the adjacent magnetic layers. Parallel alignment has lower resistance than antiparallel alignment, as shown in Figure 2.8 [10]. The GMR effect is given in [11] as:

$$\frac{\Delta R}{R_p} = \frac{R_A - R_p}{R_p} \tag{2.5}$$

where $R_p$ and $R_A$ are electrical resistances when the magnetization directions of two conductive layers are parallel and antiparallel to one another. Layer (L2) is a conductive, non-magnetic interlayer. The magnetization vectors ($M_1$ and $M_2$) in two alloy (L1) layers are aligned oppositely due to antiferromagnetic coupling. The resistance of the non-magnetic conducting layer is high in this case due to the very thin layer of material,

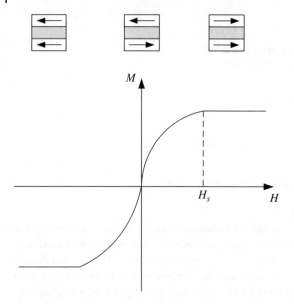

**Figure 2.7** Magnetization configuration of a multi-layer at zero field and at fields greater than the saturation field ($H_s$).

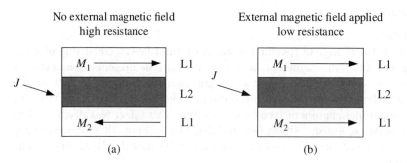

**Figure 2.8** GMR double layer: (a) layer (L1) magnetization antiparallel when no external magnetic field is applied and (b) magnetization vectors ($M_1, M_2$) parallel in an externally applied field.

usually copper, that cause electrons to scatter and hence increases the resistance. The resistance change depends on the relative orientation of electron spin surrounding the conducting layer. When an external magnetic field is applied to the device, it overcomes antiferromagnetic coupling and aligns the magnetization vectors in a parallel direction in the two alloy (L1) layers [12].

### 2.2.2.3 Sensing Module Design

GMR sensors are designed based on the requirements of the field of application. The elements involved in the sensing unit may have different effects on its action. The design of the sensing module consists of three subsystems: the GMR sensor, the signal processing unit, and the power supply unit.

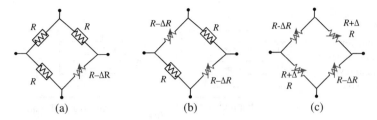

**Figure 2.9** Bridge configuration of a GMR sensor: (a) unique element, (b) half-bridge, and (c) full bridge.

#### 2.2.2.4 GMR Sensor Configuration

The GMR sensors for sensing magnetic field are best designed in full Wheatstone bridge configuration. They are also fabricated as simple GMR resistors or GMR half-bridges. Figure 2.9 [12] shows various bridge configurations used in GMR sensors. A half-bridge configuration has two active resistors and two shielded resistors patterned photolithographically. The two resistors are plated with small magnetic shields protecting them from the applied magnetic field and these resistors act as reference resistors. The remaining GMR resistors are both exposed to the external magnetic field.

The output voltage of the above given configuration (Figure 2.9) is as follows:

$$Vout_+ - Vout_- = \frac{\Delta R}{2R - \Delta R}(V_+ - V_-) \tag{2.6}$$

$$Vout_+ - Vout_- \approx \frac{\Delta R}{2R}(V_+ - V_-) \tag{2.7}$$

where $Vout_+$ and $Vout_-$ are the voltage outputs, $V_+$ and $V_-$ are the inputs connected to the power supply unit, and $R$ is the characteristic resistance of the four elements in the absence of a magnetic field. $\Delta R$ is the change in resistance of the two active elements in the presence of a magnetic field and is linearly proportional to the measured field. The MR ratio of GMR sensors is near to 7%. As $\Delta R$ is small compared to $R$, the output of the GMR sensor is also linearly proportional to the resistance change.

#### 2.2.2.5 Power Supply

The power supply unit provides a stable source of power for the GMR sensor and the instrumentation amplifier used in the process. The battery source usually fluctuates and needs to be regulated before being applied to the sensing module. The voltage regulator is used to regulate the voltages for the whole system.

#### 2.2.2.6 Applications of GMR Sensors

GMR sensors have a wide range of applications and are applied in several fields, such as physics, engineering, biology, space etc. GMR magnetic field sensors are highly sensitive and therefore can provide positional information for actuating parts in proximity detectors and linear transducers. The application examples related to measuring displacements include hydraulic/pneumatic pressure of cylinder stroke position, suspension position, fluid level, machine tool slide position, aircraft control-surface position, and vehicle detection. GMR sensors are also used as current sensors. They are deployed in industrial instrumentation, AC or DC current sensing, industrial process control, current probes etc. Furthermore, they are also used for detecting different types of magnetic

Electric current ($J$)

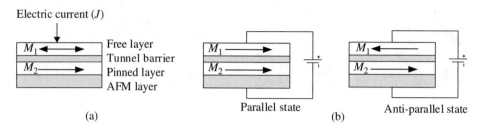

Figure 2.10 Schematic diagram of TMR: (a) MTJ element structure and (b) magnetization states based on the tunneling effect.

media, for example in magnetic ink detection, media magnetic signature detection, magnetic anomaly detection etc., where which magnetic field perturbations are detected and the desired parameters are calculated [13].

### 2.2.3 TMR Sensors

#### 2.2.3.1 TMR Effect

An MR effect that occurs in an MTJ is known as the TMR effect. This effect was first observed by Julliere in 1975. It exhibits greater resistivity change compared to AMR and GMR. Figure 2.10 shows a schematic diagram of the MTJ element. It consists of three main layers: a bottom pinned ferromagnetic layer, a middle tunnel barrier layer, and a top free ferromagnetic layer. The bottom layer is a composite of an antiferromagnetic (AFM) layer and a ferromagnetic pinned layer. Exchange coupling between the two layers tightly holds the magnetic behavior of these two layers. The middle layer is made up of MgO or Al$_2$O$_3$. The top layer is free as it is not pinned to any layer. The arrows represent the magnetization vectors ($M_1$ and $M_2$) of the layers [14].

The resistance change for the TMR effect is given by

$$\frac{\Delta R}{R_p} = \frac{R_A - R_p}{R_p} \tag{2.8}$$

where $R_p$ and $R_A$ are electrical resistances when the magnetization directions of the free and pinned layers are parallel and anti-parallel to one another.

#### 2.2.3.2 Principle of TMR Sensors

Figure 2.11 [15] represents a curve between MR and magnetic field ($R$ vs $H$). It is assumed that the response curve is linear with no magnetic hysteresis. The curve is marked with high resistance ($R_H$) and low resistance ($R_L$) states. Also, the response curve is unsymmetrical about the $H = 0$ axis. This is due to the bias or offset field ($H_o$). The offset field values range from 1 to 40 Oe. The white arrows indicate the magnetization direction in the free layer while the grey arrows show the pinned layer magnetization direction. The resistance change (MR effect) is shown as a function of the angle between $M_1$ of the free layer and $M_2$ of the pinned layer. When $M_1$ and $M_2$ are antiparallel, the resistance has maximum value ($R_H$). On the other hand, if $M_1$ and $M_2$ are parallel to one another, the resistance has a low value ($R_L$). The value halfway between $R_H$ and $R_L$ is considered to be the operating point of the sensor as it has linear behavior at that point [14].

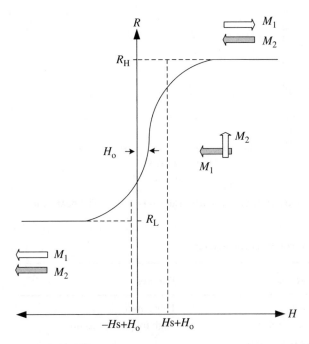

**Figure 2.11** MTJ response curve.

### 2.2.3.3 Sensing Module Design

The TMR sensing module involves the same steps as discussed previously for the GMR sensing module. It mainly consists of a TMR sensor, a signal processing unit, and a power supply unit.

### 2.2.3.4 TMR Sensor Configuration

A resistance bridge is usually used as a transducer to give a voltage signal at the output that is easily amplified. This configuration helps in reducing noise in the device, temperature drift, and other related deficiencies. Figure 2.12 [16] shows a unique push–pull Wheatstone bridge that is composed of four unshielded TMR sensor elements. The bridge circuit is connected electrically through the pads ($V_{CC}$, GND, $V_+$ and $V_-$) and a steady voltage is applied between $V_{CC}$ and the electrical ground (GND). When an external magnetic field ($H$) is applied along the positive sense axis, the resistance increases for one set of resistors and decreases for the other set of resistors and vice versa for the negative sense axis. This configuration increases the sensitivity of the sensor and is called the push–pull bridge circuit configuration. The MTJ bridge output is an analog voltage signal, as shown in Figure 2.12. Furthermore, a dedicated sensing module is designed for practical purposes that includes a signal processing unit consisting of analog signal processing and digital signal output, and a power supply unit as discussed previously for GMR sensors.

### 2.2.3.5 Applications of TMR Sensors

TMR is relatively new technology and magnetic sensors based on the TMR effect are utilized for various purposes [17]. Table 2.1 gives detailed descriptions of different types of TMR sensors and their specific application examples.

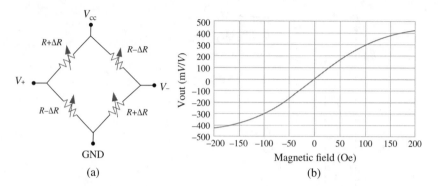

**Figure 2.12** TMR sensor configuration and its response characteristics: (a) a push–pull Wheatstone bridge and (b) the transfer curve.

**Table 2.1** TMR sensor application examples with their features.

| Product | Applications | Features |
| --- | --- | --- |
| TMR switch sensor | Flow meters | Low power |
| | Motor control | High frequency response |
| | Proximity switches | |
| TMR linear sensor | Current sensors | Low power |
| | Magnetic field sensors | High sensitivity |
| | Position sensors | Large dynamic range |
| TMR angle sensor | Flow meters | Robust output with high amplitude |
| | Rotary encoders | Allow large air gap |
| | Potentiometers | |
| | EPS motors | |
| | Steering angles | |
| | Pedal opening | |
| | Throttle valve opening | |
| TMR gear tooth sensor | Gear tooth detection | Small pitch detection |
| | Linear and rotary encoders | High sensitivity |
| | Speed sensors | Allow large air gap |

## 2.2.4 CMR Sensors

### 2.2.4.1 CMR Effect

The increased interest in the last few decades in ferromagnetic materials, particularly thin-film multi-layer MR materials (e.g. GMR and TMR) has led to the discovery of CMR. CMR is the MR related to a ferromagnetic-to-paramagnetic phase transition [18]. It has been observed that materials, particularly three-dimensional transition metal oxides, exhibit large room temperature MR associated with a paramagnetic-to-ferromagnetic phase transition. The interest has grown to produce metal oxide devices whose performance is better than GMR devices. It is now known that the large MR in these oxides is the result of a specific type of metal–insulator

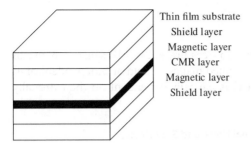

Thin film substrate
Shield layer
Magnetic layer
CMR layer
Magnetic layer
Shield layer

**Figure 2.13** Schematic representation of CMR sensors.

transition. The major compounds used in CMR studies are manganese-based perovskite oxides. CMR studies have suggested an important aspect that complements its implementation for practical applications. Its MR ratio is high, e.g. 50% at low magnetic fields of 10 Oe and room temperature. Recent studies on manganite perovskite show nearly 5% MR ratio at 1800 Oe and 247 K. Other features, such as power consumption, noise, and compatibility with established fabrication methods, need to be considered in future work.

#### 2.2.4.2 CMR Sensor Structure

The CMR sensor consists of two magnetic layers and a thin layer of CMR sandwiched between two magnetic layers. Figure 2.13 represent the structure of the CMR sensor. The two magnetic layers control the magnetic field and thus the resistance of the CMR layer [19]. When the magnetization vectors of the first and second magnetic layers are aligned parallel to one another, the fields from the two magnetic layers add up to give rise to a large magnetic field and therefore create low resistance conditions. In the case where the two magnetization vectors from the magnetic layers are anti-parallel, a small field is produced and high resistance conditions prevails.

A CMR sensor responds to the low fields produced by the recording media and therefore finds application in magnetic disk read/write heads.

## 2.3 Limitations of MR Effect Based Sensors

The discovery of MR effect based sensors has led to the replacement of more expensive wire-wound sensors in a number of different applications. MR effect ratios are the metric used for describing the MR effect, but the more useful figure of merit is related to the detectivity of the signal. It includes both the sensitivity of the MR sensor and the background noise level present in the sensed signal. Although the sensitivity of MR sensors is independent of frequency over their range, these sensors are affected by noise sources. At low frequencies, the detectivity of MR sensors is reduced considerably due to noises with 1/f characteristics. At high frequencies, the detectivity is limited by Johnson noise inherent in the resistances of the device [20]. The performance of MR sensors is therefore affected by the presence of different noises inherent to the sensor structure itself as well as the external noise present in the environment.

### 2.3.1 Noise Performance

#### 2.3.1.1 Sources of Noise
In this section we discuss the different noise sources present in MR sensors. Noise in magnetic sensors can be dependent on the applied magnetic field or independent of it. For example, low frequency noise (1/f noise) is partially dependent on the externally applied magnetic field.

#### 2.3.1.2 Frequency-independent Noise (Thermal Noise and Shot Noise)
Frequency-independent noise is also called white noise. Such noise has flat characteristics during the measurement as its correlation characteristic time is less than the minimal sampling time.

**Thermal noise** Thermal noise was first observed by Johnson and is directly related to the resistance of the sensing element (sensor). The power spectral density (PSD) of such noise is given by [21]:

$$S_V(\omega) = \frac{\gamma_H V^2}{N_c f^\beta} \tag{2.9}$$

where $N_c$ is the number of carriers, $\gamma_H$ and $\beta$ are constants, $\omega$ is the angular velocity, $f$ is the frequency of the entity under observation, and $V$ is the applied voltage.

Thermal noise is difficult to eliminate and is reduced by changing the resistance or the temperature only as it is independent of the voltage supplied to the sensor. This noise impacts the working bandwidth of the sensor and increases with the square root of the available bandwidth.

**Shot noise** This type of noise occurs due to the discrete nature of electrons. It is only present in sensors where there is a barrier in the current path. The quantum nature of electrons is only shown in TMR sensors in which junctions exhibit shot noise and dominate the thermal noise regime [13]. It is observed that shot noise is small compared to thermal noise at room temperature. Due to large low frequency noise, shot noise measurements are carried out at higher frequencies. The PSD of shot noise at $t = 0$ is given in [22] as:

$$S_I(\omega) = 2eI \tag{2.10}$$

where

$$< I > = \frac{e}{2\pi h} \int dET f \tag{2.11}$$

and $dE$ is a narrow energy interval. $T$ is the probability with which one particle that is incident on a barrier is transmitted in a fictitious experiment. $f$ is the probability with which an incident beam is occupied, $h$ is the height of the conductor, and $e$ is the unit charge.

#### 2.3.1.3 Low Frequency Noise
**Low-frequency (1/f) noise** 1/f noise, or frequency decreasing noise, occurs in almost every fluctuating system. This type of l/f noise is dominant in GMR and TMR sensors and is one of the limitations on the performance of MR sensors. It is only dependent on resistance fluctuations. Its behavior is given by the PSD by varying $V^2$ or $I^2$ and helps to differentiate between white noise and low 1/f noise. It is also affected by the size and

shape of the sensor. The PSD of the 1/f noise decreases as the volume of the sensor increases [13]. This noise can occur in both magnetic and non-magnetic field conditions. The general PSD proposed by Hooge in [23] is similar to (2.9).

**Random telegraph noise (RTN)** RTN is the noise that arises from the fluctuations of a specific source between two different levels. Such noise is more pronounced in small MR sensors and is sometimes observed in large GMR at reasonable current levels. RTN is a fluctuation between two levels with comparable energies and is dependent on the temperature, field, and applied bias current. RTN is difficult to eliminate and a sensor with such noise is usually difficult to operate, although it is possible to theoretically suppress this noise by data treatment.

### 2.3.1.4 High-frequency Noise
At high frequencies noise is usually dominated by thermal or shot noise. In the gigaHertz (GHz) range, thin magnetic films present ferromagnetic resonances with frequencies dependent on the material and the shape of the sensors. In small elements, quantization of the spin waves induces a greater number of resonances. The noise detected comes from either the GMR effect at resonance frequency or the thermal or shot noises.

### 2.3.1.5 External Noise
There are three main types of disturbances that contribute to producing magnetic noise. First, the disturbances from power supplies include a number of discrete frequencies (power supply line frequencies and higher frequencies up to MHz) and their harmonics. All these lines produce a real AC magnetic field and correspond to the bias current in MR sensors. The second source is a 1/f magnetic noise that exists almost everywhere. In a laboratory setup, this noise has an intensity of about 100 nT at 1 Hz and decreases slightly faster than 1/f [13]. The third noise is created by vibration of the MR sensors in an existing DC field. Such a noise is only detected with very sensitive sensors. This noise is easily recognized because it appears as bursts at fixed very low frequencies.

### 2.3.1.6 Noise Measurement
Noise is usually difficult to measure quantitatively, but several techniques are employed to estimate noise in measuring devices. One such method uses standard design arrangements with regard to common errors. The equipment design normally used for interfacing magnetic sensors in a particular application consists of the following parts and is shown in Figure 2.14:

- sensor
- biasing source (voltage or current)
- front end electronics for signal conditioning
- amplification, filtering, and shaping
- acquisition system.

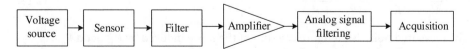

**Figure 2.14** Schematics for electronic devices for noise measurement techniques.

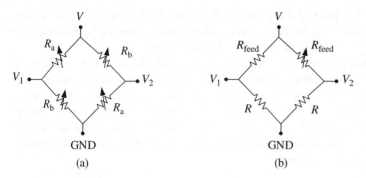

**Figure 2.15** Wheatstone bridge configurations: (a) full bridge and (b) half-bridge.

The main voltage or current applied in the process is a few orders of magnitude higher than the noise. Because of this, differential measurements are preferred. Figure 2.15 shows the full-bridge configuration that is commonly used in MR sensor applications and is very efficient for measuring noise. In order to perform such measurements, the voltage applied must be stabilized. In general, a battery with proper filtering is used for practical purposes.

Let us suppose that the MRs $R_a$ and $R_b$ shown in Figure 2.15a are identical, each having its own noise. The resistance measured between $V_1$ and $V_2$ is $R$ ($R = R_a = R_b$). The corresponding noise is the average noise of $R$ and is due to either resistance fluctuations or thermal noise. If there is some noise in the voltage supply, it is suppressed by differential measurement. It is observed that when an external field is applied, the bridge is no longer in a balanced condition and the noise proportional to this unbalanced condition may appear at the output. Sometimes it is difficult to have four matched MR elements. In this case, the half-bridge configuration shown in Figure 2.15b is used. The signal associated with the element is divided by four and the noise of this single element is also reduced by a quarter. The resistances $R_{feed}$ are usually chosen so that $R_{feed} \gg R_a$. Also, $R_a$ and $R_b$ are equal. The voltages $V_1$ and $V_2$ are given by:

$$V_1 = V \frac{R_a}{(R_{feed} + R_a)} \tag{2.12}$$

$$V_2 = V \frac{R_b}{(R_{feed} + R_b)} \tag{2.13}$$

The noise induced by the resistances can be calculated from (2.12) and (2.13). If the voltage is fixed and well filtered, the noise fluctuations on the output voltages are given in [13] as:

$$\frac{\delta V_1}{V_1} = \frac{\delta R_a}{R_a} - \frac{(\delta R_{feed} + \delta R_a)}{(R_{feed} + R_a)} \tag{2.14}$$

The measured noise is therefore the sum of the noise induced by $R_{feed}$ and $R_a$.

As shown in Figure 2.14, the MR sensor output is fed to the preamplifier block. This is usually a very low noise preamplifier. Each preamplifier has an input voltage and current noises. In the case where the resistance is connected to the input, the noise is given by the PSD as:

$$(S_V)^{\frac{1}{2}} = \sqrt{4kRT + e_n^2 + R^2 i_n^2} \tag{2.15}$$

where $e_n$ is the voltage noise and $i_n$ is the current noise of the preamplifier. $T$ is the temperature, $R$ is the resistance as shown in Figure 2.15, and $k$ is the Boltzmann constant. After this, a second amplifier is used with a total gain in the range $10^4$ to $10^5$. A low pass filter is then employed to avoid spectral aliasing for proper noise measurement. The acquisition is performed with a 16 bit acquisition card. Typical values for noise in MR sensors are below 100 kHz for low frequency and above 1 GHz for resonant noise measurements [13].

### 2.3.1.7 Sensitivity, Signal-to-noise Ratio, and Detectivity

Noise PSD is given in $V^2/Hz$ and it is easy to compare a signal that is measured in volts (V) to the square root of the PSD ($V/Hz^{1/2}$). To evaluate the signal-to-noise ratio (SNR), a reference signal with a known frequency is often used. If a signal $V_0 \cos(\omega t)$ is seen in the acquisition system, its PSD is the power associated with this signal taken on a bandwidth of 1 Hz over 1 second and corresponds to $(V_0)^2/2$. This allows direct calibration of a sensor. The sensitivity for the MR signal is usually given in V/V/T. If the sensor is linear and centered, the output voltage and the sensitivity of the sensor are given as:

$$V_{out} = (R_0 + \frac{\delta R}{\delta H} H + ...)I \tag{2.16}$$

The sensitivity is:

$$S = \frac{\frac{\delta R}{\delta H}}{R_0} \tag{2.17}$$

In order to compare different sensors, another term, detectivity, is used. Detectivity is equal to the PSD divided by the sensitivity. For example, if a sensor exhibits a thermal noise of $1nV/Hz^{1/2}$ and a sensitivity of 20 V/V/T, the corresponding detectivity will be 50 pT for 1 V of bias voltage [13].

### 2.3.2 Noise Shielding and Preventive Measures

Electrostatic shielding, magnetic shielding (magnetostatic shielding), and high-frequency electromagnetic shielding are the three main types of electromagnetic shielding [24]. Electrostatic shielding happens when a conductive enclosure, known as a Faraday cage, is employed to block the external electrostatic fields. The enclosure can reduce the influence of the external electrostatic fields on the internal field. For a magnetic field above 100 kHz, reduction of the external high-frequency magnetic field is realized by utilizing the eddy current generated by the low-resistance conductive materials to emanate the reverse magnetic flux. For a magnetic field below 100 kHz, high-permeability materials are used to draw the magnetic field into the materials themselves.The term shielding effectiveness (SE) is used to describe the ability of

**Table 2.2** The relative conductivity and relative permeability of materials.

| Materials | Low-carbon steel | Stainless steel | Aluminum | Copper | Iron |
|---|---|---|---|---|---|
| Relative conductivity | $10^6$ | $10^6$ | $3.35 \times 10^7$ | $5.8 \times 10^7$ | $9.93 \times 10^6$ |
| Relative permeability | 2000 | 1.2 | 1.000021 | 0.99999 | 200 |

*Source:* Reprinted with permission from F. Li, The design and application of the hardware of the transient magnetic field measurement device, University of Electronic Science and Technology of China, July 2012.

the material to shield magnetic fields. For static magnetic fields and low-frequency magnetic fields, the shielding effectiveness is determined by

$$SE_H(dB) = 20lg0.22\mu_r[1 - (1 - \frac{t}{r})^3] \tag{2.18}$$

where $\mu_r$ is the relative permeability of the material, $t$ is the thickness of the shielding layer, and $r$ is the equivalent radius of the shielding.

In this book, research related to magnetic fields below 100 kHz is considered. Because of the high sensitivity of MR effect based sensors, one single sensor often fails to stave off the crosstalk error contributed by the interfering magnetic field. This phenomena is especially obvious in power systems where the magnetic environment is complex. Methods have been proposed to handle this shortcoming. One effective approach is to utilize high-permeability materials to eliminate the magnetic noise when information about the noise is uncertain. These materials can be used to make closed containers to reduce the field inside. Commonly used materials include low-carbon steel, stainless steel, copper, aluminum, and iron. The relative conductivities and relative permeabilities of these materials are shown in Table 2.2 [25]. Laboratory experiments have been conducted to test the shielding effectiveness of different materials. In the experiments, boxes with different materials and thickness were made as shown in Figure 2.16 [25]. The low-carbon steel boxes had thicknesses of 1 and 2 mm, the stainless steel boxes were 1.2 mm thick, the aluminum boxes were 2.6 and 3 mm thick and the copper boxes were 1.1 and 1.5 mm thick. The iron wire net was made with wire of diameter 0.3 mm. The dimensions of the mesh were $3 \times 3$ mm. The magnetic field of 50 Hz with sinusoidal waveforms was generated by the current-carrying coil. The magnetic fields detected by the MR sensor with the shielding boxes were recorded. Figure 2.17 shows the waveform detected by the MR sensor without shielding boxes [25]. The strength of the magnetic field is 1.402 Gs. Results with shielding boxes are shown in Figure 2.18 [25]. The strengths of the magnetic fields are presented in Table 2.3 [25]. It should be noted that higher permeability materials have better shielding effectiveness and the thicker the shielding is, the better the shielding effectiveness.

Figure 2.19a shows the simulated shielding effect of a cylinder-shaped container made of mu-metal (relative permeability $\mu_r = 1 \times 10^4$). While the external magnetic field is $1 \times 10^{-5}$ T, the internal magnetic field is reduced to $1 \times 10^{-11}$ T. A multi-layer shielding with smaller thickness can be used to obtain a comparable shielding effect. Figure 2.19b shows the simulation result of a double-layer cylinder-shaped mu-metal container. With the same external magnetic field, $1 \times 10^{-5}$ T, the magnetic field is reduced to $1 \times 10^{-8}$ T by the outer layer and then to $1 \times 10^{-5}$ T by the inner layer.

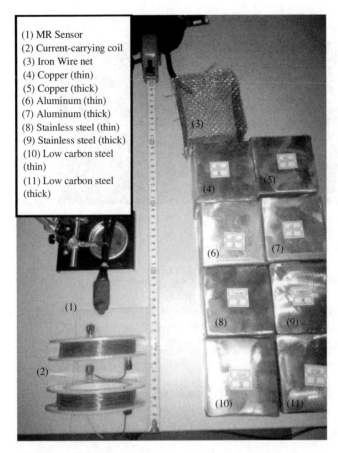

(1) MR Sensor
(2) Current-carrying coil
(3) Iron Wire net
(4) Copper (thin)
(5) Copper (thick)
(6) Aluminum (thin)
(7) Aluminum (thick)
(8) Stainless steel (thin)
(9) Stainless steel (thick)
(10) Low carbon steel (thin)
(11) Low carbon steel (thick)

**Figure 2.16** Shielding boxes made of different materials. (*Source:* Reprinted with permission from F. Li, The design and application of the hardware of the transient magnetic field measurement device, University of Electronic Science and Technology of China, July 2012.)

### 2.3.3 Cross-axis Noise

MR sensors are manufactured as resistances in Wheatstone bridge configuration to form magnetic field sensors. These sensors are designed to sense magnetic fields along the sensitivity direction. However, there is some interference from off-axis fields that are orthogonal to the sensitive axis fields. Such an effect is known as the cross-axis effect or cross-axis noise and is characterized by the voltage error due to the cross-field intensity [26].

**Cross-axis effect for AMR magnetic sensors** Figure 2.20 [27] shows the AMR thin ferromagnetic film element with magnetic field vectors and the direction of current inside the element. When a magnetic field $(H_c)$ is applied perpendicular to the easy axis $(H_a)$, the film domains rotate towards the sensitive axis $(H_s)$. A linear resistance change is observed when the current is applied at $45°$ from both the axes. Generally, the cross-axis effect is caused by the dimensional characteristics of the AMR element design. Magnetic fields in the $H_a$ direction give rise to small magnetic disturbances.

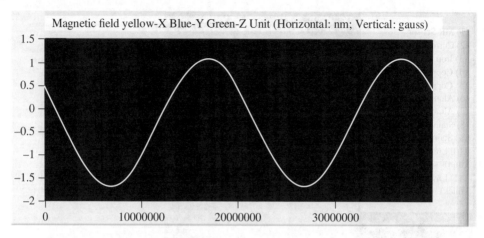

**Figure 2.17** Magnetic waveform of 50 Hz without shielding.

These magnetic distortions are increased by reducing sensor size. Typically, the cross-field error accounts to tenths of a percentage point.

**Impact on measurements** In practice, the cross-field errors are ignored in output voltage $V_{out}$ calculations and the voltage is given by the bridge sensitivity ($S_B$), voltage ($V_B$), and the field along the sensitive axis ($H_S$):

$$V_{out} = V_B \times S_B \times H_S \tag{2.19}$$

To include the effect of cross-axis fields, the orthogonal components of the field ($H_c$) are added to the equation and these impact the sensitivity and offset voltage:

$$V_{out} = V_B[S_B \times (1 + C \times (H_c)^2) \times H_S] + [D \times H_c] \tag{2.20}$$

where $S_B$ is the sensitivity along the designated path, and $C$ and $D$ are the cross-field sensitivity and offset values, respectively. The major contribution to the error voltage is due to $D$, therefore (2.20) becomes:

$$V_{out} = V_B[S_B \times H_S] + [D \times H_c] \tag{2.21}$$

With $D$ dependent on the shape and size of the sensor, there is reasonable consistency of $D$ for each design of sensor [13]. Typical values of $D$ maybe very low, such as 0.003 (0.3%) or as high as 0.1 (1%). GMR sensors have very little off-axis sensitivity because they are designed as very thin films where demagnetizing fields are very strong along the off-axis sensitivity axes. A flux concentrator is usually used to keep the off-axis sensitivity in the MR sensor below 1% [28].

## 2.4 Sensor Circuitry Design and Signal Processing

When the sensor has been manufactured, the performance of the measurement system is dependent on the circuit and the sequential signal processing since it is necessary to transmit and extract useful information from the electrical signal containing the magnetic field parameters.

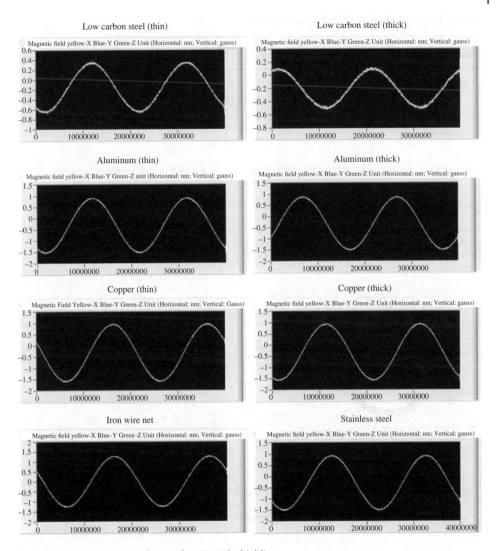

**Figure 2.18** Magnetic waveforms of 50 Hz with shielding.

### 2.4.1 AMR: Set/reset Pulse

In AMR sensors, the MR is arranged in the Wheatstone bridge configuration. At the beginning, when no magnetic field is applied to the sensing element, the magnetic domains inside the permalloy (NiFe) are oriented randomly. When an external magnetic field is applied to the AMR element, these magnetic domains are aligned either parallel or anti parallel, depending on the direction of the applied field. Care must be taken when exposing the sensor to the external field. When the AMR sensors are exposed to a field greater than the particular threshold value (more than 20 Gs), the magnetic domains cannot recover their original positions where maximum resistance change is expected to be detected. This affects the performance of the AMR sensor and in some

**Table 2.3** Magnitudes with different shielding boxes.

| Materials | Thickness (mm) | Magnitude (Gs) | SE (dB) |
|---|---|---|---|
| Low-carbon steel | 1 | 0.491 | 9.1 |
| | 2 | 0.3 | 13.4 |
| Stainless steel | 1.2 | 1.213 | 1.26 |
| Aluminum | 2.6 | 1.223 | 1.19 |
| | 3 | 1.183 | 1.48 |
| Copper | 1.1 | 1.263 | 0.9 |
| | 1.5 | 1.263 | 0.9 |
| Iron wire net | | 1.273 | 0.84 |

*Source:* Reprinted with permission from F. Li, The design and application of the hardware of the transient magnetic field measurement device, University of Electronic Science and Technology of China, July 2012.

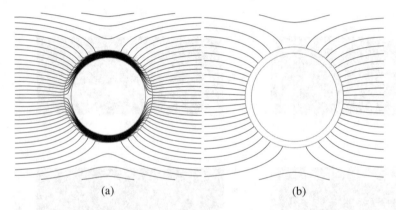

(a)        (b)

**Figure 2.19** Simulation results of mu-metal cylindrical magnetic shielding: (a) single layer of 1 cm thickness and (b) double layers of 1.5 mm thickness each. (*Source:* Reprinted with permission from Q. Huang et al., Magnetics in Smart Grid, *IEEE Transactions on Magnetics*, IEEE, July 2014.)

cases may result in measuring erroneous data. To solve this, AMR sensors are provided with a set–reset pulse circuit.

A set–reset function is used in AMR sensors to:

- balance the effect of the strong magnetic field applied to the sensor
- align the magnetic domain in optimum positions to perform sensitive functions
- flip the magnetic domain to find the bridge offset value under varying temperature situations.

**Set–reset operation** To implement the set and reset functions on AMR sensor elements, a strap or planar coil is provided on the sensor device. The strap creates an electric pulse (minimum peak current value) that can generate the magnetic field imposed on the sensor bridge elements (>40 Gs). This help to realign the magnetic domains of the sensor elements. Figure 2.21 [29] shows the schematics of a simple set–reset circuit. Set

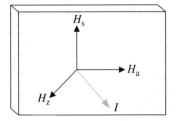

**Figure 2.20** AMR thin-film element with cross-axis effect.

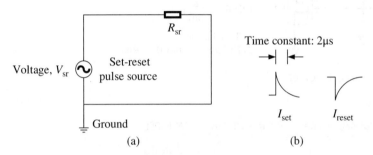

**Figure 2.21** Set–reset circuit: (a) circuit diagram and (b) exponential set–reset pulse.

pulses are pulsed currents that enter the positive pin of the set–rest strap. Similarly, reset pulses are the pulsed currents that enter the negative pin of the given strap.

The pulses can be exponential or squared waveforms. The important parameter to consider is the duration of peak electric pulses. The duration should be kept to a minimum to avoid heating the sensor bridge elements. In an exponential waveform, short duration pulses are provided through a capacitive "charge and dump" circuit. The resulting waveform is the dampened exponential pulse waveform. The value of the capacitor in this case is in the range of hundreds of nano-Farads to a few micro-Farads, depending on the strap resistance included. The decay of the exponential waveform is governed by a time constant, i.e. the capacitance multiplied by the resistance.

A detailed set–reset pulse circuit for magnetic sensor HMC2003 is shown in Figure 2.22 [30]. Initially, the status of SET is 0 and RESET is 1. At that moment, the voltages across C1 are high. The transistor PMOS will not conduct and the transistor NMOS will conduct. The voltage across C2 is 0. When RESET becomes 0, NMOS will not conduct and PMOS will still conduct. The voltage across C2 is still 0. When RESET is zero, SET becomes 1 and the triode will conduct. The waveform of the SET pulse is as shown in Figure 2.23 [30].

When SET is 1 and RESET is 0, PMOS conducts. When C2 completes charging, the voltage at the left end of C2 remains at 20 V while at the right end it is 0 V. Then RESET becomes 0 while SET changes from 1 to 0. When SET changes from 1 to 0, PMOS will not conduct but the voltage across C2 remains the same. RESET then changes from 0 to 1, SET becomes 0, and NMOS conducts. The RESET pulse waveform is shown in Figure 2.24.

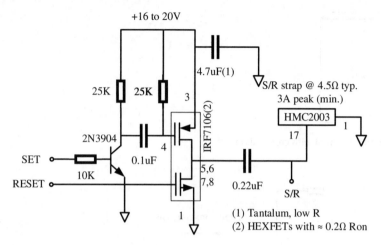

**Figure 2.22** Set–reset circuit of HMC2003.

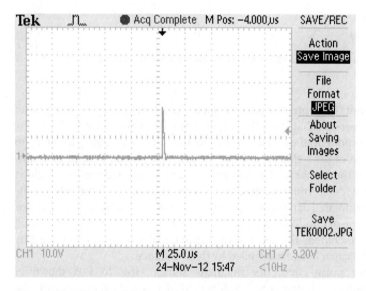

**Figure 2.23** SET pulse of the set–reset circuit of HMC2003.

### 2.4.2 GMR: Temperature Compensation and Unipolar Output

**Signal processing** The signal processing unit is used to process the voltage signal from the GMR sensor and calculate the desired parameter. The steps involved in signal processing are shown in Figure 2.25 [31]. A low-pass filter is used to remove interfering signals, particularly high frequency noise from the voltage signal coming from the GMR sensor. The output of the GMR sensor is a common-mode signal and is about half of the power supply. A differential amplifier, such as an instrumentation amplifier, is used to remove useless common-mode noise and the signal is amplified to a useful differential

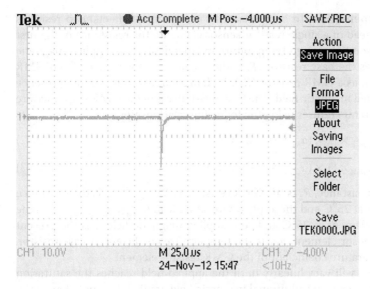

**Figure 2.24** RESET pulse of the set–reset circuit of HMC2003.

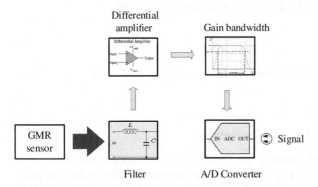

**Figure 2.25** Schematics of a signal processing system.

signal. The bandwidth gain is used to improve the frequency response of the GMR sensor output. Finally, the analog-to-digital (A/D) converter is used to convert the analog signal from the GMR sensor to a digital signal.

**Temperature characteristics** Temperature deviations often cause changes to the behavior of GMR sensors, for example:

- Changes to the base resistance of the GMR sensor element, i.e. when no external magnetic field is acting on the sensor element, the difference in temperature causes the temperature coefficient of the resistance to change.
- When an external magnetic field is applied, the percentage change of MR (% GMR) of the sensor element will change. In general, the % GMR decreases with increase in temperature.
- The maximum output of the sensing element is also affected by a change in temperature. The saturation field usually decreases with an increase in temperature.

**Temperature compensation** To compensate for temperature effects, two resistors are used in series instead of a single resistor sensor element. The resistance of the first resistor remains constant at a given temperature independent of the applied magnetic field. This is known as the base resistor ($R_1$). The second resistor is made from GMR material and is known as a GMR resistor ($R_2$). It changes in resistance in the presence of the externally applied magnetic field. The base resistance and the GMR resistance are computed at various temperatures using the following relationships:

$$R_1 = R_{1,baseresistance} \times [1 \times (\alpha_{R1} \times (t - 25°C))] \tag{2.22}$$

$$R_2 = R_{2,baseresistance} \times [1 \times (\alpha_{GMR} \times (t - 25°C))] \tag{2.23}$$

where $R_1$ is the base resistance, $R_2$ is the GMR resistance, $R_{1,baseresistance}$ is the resistance of the sensor element at 25°C and the applied saturating field ($H_{sat}$), $R_{2,baseresistance}$ is the resistance of sensor element at 25°C and zero applied magnetic field, $t$ is the present temperature of the sensor element, $a_{R1}$ is the temperature coefficient of base resistor, and $a_{GMR}$ is the temperature coefficient of the GMR resistor element.

The GMR resistance will vary linearly until the applied field reaches the saturation field ($H_{sat}$). At this $H_{sat}$ field, any additional magnetic field applied does not change the resistance of the device. The equation of the GMR resistance, taking into account both the % GMR and the changes in saturation field with temperature, is:

$$R_2 = R_{02} \times [1 - \frac{H_{app}}{H_{satT}}] \tag{2.24}$$

where $H_{app}$ is the applied magnetic field and $H_{satT} = H_{sat}$ at 25°C $\times [1 - (\alpha_{Hsat} \times (t - 25°C))]$.

### 2.4.3 TMR: Higher Noise Level at Low Frequencies

Compared with AMR and GMR sensors, the TMR sensor circuit does not suffer from the problem caused by the set–reset pulse or temperature compensation and unipolar output. Therefore this circuit is simple, as shown in Figure 2.26, where U1 is the TMR sensor and U2 is a differential amplifier. However, there are still some problems related to TMR sensors to be considered. The 1/f noise in the MTJ is the largest of all the MR-based devices and is divided into two types: external and magnetic.

**External 1/f noise** The electronic 1/f noise is created when the junction of the TMR is in a saturated state. It is represented by the modified Hooge formula as:

$$S_V(\omega) = 2\pi \frac{\alpha V^2}{A\omega} = \frac{\alpha V^2}{Af} \tag{2.25}$$

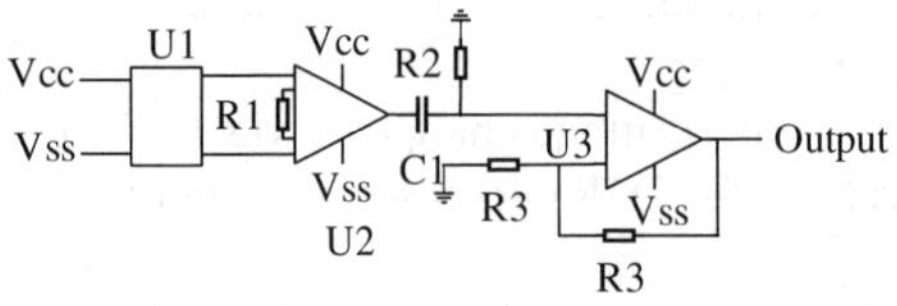

**Figure 2.26** Circuit design of a TMR sensor.

where $\alpha$ is the parameter representing the dimension of the surface, $V$ is the voltage of the sensor, $A$ is the active surface of the device, and $f$ is the frequency of the noise.

This noise decreases when the polarization increases. This is a measure of a better barrier and more controlled interfaces. The electronic noise also decreases when the barrier becomes thinner. The resistance variation in this case is observed compared to the roughness of the thickness barrier and is measured as $\alpha$ in (2.25). In the saturation condition, $\alpha$ is increased by a factor of 2 or more because the number of conducting channels are near in parallel condition that corresponds to the reduction of the effective surface of the junction.

**Magnetic 1/f noise** The magnetic 1/f noise in TMR is mainly due to the fluctuations in domain. This type of noise can be reduced by shaping the free layer of the MTJ, which will reduce the domain walls. A yoke-shaped free layer electrode in MTJs can decrease the magnetic noise by more than an order of magnitude compared to rectangular MTJs. An increase in the number of junctions also increases the voltage biasing, with a decrease in TMR factor.

## 2.5 Overview of Established Magnetic Field Sensing Technologies

Magnetic field based MR sensors are used in a wide range of applications to detect current, rotation, motion, position, and other parameters. Different sensor types are utilized for sensing magnetic fields. In order of technical progress, the important MR sensors are AMR, GMR, and TMR. Sensors based on the Hall effect are used in clamp ammeters. These were some of the first magnetic sensors used but they suffer from the disadvantages of low sensitivity and nonlinearity. Recently, CMR technology has been discovered that is related to thin-film MR material. It was found that CMR sensors have higher MR characteristics than GMR sensors. However, these sensors are not commercialized due to their low temperature and large equipment requirement. The relatively new TMR sensor technology has better temperature stability, high sensitivity, a wide linear range, and does not require a set–reset coil, unlike AMR and GMR technologies. Table 2.4 shows a summary of the characteristic features of commonly used MR sensor technologies.

**Table 2.4** Different features of MR sensor technologies.

| Technology | Hall effect | AMR | GMR | TMR |
|---|---|---|---|---|
| Resolution (mOe) | ~500 | ~0.1 | ~3 | ~0.1 |
| Field sensitivity (mV/V/Oe) | ~0.05 | ~1 | ~2 | ~0.1 |
| Dynamic range (Oe) | ±1000 | ±10 | ±20 | ±150 |
| Power consumption (mA) | 5–20 | 1–10 | 1–10 | 0.001–0.01 |
| Die size (mm²) | 1 × 1 | 1 × 1 | 1 × 1 | 1 × 1 |
| Temperature performance(°C) | <500 | <500 | <500 | <200 |

## Bibliography

1 M. J. Caruso, T. Bratland, C. H. Smith, and R. Schneider, "A new perspective on magnetic field sensing," *Sensors*, vol. 15, pp. 34–47, 1998.

2 J. E. Lenz, "A review of magnetic sensors," *Proceedings of the IEEE*, vol. 78, no. 6, pp. 973–989, 1990.

3 H. Nalwa, *Handbook of Thin Films*. Cambridge, MA: Elsevier, 2001.

4 S. Tumanski, *Thin Film Magnetoresistive Sensors*. Boca Raton, FL: CRC Press, 2001.

5 Y. Cai, Y. Zhao, X. Ding, and J. Fennelly, "Magnetometer basics for mobile phone applications," *Electronic Products*, vol. 54, no. 2, 2012.

6 D. He, *AMR Sensor and its Application on Nondestructive Evaluation*. InTech, 2017.

7 KOHDEN. (2018) The NIST definition of cloud computing. [Online]. Available: http://www.hkd.co.jp/english/amr_tec_amr/

8 P. Ripka and M. Janosek, "Advances in magnetic field sensors," *IEEE Sensors Journal*, vol. 10, no. 6, pp. 1108–1116, 2010.

9 D. Samal and A. Kumar, *Giant Magnetoresistance*. Springer-Verlag, 2008.

10 NVE. (2018) How GMR works. [Online]. Available: https://www2.nve.com/gmrsensors/gmr-operation.htm

11 I. Ennen, D. Kappe, T. Rempel, C. Glenske, and A. Hutten, "Giant magnetoresistance: Basic concepts, microstructure, magnetic interactions and applications," *Sensors*, vol. 16, no. 6, p. 904, 2016.

12 C. Reig, M.-D. Cubells-Beltran, and D. Ramirez Munoz, "Magnetic field sensors based on giant magnetoresistance GMR technology: Applications in electrical current sensing," *Sensors*, vol. 9, no. 10, pp. 7919–7942, 2009.

13 C. Reig, S. Cardoso, and S. C. Mukhopadhyay, "Giant magnetoresistance (GMR) sensors," *Smart Sensors, Measurement and Instrumentation*, vol. 1, pp. 157–80, 2013.

14 M. Julliere, "Tunneling between ferromagnetic films," *Physics Letters A*, vol. 54, no. 3, pp. 225–226, 1975.

15 MultiDimension. (2015) Introduction to TMR magnetic sensors. [Online]. Available: http://www.dowaytech.com/en/1776.html

16 MultiDimension. (2018) Large dynamic range TMR linear sensor. [Online]. Available: http://www.dowaytech.com/en/1800.html

17 J. Paul, "Sensors based on tunnel magnetoresistance – new technology, new opportunities," in 17th International Conference on Sensors and Measurement Technology, Nurnberg, 2015.

18 A. Ramirez, "Colossal magnetoresistance," *Journal of Physics: Condensed Matter*, vol. 9, no. 39, p. 8171, 1997.

19 J. Nickel and S. Zhang, "Colossal magnetoresistance sensor," Patent, Nov. 10, 1998, US Patent 5,835,003.

20 A. Jander, C. Smith, and R. Schneider, "Magnetoresistive sensors for nondestructive evaluation," *10th SPIE International Symposium, Nondestructive Evaluation for Health Monitoring and Diagnostics*, vol. 5770, pp. 1–14, 2005.

21 J. B. Johnson, "Thermal agitation of electricity in conductors," *Physical Review*, vol. 32, no. 1, p. 97, 1928.

22 Y. M. Blanter and M. Büttiker, "Shot noise in mesoscopic conductors," *Physics Reports*, vol. 336, no. 1-2, pp. 1–166, 2000.

23  F. Hooge and A. Hoppenbrouwers, "1/f noise in continuous thin gold films," *Physica*, vol. 45, no. 3, pp. 386–392, 1969.

24  G. Lovat, *Electromagnetic Shielding*. Wiley, 2007.

25  F. Li, "The design and application of the hardware of the transient magnetic field measurement device," Masters Dissertation, University of Electrical Science and Technology of China, 6, 2012.

26  M. Caruso, "Cross axis effect for AMR magnetic sensors," *Application Note AN215*, 2003.

27  Honeywell. (2018) Cross axis effect for AMR magnetic sensors. [Online]. Available: http://c1233384.r84.cf3.rackcdn.com/UK_HMP_HMC1052L_4AN.pdf

28  P. Ripka, *Magnetic Sensors and Magnetometers*. Norwood, MA: Artech House, 2001.

29  Honeywell. (2018) Set/Reset function for magnetic sensors – AN213. [Online]. Available: http://www.seraphim.com.tw/upfiles/c_supports01284968029.pdf

30  ——. (2018) Set/Reset function for magnetic sensors – AN201. [Online]. Available: https://neurophysics.ucsd.edu/Manuals/Honeywell/AN-201.pdf

31  O. Yong, J. He, J. Hu, and S. X. Wang, "A current sensor based on the giant magnetoresistance effect: Design and potential smart grid applications," *Sensors*, vol. 12, no. 11, pp. 15 520–15 541, 2012.

# 3

# Magnetic Field Measurement for Power Transmission Systems

## 3.1 Introduction

In recent times, traditional power grid installations have been rapidly transformed to incorporate the outcomes of research and development to form modern grids, commonly termed smart grids. A number of aspects of future smart grids were declared in the Standardization Management Board Smart Grid Strategic Group (SG3) standardization road map, drafted by International Electro Technical Commission. The road map emphasizes efficient electric power transmission and distribution infrastructures as its key pillars [1]. It urges the cultivation of a resilient communication link amongst the different building blocks of a power grid to perform condition monitoring and troubleshooting at control stations. For power delivery in conventional grids, a number of factors play an important role, such as timely and low cost maintenance, monitoring in real time, and optimized utility of available infrastructure. In existing grids, power is transmitted by means of underground and overhead high-voltage transmission lines (HVTLs), which incur significantly reduced costs compared with existing technologies for deployment and maintenance. For this reason, overhead HVTLs are utilized for power delivery in most of the world.

As a result, overhead HVTLs act as primary medium, connecting generation points to consumers spread across wide areas. These overhead power lines are installed in a range of climatic conditions where they are continuously challenged by forces of nature. At one extreme, their resilience is tested by snow, ice, rainstorms and wind gusts, and at the other hot and humid weather tests their survival. Thunderstrikes on support towers damage insulation and travel as electrical surges in the network. In the developed world, where installation of these overhead lines meets and surpasses the load requirements, efforts are now being made to develop autonomous state-of-the-art monitoring solutions to avoid power outages. Another objective of monitoring is to anticipate load requirements for critical lines during peak season.

An important design parameter for interconnected power transmission and distribution systems is the development of a real-time monitoring and predictive fault analysis scheme for overhead HVTLs. This argument is even more valid for enforcement of smart grid objectives where sensing and actuation are time-critical for efficient and uninterrupted operation of the network. On the other hand, while the importance of overhead power lines is evident, difficulties in deploying condition monitoring systems are not visible to common users. Service staff require safe and established equipment

*Magnetic Field Measurement with Applications to Modern Power Grids*, First Edition.
Qi Huang, Arsalan Habib Khawaja, Yafeng Chen and Jian Li.
© 2020 John Wiley & Sons Ltd. Published 2020 by John Wiley & Sons Ltd.

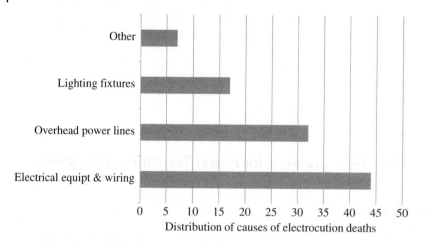

**Figure 3.1** Average deaths caused by contact with electricity among electrical workers in construction.

for troubleshooting and general maintenance. Since these power lines are a key player in the reliability of a power network, it is desirable to monitor the operating conditions of HVTLs in modern power grids for implementation in emerging smart grid scenarios.

Information obtained from the US Bureau of Labor Statistics (BLS) and shown in Figure 3.1 [2] reveals that from 2003 to 2005, the main contributors to the death toll of electrical workers was overhead power lines and electrical equipment. Electrocution is the fourth leading cause of death in construction and causes 9% of worker deaths on average. Electrical power installers/repairers have the highest death toll due to electrocution of 31.8% when statistics were collected for deaths per 100,000 full-time workers. Non-contact sensor-based power line diagnosis solutions have the potential to solve complex problems and greatly facilitate our lives.

Overhead HVTLs are a vital mode of transportation of power for a widely distributed transmission network. The overhead lines serve to connect generation units to distribution sites via HVTLs terminating at substations and then to end users by relatively low voltage distributed lines. The overhead lines remain exposed to extreme conditions at all times. Such exposure commonly results in reduction of ground clearance in the form of conductor elongation (sag). Another effect of extreme weather is wind-induced conductor motion, referred to as aeolian vibration and conductor galloping. Outdoor conditions are not only limited to environmental factors, but may also include line faults. Altogether, this hinders the efficient operation of the power grid and may result in permanent damage to the network. Thus, a timely estimate of phase current, conductor elongation, and wind-induced motion can dramatically improve the security of the power grid and reduce outage times.

In recent years, many researchers have proposed methods for operation state monitoring of overhead transmission lines [1; 3; 4; 5]. These methods have involved both contact and non-contact based retrieval of spatial parameters and line current. In industry, non-contact measurement products, such as Sagometer by Engineering Data Management, have recently been commercialized. This product measures sag in transmission lines using a camera as a sensor and digital image processing techniques. Another device, developed by Electric Research Power Institute, is attached with live current carrying conductors to monitor conductor temperature, sag, current, and vibration. One research group has proposed the use of non-contact magnetic field sensors for estimation of the spatial and electrical parameters of the transmission line. However, this method uses a large number of sensing units and computationally expensive stochastic optimization techniques, which restricts the practical use of the approach.

Recently, magnetic field based condition monitoring has been shown to be useful in overhead power transmission line condition monitoring. These recent advances in non-contact based sensing technologies have greatly transformed our lives. Using such sensing technologies in monitoring and diagnosis is of utmost importance for high power electrical installations.

## 3.2  Electric Current Reconstruction

### 3.2.1  Reconstruction with Stochastic Optimization Techniques

Based on the Biot–Savart law, the phase currents and source positions determine the distribution of the magnetic field emanated. With the magnetic field data measured at a number of field points, an inverse problem must be solved to derive the phase currents as sources of the magnetic field. When the relative positions of field points and current sources are fixed, the source currents can be straightforwardly obtained by using deterministic optimization strategy.

However, the spatial parameters of transmission lines can change due to the conductor sagging effect or galloping. As a result, both the spatial and electrical parameters of current sources are not known. In order to solve the inverse problem, a stochastic optimization technique based on an artificial immune system (AIS) algorithm is used to reconstruct both the electrical and spatial parameters from the magnetic field data. The source reconstruction process is shown in Figure 3.2. It starts with a group of default position parameters $P_0$ of the transmission line configuration, which are recorded under the normal operation state. Based on the measured magnetic field $B_{mea}$, phase currents $I_p$ are estimated by an inverse current program (ICP), which is based on least squares estimation by this equation

$$I_p = (A^T A)^{-1} A^T B_{mea} \qquad (3.1)$$

where $\mathbf{A}$ is the coefficient matrix, which depends on the position parameters of the current sources. The magnetic field $B_{cal}$ is obtained by using $I_p$ and matrix $\mathbf{A}$ in a magnetic field evaluation (MFE) module as

$$B_{cal} = A I_p \qquad (3.2)$$

A defined minimum threshold value of the Euclidean distance $\| \mathbf{B}_{cal} - \mathbf{B}_{mea} \|$ is defined as the end condition for terminating the reconstruction process. If the $B_{cal}$ generated by default position parameters $P_0$ does not meet the end condition, the AIS algorithm randomly generates new position parameters $P_s$ in the source position optimization (SPO) module. With reference to the $B_{mea}$ and the newly generated $P_s$, the new $I_p$ is computed by the ICP again. The new $B_{cal}$ is calculated using the $P_s$ and new $I_p$ in the MFE module and compared with the $B_{mea}$ again. The Euclidean distance between

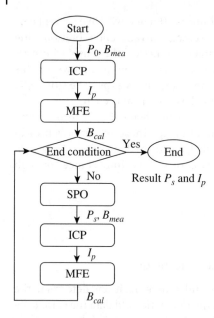

**Figure 3.2** Flowchart of current source reconstruction based on magnetic field sensing. (*Source:* Reprinted with permission from X. Sun et al., Overhead high-voltage transmission-line current monitoring by magnetoresistive sensors and current source reconstruction at transmission tower, *IEEE Transactions on Magnetics*, vol. 50, no. 1, pp. 1–5, Jan. 2014.)

them is found. When the Euclidean distance is less than the minimum threshold value, the optimizing process finishes. The resulting $P_s$ and $I_p$ are saved as the true values of the transmission line parameters; otherwise, the iteration continues. This reconstruction process is repeated multiple times ($N$) in order to obtain the final results of $P_s$ which are the averages of these $N$ optimizations. Accordingly, the final $I_p$ is obtained from the optimized $P_s$ and the measured magnetic field [6].

### 3.2.1.1 Experimental Proof
In order to verify this current source reconstruction technology for HVTL monitoring, a laboratory setup including an MR three-axis sensor (Honeywell HMC2003) array and three-phase straight transmission power lines was established to act as the test bed. This setup is shown in Figure 3.3. The 11 MR sensors of the array are evenly spaced. The vertical distance between the sensor array and transmission power lines is 26.0 cm. In the normal operation state, the heights of the three power lines are 68.0 cm from the ground. When the power supply is connected to the power lines, the measured phase current amplitudes are 16.2, 16.4, and 16.2 A in phases A, B, and C, respectively. The emanated magnetic field was measured by the sensor array. As shown in Figure 3.4a, the measured magnetic field values by the MR sensor array coincide well with the calculated values based on the Biot–Savart law. Based on the measured magnetic field values, the operation states of the power lines were reconstructed. The reconstruction result of the phase currents is shown in Figure 3.4b. The reconstructed amplitudes of the phase currents are 16.12, 16.40, and 16.11 A in phases A, B, and C. The average error for the actual amplitude value is 0.35%. The current cycle is found to be 19.96 ms, corresponding to

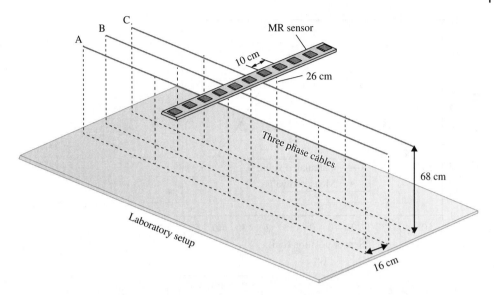

**Figure 3.3** Experimental setup for verifying the current source reconstruction method for transmission line monitoring. (*Source:* Reprinted with permission from X. Sun et al., Overhead high-voltage transmission-line current monitoring by magnetoresistive sensors and current source reconstruction at transmission tower, *IEEE Transactions on Magnetics*, vol. 50, no. 1, pp. 1–5, Jan. 2014).

a system frequency of 50.1 Hz with an error of 0.2%. In order to verify this technology with abnormal operation states, tests were carried out in the four cases as shown in Table 3.1. Cases 1 and 2 mimic the occurrence of cable sagging while cases 3 and 4 mimic the occurrence of current imbalance. The phase current reconstruction results for these cases are shown in Figure 3.5. In these results, the errors of phase current amplitudes are all less than 0.4%. The error of the reconstructed system frequency compared to the actual value is less than 0.3%. These experiments performed with the test bed prove the principle of the current monitoring technology based on magnetic field measurement and current source reconstruction. The technology can function properly regardless of whether the transmission lines are in the normal operation state or abnormal conditions.

### 3.2.1.2 Application in 500 kV Transmission Line Model

This reconstruction method was tested with the simulation model of the 500 kV transmission lines in Figure 3.6. A typical HVTL emanates a magnetic field in the amplitude of several hundred microTesla at the top level of the transmission tower. The magnetic field can be accurately measured by commercially available MR sensors that can provide sensitivity down to around $10^{-9}$ Tesla and spatial resolution of 0.9 mm [7]. Figure 3.6 shows a transmission tower with three-phase 50 Hz 500 kV (maximum load current is 3.75 kA per phase) transmission lines [8]. The configuration of the three-phase conductors is a flat formation, as shown on the right of the figure. A magnetic sensor array is installed on the top level of the tower and used to measure the magnetic field emanated by the phase conductors. The array is composed of 11 MR sensors with 1.0 m spacing between them. finite element analysis (FEA) simulation was conducted to investigate the influence of the steel structure of the transmission tower on the magnetic field measurement. Only the upper framework of the transmission tower is considered because the lower

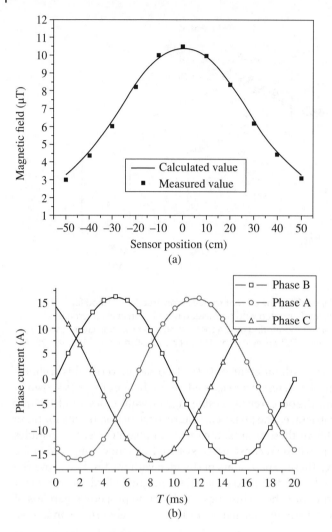

**Figure 3.4** Magnetic field values in the experiment and phase current reconstruction results by the current source reconstruction method: (a) magnetic field values measured by MR sensors and calculated values based on Biot–Savart law and (b) the reconstructed phase current curve of each phase conductor. (*Source:* Reprinted with permission from X. Sun et al., Overhead high-voltage transmission-line current monitoring by magnetoresistive sensors and current source reconstruction at transmission tower, *IEEE Transactions on Magnetics*, vol. 50, no. 1, pp. 1–5, Jan. 2014).

part is relatively far away from the region of interest. The meshed FEA model of the transmission tower upper framework is shown in Figure 3.7a. In this FEA model, there is a magnetic field emanating from the transmission lines operating in the normal state with 50 Hz 3.75 kA phase currents. The magnetic field distribution in the sensing zone where the magnetic sensor array is placed is simulated with the tower framework effect considered. On the other hand, the magnetic field distribution in the sensing zone is calculated analytically without considering the effect of the steel tower structure. The FEA simulation results with the steel tower structure considered and the calculation results

**Table 3.1** Abnormal operation states in tests.

| Case | Current amplitudes (A) | Power lines height (cm) |
|------|------------------------|-------------------------|
| 1 | 16.8, 16.5, 16.2 | 68.0, 48.0, 68.0 |
| 2 | 16.8, 16.5, 16.6 | 48.0, 68.0, 68.0 |
| 3 | 16.6, 8.1, 16.6 | 68.0, 68.0, 68.0 |
| 4 | 16.5, 16.3, 8.3 | 68.0, 68.0, 68.0 |

*Source:* Reprinted with permission from X. Sun et al., Overhead high-voltage transmission-line current monitoring by magnetoresistive sensors and current source reconstruction at transmission tower, *IEEE Transactions on Magnetics*, vol. 50, no. 1, pp. 1–5, Jan. 2014.

without considering the framework of the tower are shown in Figure 3.7b. The Euclidean distance between them is determined to be a very small value. It indicates that the influence of the upper tower framework on the magnetic field distribution in the sensing zone is negligible (approximately less than 1%). Thus the magnetic field emanated by the HVTL can be measured by the magnetic sensor array without distortion from the steel structure. Based on our simulation results, the influence of the tower frameworks is negligible on the magnetic field distribution in the vicinity of the lines. The influence of charges and currents induced in the tower can also be neglected. This fact has been verified theoretically and experimentally by the literature [8]. Thus in this work we only need to consider the effects of transmission line conductor sagging and image current in the conducting ground when calculating the magnetic field distribution.

The magnetic field from the transmission lines can be accurately calculated by an analytical method [9; 10]. The resulting magnetic field of multi-conductor power lines can be evaluated by superimposing the contribution from each phase current flowing in the conductors and the image currents. The detailed calculation of the magnetic field of the transmission line with sagging is described in [9].

Figure 3.8a shows the magnetic field distribution of the 500 kV transmission lines in Figure 3.6 on the top level of the tower. The magnetic field is simulated with conductor sag of 10 m. We assume that every conductor suffers the same sag. The maximum magnetic flux density of the resulting magnetic field is 292 $\mu$T obtained at the center sensor position. Figure 3.8 shows the phase current reconstruction results from the measured magnetic field. It is found that the amplitudes of phase currents are reconstructed with an average error of 0.13% compared to the actual value. The current cycle is found to be 20.09 ms, corresponding to a system frequency of 49.78 Hz with an error of 0.4% compared to the actual mains frequency.

Figure 3.9a shows the magnetic field distribution when there is current amplitude imbalance in phase B. Accordingly, the conductor sag of this transmission line extents from 10 to 12 m due to the increased cable temperature induced by the current. In this case, the complete phase current curves for the three phases can also be reconstructed. It can be seen in Figure 3.9b that each reconstructed phase current matches well with the actual phase current and the current amplitude imbalance (normal current amplitude is 3.75 kA) in phase B is also deduced. The reconstructed current amplitude of 4.124 kA accurately reflects the amplitude imbalance in phase B.

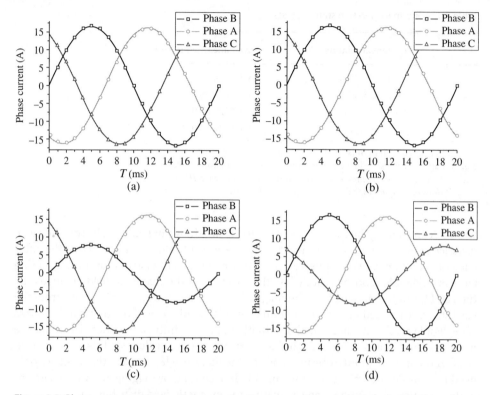

**Figure 3.5** Phase current reconstruction results for the abnormal cases experiment using the laboratory setup: (a) reconstruction results for case 1, (b) reconstruction results for case 2, (c) reconstruction results for case 3, and (d) reconstruction results for case 4. (*Source:* Reprinted with permission from X. Sun et al., Overhead high-voltage transmission-line current monitoring by magnetoresistive sensors and current source reconstruction at transmission tower, *IEEE Transactions on Magnetics*, vol. 50, no. 1, pp. 1–5, Jan. 2014.)

Figure 3.10a shows the magnetic field distribution where there is 10% imbalance in phase angle (the phase angle changes from 0 to 12°) in phase B. From Figure 3.10b, the phase imbalance is reflected in the 0.67 ms delay in the phase B current curve along the time axis. The error compared to the actual value of 0.673 ms is 0.4%.

In the above cases, the reconstructed electrical values agree with the actual values with very small errors. The current monitoring technology based on current source reconstruction and magnetic field measurement can successfully provide the electrical parameters of the transmission lines, including current amplitude, phase, and frequency, accurately with an error of less than 1%. The technology can function in the normal operation state of the transmission lines and also in abnormal situations, including current amplitude imbalance and phase imbalance.

However, practicality can be limited by employing such a large number of sensors as simultaneous data acquisition from 11 three-axis sensors (requiring 33 channel simultaneous sampling analog-to-digital converters) and power consumption of this detailed circuitry are other bottlenecks. Sun et al. also proposed the estimation of the spatial parameters of each phase conductor with a number of magnetic field sensors, deployed

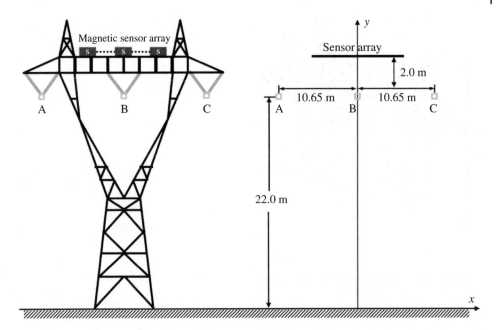

**Figure 3.6** Simulation model of 500 kV overhead HVTLs. Squares marked "s" denote the MR sensor array. (*Source:* Reprinted with permission from X. Sun et al., Overhead high-voltage transmission-line current monitoring by magnetoresistive sensors and current source reconstruction at transmission tower, *IEEE Transactions on Magnetics*, vol. 50, no. 1, pp. 1–5, Jan. 2014.)

along the span length of HVTLs between support towers. However, all these methods utilized a stochastic optimization algorithm to re-estimate the spatial parameters of all the phase conductors for every reconstructed value. In reality, these parameters remain same for a considerable period of time. In addition, computationally expensive functions for convergence are inherent in the stochastic optimization technique. This implies that the number and positioning of sensors in the optimization process need further investigation. Furthermore, a simple and robust optimization algorithm is essential for a real-time monitoring system.

### 3.2.2 Reconstruction with Optimal Placement of Minimum Sensing Nodes

Motivated by the underlying research and practical implications of the existing methods, here we present an efficient solution to accelerate the development of a practical system. For our method, dual-axis magnetic field sensors equivalent to the number of phase conductors are sufficient. This method can reconstruct the electric current and estimate sag in conductors at a support tower where the proposed monitoring system is mounted. The sensors are arranged as an array and can be placed at any height on the vertical plane of the support tower, above or beneath the conductors. They work by sensing only the vertical and horizontal components of the magnetic field generated from current-carrying conductors. Since current estimation from magnetic field measurement requires solving an inverse reconstruction problem which is generally ill-posed, in our case small noise in the sensor output measuring the magnetic field will have an

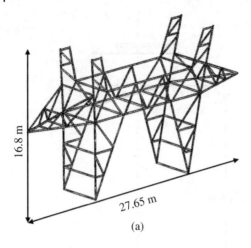

(a)

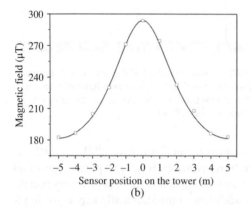

(b)

**Figure 3.7** Transmission tower simulation model and magnetic field distribution in the sensing zone where the magnetic sensor array would be placed. (a) Transmission tower upper framework. (b) The curve denotes the magnetic field distribution calculated without considering the tower framework effects and squares denote the magnetic field computed by FEA with the tower framework effect considered. (*Source:* Reprinted with permission from X. Sun et al., Overhead high-voltage transmission-line current monitoring by magnetoresistive sensors and current source reconstruction at transmission tower, *IEEE Transactions on Magnetics*, vol. 50, no. 1, pp. 1–5, Jan. 2014.)

amplified impact on the reconstructed current. To handle this, the condition number of the position matrix used to solve the inverse problem is kept close to unity, which reduces the impact of noise on the current reconstruction results. It is observed that placing the sensors away from each other and parallel to conductors not only improves the condition number but also aids in convergence of the algorithm for sag estimation. The previous research on calculation of the magnetic field around power lines proved that the impact of sag, span length, and neighboring span contribute to calculation error. Therefore, for accurate sag and current estimation, all such practical factors have been considered in our solution. With our methodology, the position of the sensor array can be adjusted on the support tower according to the circuit configuration. The method

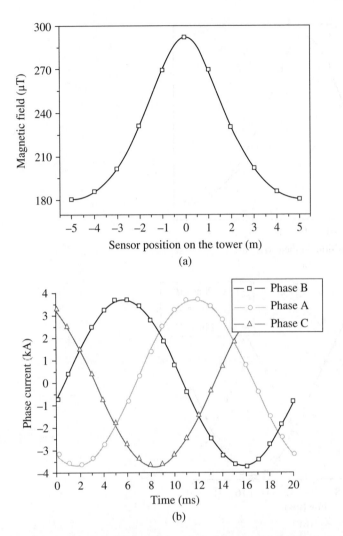

**Figure 3.8** Magnetic field at the top level of the transmission line tower under the normal operation state and the corresponding phase current reconstruction results: (a) squares denote the magnetic field values at the positions of the magnetic sensors and (b) the reconstructed phase current curve of each phase conductor. (*Source:* Reprinted with permission from X. Sun et al., Overhead high-voltage transmission-line current monitoring by magnetoresistive sensors and current source reconstruction at transmission tower, *IEEE Transactions on Magnetics*, vol. 50, no. 1, pp. 1–5, Jan. 2014.)

has been verified with numerical simulations of a typical tower configuration, simulating scenarios of symmetrical and unsymmetrical sag. A scaled laboratory setup is then arranged in which the reconstructed current using a magnetic field sensor array is compared with current measurements from the ammeter. Similarly, estimated sag is compared with the actual sag measured with a calibrated scale in the laboratory. This detailed account is presented in this chapter to demonstrate a working model of the proposed system.

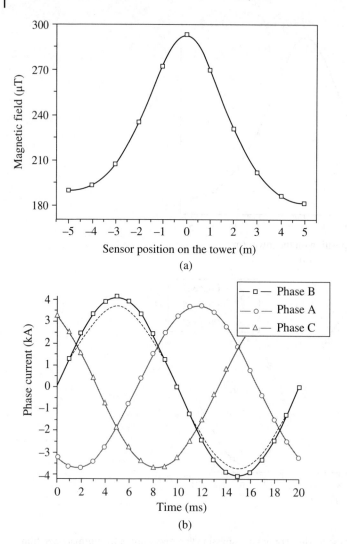

**Figure 3.9** Magnetic field at the top level of the transmission line tower when there is current amplitude imbalance in phase B and the corresponding phase current reconstruction results: (a) squares denote the magnetic field values at the positions of the magnetic sensors and (b) the reconstructed phase current curve of each phase conductor. (*Source:* Reprinted with permission from X. Sun et al., Overhead high-voltage transmission-line current monitoring by magnetoresistive sensors and current source reconstruction at transmission tower, *IEEE Transactions on Magnetics*, vol. 50, no. 1, pp. 1–5, Jan. 2014.)

### 3.2.2.1 Problem Formulation for Sensor Placement

We present an innovative method for electric current and sag estimation for HVTLs with a magnetic field sensor array mounted on a support tower. It is achieved by first interpreting the sensed magnetic field magnitude for one component (e.g. horizontal) into an electric current by inverse reconstruction. The results are then validated by solving a forward problem, and the difference between the calculated and measured

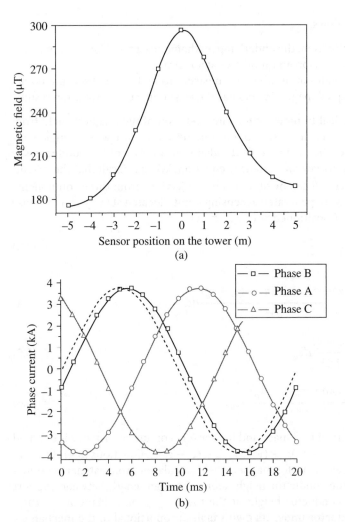

**Figure 3.10** Magnetic field at the top level of the transmission line tower when there is phase angle imbalance in phase B and the corresponding phase currents reconstruction results: (a) squares denote the magnetic field values at the positions of the magnetic sensors and (b) the reconstructed phase current curve of each phase conductor. (*Source:* Reprinted with permission from X. Sun et al., Overhead high-voltage transmission-line current monitoring by magnetoresistive sensors and current source reconstruction at transmission tower, *IEEE Transactions on Magnetics*, vol. 50, no. 1, pp. 1–5, Jan. 2014.)

magnetic field for the the other component (e.g. vertical) is iteratively reduced. Inverse reconstruction problems are frequently encountered in electrical engineering applications, mainly for the following:

1) Source reconstruction in cases where the system model and effect are known.
2) System model identification when cause and effect are known. Our application can be considered as a special case where source reconstruction has to be performed as well as determining the system model, entirely with a perceived magnetic field.

We dealt with this as follows:

1) For sag and current estimation, the underlying mathematical model for magnetic field components is used to develop an innovative algorithm.
2) We then applied the condition number as a measure to find optimal sensing points so the error in the sampled magnetic field has the least impact on estimation results.

To develop a mathematical framework for interpretation of the magnetic field from a number of conductors at each sensing point, one needs to start with a single conductor and sensing point. The magnetic field radiated from each of the conductors is governed by the Biot–Savart law and catenary equation. We can recall that the magnitude of magnetic field vector $\vec{B}$ can be obtained by projections from three components $(X, Y, Z)$. Therefore, the vector generated at sensing point $s$ located at $(x_s, y_s, z_s)$ by phase conductor $p$ with current $I$ positioned at $(x_p, y_p, z_p)$ is

$$\vec{B} = \vec{B}_{X_s} + \vec{B}_{Y_s} + \vec{B}_{Z_s} \tag{3.3}$$

where

$$\vec{B}_{X_s} = \hat{i}_x \frac{\mu_0 I_p}{4\pi} \int_{-\frac{L}{2}}^{\frac{L}{2}} \frac{(z_s - z_p)\sinh \alpha z_p - (y_s - y_p)}{|\vec{r}|^3} dz_p \tag{3.4}$$

$$\vec{B}_{Y_s} = \hat{i}_y \frac{\mu_0 I_p}{4\pi} \int_{-\frac{L}{2}}^{\frac{L}{2}} \frac{x_s - x_p}{|\vec{r}|^3} dz_p \tag{3.5}$$

$$\vec{B}_{Z_s} = \hat{i}_z \frac{\mu_0 I_p}{4\pi} \int_{-\frac{L}{2}}^{\frac{L}{2}} \frac{-\sinh \alpha z_p (x_s - x_p)}{|\vec{r}|^3} dz_p \tag{3.6}$$

Here, $\mu_0$ is the permeability of free space and the sensor–conductor distance is denoted by the distance vector, $\vec{r} = (x_s - x_p)\hat{i}_x + (y_s - y_p)\hat{i}_y + (z_s - z_p)\hat{i}_z$. Equations (3.4)–(3.6) are deduced using the Biot–Savart law for magnetic field calculation from catenary conductors, where $y_p$ refers to the conductor height along the span length between support towers. It is the sum of conductor height at the tower $y_{tower}$ and shape of the catenary adopted by the conductor under its own weight, proportional to the mechanical parameter $\alpha_p$:

$$y_p = y_{tower} + \frac{1}{\alpha_p}(\cosh(\alpha_p z_p) - 1), -\frac{L}{2} \le z_p \le \frac{L}{2} \tag{3.7}$$

where $L$ denotes the length of span between adjacent towers.

For practical measurements, magnetic field sensors are available in one- to three-axis arrangements to sense corresponding projections of magnetic field vectors. It is evident from (3.4) to (3.6) that all three components of the magnetic field are proportional to the electric current $I_p$. Therefore, a mono-axial sensor is sufficient to solve the current reconstruction problem. However, in our case sag is considered a dynamic entity, which when changed can distort the reconstruction results. We therefore propose to use two-axis magnetic field data, with one-axis readings for reconstruction and the other axis for comparison in cases when the sag is changed. For instance, we require to measure magnetic field along $x$ and $y$ axes, exploiting the fact that the magnetic field component for the vertical component will be zero when the sensor is placed underneath or above the phase conductor, i.e. term $(x_s - x_p)$ in (3.5). This will provide a search direction in the algorithm detailed later in the chapter. In contrast to the three-axis measurements in [6] and [4] with a large number of sensors, dual-axis measurements from sensors equal to the number of conductors will not only reduce the hardware complexity, setup cost, and power requirements of the system, but also are a significant leap towards a practical design.

Complete system realization using a three-dimensional model designed in Google Sketchup, is given in Figure 3.11. A proposed design is shown for a typical one-circuit tower configuration. The tower is equipped with dual-axis sensors equal to the number of phase conductors. For best positioning of the sensor array to reduce error in inverse current reconstruction, its condition number is tested by varying the height of the array and the sag in the conductors. A detailed account of the condition and sensitivity of the proposed system is included in Section 3.1.1. The sensing system requires simultaneously sampling along the horizontal and vertical axes. Once sensor analog output has been conditioned and sampled, a micro-controller performs an estimation algorithm to calculate the electric current and sag. The estimated results are then transmitted to a relay node, cloud or substation.

One needs to consider the magnetic field of conductors from the neighboring span when the sensor is placed at the support tower. As the magnetic field decays significantly with distance $|\vec{r}|^3$, so we can consider the magnetic field radiated only from half spans on both sides of tower. For the same effect, integration for two half span lengths is sufficient, contrary to the integral calculation for two full span lengths found in literature. This significantly reduces the computational cost in terms of time and energy. The mechanical tension parameter responsible for sag, $\alpha_p$, can be considered the same for the conductor in the neighboring span due to the fact that conductor elongation is roughly uniform for the same phase conductor in neighboring span lengths. Since each span length terminates at the tower, we need to integrate both of them separately. The first integral calculation considers the half span $(s_1)$ from the maximum sag point ending at the tower whereas the second integral contains the half span $(s_2)$ from the tower ending at the maximum sag point span away from the tower. For the same effect, the horizontal and vertical components of the magnetic field become

$$\vec{B}_{X_s} = \hat{i}_x \frac{\mu_o I_p}{4\pi} \left\{ \begin{array}{l} \int_{-\frac{L}{2}}^{0} \left[ \frac{(z_s - z_p)\sinh \alpha_p z_p - (y_s - y_p)}{|(x_s - x_p) + (y_s - y_p) + (z_s - z_p)|^3} \right] dz_{p_{s1}} \\ + \int_{0}^{\frac{L}{2}} \left[ \frac{(z_s - z_p)\sinh \alpha_p z_p - (y_s - y_p)}{|(x_s - x_p) + (y_s - y_p) + (z_s - z_p)|^3} \right] dz_{p_{s2}} \end{array} \right\} \tag{3.8}$$

$$\vec{B}_{Y_s} = \hat{i}_y \frac{\mu_o I_p}{4\pi} \left\{ \begin{array}{l} \int_{-\frac{L}{2}}^{0} \left[ \frac{(x_s - x_p)}{|(x_s - x_p) + (y_s - y_p) + (z_s - z_p)|^3} \right] dz_{p_{s1}} \\ + \int_{0}^{\frac{L}{2}} \left[ \frac{(x_s - x_p)}{|(x_s - x_p) + (y_s - y_p) + (z_s - z_p)|^3} \right] dz_{p_{s2}} \end{array} \right\} \tag{3.9}$$

Here, $-\frac{L}{2}$ and $\frac{L}{2}$ are the distances of the maximum sag point of each phase conductor towards and away from the sensing units, respectively. By integrating the variables in (3.8) and (3.9), we get magnetic field coefficients $(A_X$ and $A_Y)$ for the phase conductor with current $(I_p)$

$$\vec{B}_{X_s} = \hat{i}_x \frac{\mu_o I_p}{4\pi} A_X \tag{3.10}$$

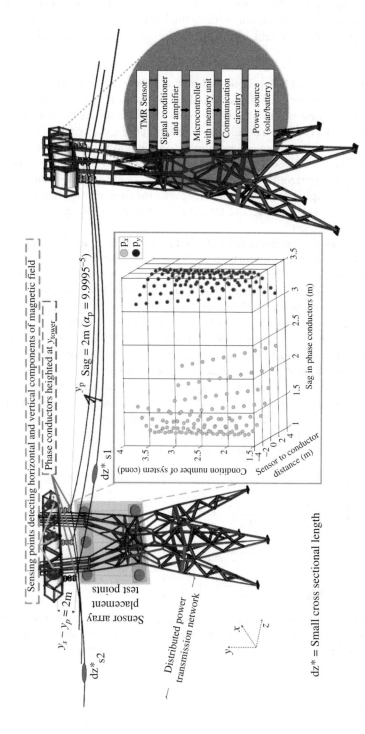

**Figure 3.11** Current reconstruction and sag estimation system realization. Inset: Condition number of position matrices ($P_x$ and $P_y$) at various sensor heights and different sag conditions. (*Source:* Reprinted with permission from A.H. Khawaja et al., Estimation of current and sag in overhead power transmission lines with optimized magnetic field sensor array placement. *IEEE Transactions on Magnetics*, vol. 53, no. 5, pp. 1–10, May 2017.)

**Labels within figure:**

TMR Sensor

Signal conditioner and amplifier

Microcontroller with memory unit

Communication circuitry

Power source (solar/battery)

Sensing points detecting horizontal and vertical components of magnetic field

Phase conductors heighted at $y_{tower}$

$y_p$ Sag = 2m ($\alpha_p$ = 9.9995$^{-5}$)

$y_s - y_p$ = 2m

$dz^*$ s1

$dz^*$ s2

Sensor array placement test points

Distributed power transmission network

$dz^*$ = Small cross sectional length

Condition number of system (cond)

Sag in phase conductors (m)

Sensor to conductor distance (m)

$P_x$

$P_y$

$$\vec{B}_{Y_s} = \hat{i}_y \frac{\mu_o I_p}{4\pi} A_Y \tag{3.11}$$

A number of HVTLs are suspended at the support tower, therefore one needs to determine the magnetic field contribution from each of the conductors at the sensor head. As of linearity of Biot–Savart law, we can apply the superposition principle to generalize the equations and calculate the magnetic field vector projections (horizontal and vertical) at sensing point $s$ from $j$ the phase conductors:

$$B_{X_s} = \sum_{p=1}^{j} \frac{\mu_o I_p}{4\pi} A_{P_{(X)}} \tag{3.12}$$

$$B_{Y_s} = \sum_{p=1}^{k} \frac{\mu_o I_p}{4\pi} A_{P_{(Y)}} \tag{3.13}$$

Using the same analogy, we can define a linear system adapted for any tower configuration with $i$ sensors and $j$ phase conductors. The coefficient matrix for this is based on the position of the conductors and the conductor to sensor distance. This will take the shape of a square matrix as the number of sensing points is equal to the number of phase conductors $(i = j)$. The magnetic field vector projection sensed at $i$ sensing points along the horizontal axis becomes

$$B_{X_i} = \left[ \frac{\mu_o}{4\pi} \begin{pmatrix} A_{11} & \cdots & A_{1m} \\ \vdots & \ddots & \vdots \\ A_{m1} & \cdots & A_{mm} \end{pmatrix} \right] * I_j \tag{3.14}$$

From (3.5), we can deduce that the term $A_{mm} = 0$ when $[m,m] = [(1,1), (2,2),\ldots]$. These are the sensing points on the same vertical plane, above or beneath the conductor where $x_s - x_p = 0$. Consequently, the vertical component of magnetic field vector becomes

$$B_{Y_i} = \left[ \frac{\mu_o}{4\pi} \begin{pmatrix} 0 & \cdots & A_{1n} \\ \vdots & \ddots & \vdots \\ A_{n1} & \cdots & 0 \end{pmatrix} \right] * I_j \tag{3.15}$$

The product of the magnetic field coefficients and constant $\left( \frac{\mu_o}{4\pi} \right)$ can be denoted as position matrix $P_{cmpt}$ for the component $(cmpt)$. We can generalize (3.14) and (3.15) to formulate a linear system. Since $P_{cmpt}$ is a non-singular matrix, the solution of the inverse problem exists in all such cases. The linear system and inverse current reconstruction problem becomes

$$B_{cmpt_i} = P_{cmpt} I_j \tag{3.16}$$

$$I_j = P_{cmpt}^{-1} B_{cmpt_i} \tag{3.17}$$

### 3.2.2.2 Sensitivity of the System for Source Current Reconstruction

In practical scenarios, any measurement system is prone to uncertainty, mainly from noise sources. The impact of noise on inverse reconstruction problems can be referred to as the sensitivity of the linear system. In solving an inverse matrix using the system model $P_{cmpt}$ and measured magnetic field projections, any noise in the magnetic field measurement will be amplified based on the condition of $P_{cmpt}$. For inverse reconstruction, one may use knowledge of matrix theory to understand the impact of noise on results.

For our application, $P_{cmpt}$ largely remains fixed, but errors in sensor array readouts are anticipated. As an example, noise for a magnetic field measurement system with a TMR sensor can be extrinsic, from amplifier and power supply noise, or intrinsic, from thermal, shot, electronic 1/f, and random telegraphic noise [11]. Extrinsic noise can be curbed with low noise amplifiers and a DC power source, whereas intrinsic noise remains in the system output due to the fact that it arises from the physical characteristics of the sensor itself [11]. If a measurement error $\delta$ is expected in $B_{cmpt_i}$ in (3.17), $I_j + \delta I_j$ denotes the exact solution:

$$I_j + \delta I_j = P_{cmpt}^{-1}(B_{cmpt_i} + \delta_{B_{cmpt_i}}) \tag{3.18}$$

To determine the error bound of the linear system, it is suitable to use the *2-norm* (Euclidean norm) of the matrix. The Euclidean norm is the ratio of the maximum and minimum eigenvalues of a matrix [12]. Replacing (3.17) and (3.18) using the Euclidean norm, we can calculate the error bound in $I_j$ as

$$\|\delta I_j\|_2 \leq \|P_{cmpt}^{-1}\|_2 \|\delta_{B_{cmpt_i}}\|_2 \tag{3.19}$$

Similarly, defining a bound, multiplying and rearranging each other, relative error becomes

$$\frac{\|\delta I_j\|_2}{\|I\|_2} \leq \|P_{cmpt}^{-1}\|_2 \|P_{cmpt}\|_2 \frac{\|\delta_{B_{cmpt_i}}\|_2}{\|B_{cmpt_i}\|_2} \tag{3.20}$$

The term $\|P_{cmpt}^{-1}\|_2 \|P_{cmpt}\|_2$ represents the maximum deviation in the exact solution. This term, referred to as a condition of a linear system, is a well-established measure of the sensitivity of the solution of an inverse problem. It is often called the condition number (*cond*) and is independent of round-off and computation errors. Researchers have extensively employed *cond* as a measure to find the optimal sensing positions for electromagnetic source reconstruction in current measurement, biomagnetics, shape estimation, retrieve model characteristics, and non-destructive testing.

As is evident from previous research, it is critical to minimize the *cond* of $P_{cmpt}$. This ensures that the linear system remains well conditioned. It is ascertained that array placement in the vertical plane of the tower, with each sensor at zero horizontal distance from the phase conductor, sufficiently reduces *cond* for (3.14) and (3.15), the aids in convergence of the estimation algorithm detailed in the following sections. For array placement at optimal height, we calculated the *cond* for position matrices $P_x$ and $P_y$ for a number of system models. For the configuration illustrated in Figure 3.11, different models are simulated for array placement at heights between 4 m towards and away from the conductors on vertical plane, and for symmetric sag ranging from 1 to 3.5 m. As shown inset in Figure 3.11, *cond* remains close to unity for the aforementioned models.

This analysis affirms that our method can be adopted for any tower configuration using the derived equations; the sensor array needs to be adjusted at a location where *cond* is close to unity and the magnitude of the magnetic field from phase conductors is in the sensor detection range.

### 3.2.2.3   Innovative Method for Sag and Phase Current Reconstruction

The novelty of our work lies in the fact that we can estimate electric current and conductor sag simultaneously with dual-axis sensors equal to the number of phase conductors. We have considered the effect for various important factors in magnetic field distribution in our algorithm. Those factors are span length between towers, the impact of sagged conductors, and the magnetic field from neighboring span. For sag estimation, the sensor to conductor distance remains fixed while sag for a line suspended between two towers will change, predominantly increasing with a rise in conductor temperature from time to time. After the sensors have been placed at sensing points, the magnetic fields for the horizontal and vertical components are separately represented by coefficient matrices. On installation of the array at the support tower, position matrices $P_x$, $P_y$ are determined based on mechanical parameter $\alpha_p$ for sagged conductors and conductor–sensor distances at the support tower, and then stored in memory. The proposed method can detect any change in sag whether in all (symmetrical) or in one or more (unsymmetrical) phase conductor. It also estimates the electric current by solving an inverse reconstruction problem. The algorithm is designed based on three important observations:

1) Electric current is equivalent whereas position matrices $P_{cmpt}$ differ for all magnetic field components, hence if the underlying $P_{cmpt}$ and only one magnetic field component is known one can reconstruct the electric current.
2) Magnetic field components are proportional to sag and conductor–sensor distance, implying that the product of electric current reconstruction and $P_{cmpt}$ along one axis can be weighted at other axes to verify the reconstruction results.
3) The zero diagonal term $A_{mm}$ in (3.15) indicates that change of sag in phase conductor $j$ does not contribute to vertical magnetic field magnitude at sensor $i$ closest to the victim conductor $j$, whereas the non-zero diagonal entity in (3.14) shows that a change of sag varies the horizontal magnetic field magnitude of sensor $i$ closest to victim conductor $j$. Thus, the reconstruction results using $P_y$ multiplied by $P_x$ for observations 1 and 2 can be compared with the magnetic field along the $x$ axis iteratively to determine any change in sag.

The complete process is illustrated in Figure 3.12. The algorithm starts by fetching $P_y$ from memory, stored at installation or during the last memory update, and using the magnetic field sampled along the $y$ axis to perform source current reconstruction $I_j$. Then, the product of reconstructed electric currents $I_j$ and $P_x$ is compared with the sampled magnetic field along the $x$ axis and measured against the threshold. If the difference $Diff_n$ for comparison exceeds the threshold, the comparison is carried out iteratively for updated values of $P_x$ and $P_y$ obtained by incrementing $\alpha_p$. With each iteration results are analyzed to determine the point where the difference between measured and computed $B_{xi}$ returns a small $Diff_n$. Once this criterion has been fulfilled, the algorithm terminates by returning the corresponding sag in all conductors. The system keeps on computing $I_j$ until $Diff_n$ remains within the threshold.

#### Numerical Simulations

We tested the designed algorithm to perform current reconstruction and sag estimation for a three-phase typical support tower configuration with numerical simulations processed in MATLAB, then experimentally verified the findings with a scaled laboratory setup. For numerical simulations, the complete arrangement shown in Figure 3.11

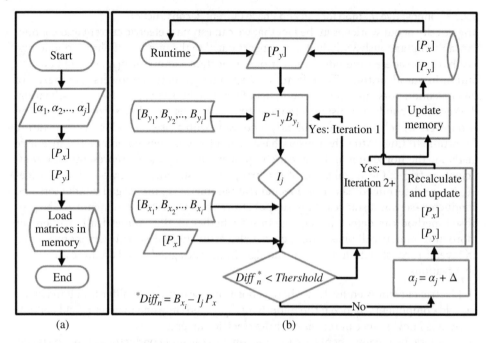

**Figure 3.12** Sag estimation and current reconstruction algorithm: (a) one-time execution at installation and (b) run-time execution. (*Source:* Reprinted with permission from A.H. Khawaja et al., Estimation of current and sag in overhead power transmission lines with optimized magnetic field sensor array placement. *IEEE Transactions on Magnetics*, vol. 53, no. 5, pp. 1–10, May 2017.)

was utilized. A sensor array was assumed to be placed at a height of 2 m above the phase conductors. The array consists of three dual-axis sensors sensitive along the horizontal and vertical axes. The sensor–sensor distance was kept the same as the conductor–conductor distance, i.e. 11.8 m. We used the linear system given above to simulate horizontal and vertical projections of the magnetic field vector at each of the sensing points. The effect of span length and sag is accounted for in position matrices $P_x$ and $P_y$ during the simulations. With each iteration, sag is redefined with an increment of 0.02 m and then used in the algorithm. This ensures that our algorithm can resolve a change of up to 0.02 m in sag. Reducing this resolution will increase computation time. However, for practical implementation, it is capped by the resolution of the sensor and the strength of the magnetic field radiated by the conductors, therefore the step size needs to be adjusted accordingly. The algorithm is tested for three different sag scenarios.

Table 3.2 elaborates different test scenarios and summarizes results for identification of the corresponding sag for respective conductors and reconstructed current. The results show that sag and electric current are correctly estimated, with error less than 1% in all cases. Graphically, $Diff_n$ for all three test scenarios can be seen in Figure 3.13 and given numerically in Table 3.2.

### Proof of Concept Experiment

To demonstrate conformity with simulation results, we installed a scaled setup of a typical one-circuit configuration in the laboratory. The arrangement consisted of two

**Table 3.2** Algorithm output for different sag scenarios in one circuit configuration.

| Test conditions/ conductor (C) | Difference between measured and calculated entities with each iteration | | | | | | | | |
| | Scenario 1: Unsymmetrical sag | | | Scenario 2: Unsymmetrical sag | | | Scenario 3: Symmetrical sag | | |
| | C 1 | C 2 | C 3 | C 1 | C 2 | C 3 | C 1 | C 2 | C 3 |
|---|---|---|---|---|---|---|---|---|---|
| **Iterations** | Diff1 | Diff2 | Diff3 | Diff1 | Diff2 | Diff3 | Diff1 | Diff2 | Diff3 |
| 1 | −1.28 | 3.599 | 0.413 | −1.287 | 3.591 | 6.005 | −9.568 | 4.853 | 5.857 |
| 30 | −0.932 | 3.405 | 0.1334 | −0.939 | 3.397 | 5.778 | −8.391 | 4.549 | 5.576 |
| 59 | −0.469 | 3.14 | −0.264 | −0.477 | 3.132 | 5.466 | −6.336 | 4.094 | 5.178 |
| 88 | 0.1727 | 2.754 | −0.876 | 0.1645 | 2.747 | 5.013 | −1.826 | 3.342 | 4.566 |
| 117 | 1.1265 | 2.144 | −1.934 | 1.1173 | 2.137 | 4.296 | 16.07 | 1.858 | 3.507 |
| 146 | 2.6906 | 1.031 | −4.206 | 2.6797 | 1.026 | 2.989 | 51.314 | −2.44 | 1.232 |
| 175 | 5.728 | −1.63 | −12.6 | 5.7137 | −1.63 | −0.14 | 2.1652 | −169 | −7.16 |
| 204 | 14.178 | −16.6 | −42.48 | 14.154 | −16.6 | −17.7 | −4.73 | −3.95 | −37 |
| 233 | 161.27 | −12.3 | −7.089 | 161.08 | −12.3 | −12.8 | −7.48 | 1.366 | −1.63 |
| 262 | 20.393 | −1.39 | −3.054 | 20.365 | −1.39 | 0.131 | −8.96 | 3.076 | 2.406 |
| **Min (Diffn)** | 0.0043 | 0.001 | 0.001 | 0.012 | 0.003 | 0.033 | 0.0011 | 0.006 | 0.007 |
| **min@Iteration** | 81 | 161 | 41 | 82 | 161 | 261 | 181 | 221 | 241 |
| **Sag (memory)** | 1.2 | 1.4 | 1.6 | 1.2 | 1.4 | 1.4 | 2.2001 | 1.8 | 1.6 |
| **New sag** | 2 | 3 | 2 | 2 | 3 | 4 | 4 | 4 | 4 |
| **Sag calculated** | 2.0101 | 3.01 | 2.01 | 2.0201 | 3.01 | 4.01 | 4.0105 | 4.01 | 4.01 |
| $I_r$ | 845.3 | −389 | -454.1 | 844.1 | −389 | −457 | 845.23 | −389 | −455 |

Sag at installation: 1 m (symmetrical), sag increment/iteration = 0.01 m, test current $I_1$ = 845.3204, $I_2$ = −389.8153, $I_3$ = −455.5052.

*Source:* Reprinted with permission from A.H. Khawaja et al., Estimation of current and sag in overhead power transmission lines with optimized magnetic field sensor array placement. *IEEE Transactions on Magnetics,* , vol. 53, no. 5, pp. 1–10, May 2017.

support towers and three phase conductors placed in flat formation configuration. For this experimental arrangement, due to non-availability of a support structure above the conductors, contrary to the system arrangement in Figure 3.11, it was convenient to place the array underneath the conductors on the support tower. However, according to our method, a position on the vertical plane is selected where the *cond* of the system model is close to unity. Consequently, each sensor is placed below the phase conductor at a distance of 27 cm and the sensor–sensor distance was kept same as the conductor–conductor distance, i.e. 30 cm. We adjusted the conductors so the phase conductors had unsymmetrical sag of 29, 30.5, and 32 cm, respectively, which was measured with a calibrated Vernier caliper. Three resistive loads each of 14.6 Ohms with maximum power rating of 4 kW were used to emulate a balanced load state whereas load current imbalance was generated by increasing the resistive load for one conductor. The condition number *cond* of the system models $P_x$ and $P_y$ under these dimensions were 2.38 and 2.67, respectively.

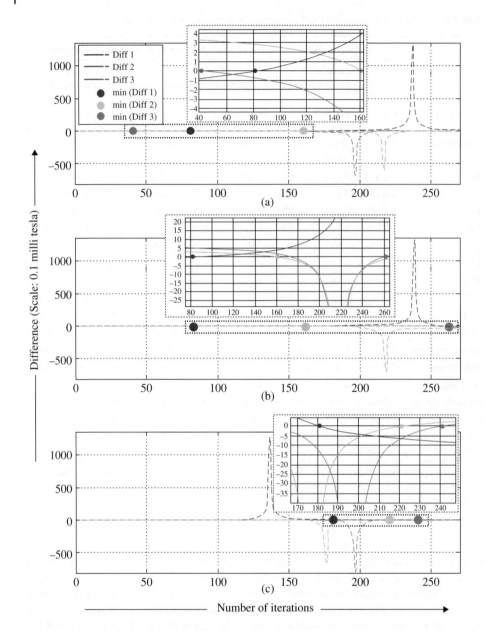

**Figure 3.13** Graphical representation of *Diff_n*, numerically described in Table 3.2. (*Source:* Reprinted with permission from A.H. Khawaja et al., Estimation of current and sag in overhead power transmission lines with optimized magnetic field sensor array placement. *IEEE Transactions on Magnetics*, vol. 53, no. 5, pp. 1–10, May 2017.)

The choice of a suitable sensor is vital for accurate results. Magnetoresistance (MR) magnetometers are considered to be one of the most common magnetic field sensors, the resistance of which varies proportionally to the applied magnetic field along the sensitive axis determined by fabricated MR layers [13]. For our application, we targeted linear magnetic field sensors based on MR. They are commercially available configured in a balanced Wheatstone bridge arrangement, where a change in magnetic field varies the resistance and yields an output voltage proportional to the applied magnetic field in the sensing direction. The work demonstrated in Chapter 2 confirms that TMR sensors are more appropriate than AMR sensors based on low power consumption and straightforward circuitry without any set/reset pulse. TMR sensors also show a better time response at exposure to a frequency that varies the uniform magnetic field and high strength transients. In addition, it has been verified that TMR sensors are low cost, have a large linear range, can work in range of temperatures, and consume far less power compared to AMR and Hall-effect magnetic field sensors [14]. We therefore used a recently developed tri-axis TMR sensor [15] in which the samples are captured only for the two components required for our method.

Another objective during experimental setup was design of a suitable instrumentation chain. As the magnetic field around the power transmission lines is in order of microTesla, it requires a large amplification according to sensor sensitivity, supply voltage, and analog to digital converter (ADC) range. For detection of such a weak magnetic field, it is necessary to filter out the Earth's magnetic field. The Earth's magnetic field remains constant at a point in space and results in a voltage bias at bridge output. The presence of the Earth's magnetic field saturates the high gain amplifier for the measurement system. We also aimed to measure the alternating magnetic field from AC power lines, so the voltage bias due to the Earth's magnetic field in output needs to be removed. This is done by connecting both ends of the sensor bridge to high-pass filters. The filters remove the offset in differential output from the Earth's magnetic field as well as any offset from resistance imbalance in the Wheatstone full bridge configuration. The differential signal from the Wheatstone bridge is amplified in two phases: first with an instrumentation amplifier (INA333) on a designed printed circuit board (PCB), then with an operational amplifier (AD8597) available on an evaluation board. The gain of INA333 is set to 23 dB, whereas the unity gain of AD8597, pre-configured on the evaluation board, is changed to 3 dB by soldering a resistor. A stable power supply drives all three sensors with a bipolar supply voltage of 5 V. A simultaneous six-channel ADC on the AD7656 evaluation board samples the analog measurements and then transmits the data to a laptop by an evaluation board. The complete experimental setup is shown in Figure 3.14. Once the phase conductors have been connected to the resistive loads, the experiment is completed in two steps:

1) The magnetic field component along the $x$ axis and current measurement using an ammeter are sampled simultaneously.
2) This is repeated for the $y$ axis magnetic field component. The sensing direction is selected with jumpers to choose Wheatstone bridge connection to instrumentation amplifier. The measurements from both sensing directions ($x$, $y$) were synchronized by current measured with an ammeter.

To test for accurate sag estimation and current reconstruction, a symmetrical sag of 20 cm (different from the real sag) is loaded into the memory. Incremental step size is kept at 0.1 cm to cater for small changes in sag. The algorithm accurately estimates the

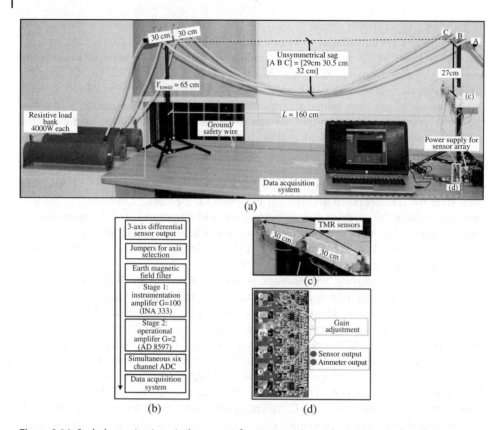

**Figure 3.14** Scaled one-circuit typical tower configuration with attached magnetic field sensor array: (a) complete experimental setup, (b) block diagram of the instrumentation chain for sensor array data processing, (c) sensor array with first-stage amplifier, and (d) second-stage amplification achieved by adding a resistor to the AD8597 unity gain configuration on an ADC evaluation board. (*Source:* Reprinted with permission from A.H. Khawaja et al., Estimation of current and sag in overhead power transmission lines with optimized magnetic field sensor array placement. *IEEE Transactions on Magnetics*, vol. 53, no. 5, pp. 1–10, May 2017.)

unsymmetrical sag of 29, 30.5, and 32 cm in conductors A, B, and C, respectively. This is calculated with an error less than 1% by a process identical to the one shown in Table 3.3. Once the algorithm has identified the sag, it proceeds with electric current reconstruction, as shown in Figure 3.15. The measured magnetic field from three sensors along the $y$ axis is compared with the calculated magnetic field, obtained by multiplying current $I_j$ measured with an ammeter and position matrix $P_y$. The current measured with three clamp ammeters is used as a reference to determine error in the reconstructed current. The results confirm that the measured magnetic field complies with the calculated magnetic field, and the reconstructed current waveform follows the ammeter waveform closely. The root mean square error between the reference and reconstructed waveforms remains less than 1%. As the Euclidean distance is the measure of similarity between two vectors, in our case it is employed to measure the dissimilarity, i.e. the percentage difference between the reference and reconstructed waveforms using magnetic field sensor data. The percentage difference remains less than 2.6% for load balanced and imbalanced states (Table 3.3).

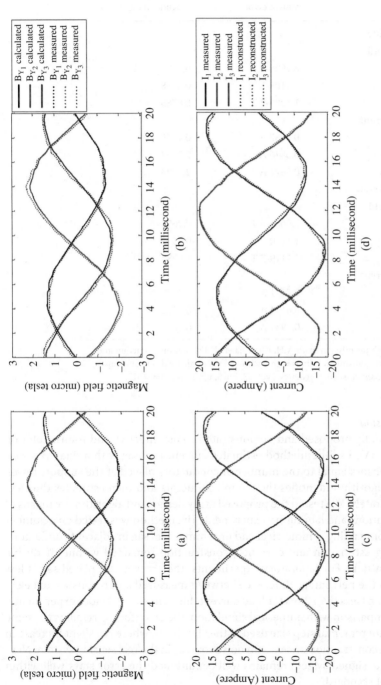

**Figure 3.15** Sensed magnetic field and reconstructed current waveforms for balanced and imbalanced load currents for one circuit configuration, referenced with an ammeter. (a) and (c) Magnetic field profile for balanced and imbalanced current in phase conductors, respectively. Here, $B_{y_i}$ calculated = $P_y * I_j$ (current sensed by ammeter) and $B_{y_i}$ measured represents sensor data. (b) and (d) Current reconstruction for balanced and imbalanced current. $I_j$ measured is ammeter readouts, $I_j$ reconstructed = $P_y^{-1} B_{y_i}$. (*Source*: Reprinted with permission from A.H. Khawaja et al., Estimation of current and sag in overhead power transmission lines with optimized magnetic field sensor array placement. *IEEE Transactions on Magnetics*, vol. 53, no. 5, pp. 1–10, May 2017.)

**Table 3.3** Root mean square error and percentage difference between reference and calculated measurements for experimental validation.

| Subfigures in figure | Root mean square error | Percentage difference (Euclidean distance) |
|---|---|---|
| **Load balanced state** | | |
| **(a) Magnetic field** | | |
| Conductor 1 | 6.9926-8 T | 0.242 |
| Conductor 2 | 9.4189-8 T | 0.3268 |
| Conductor 3 | 1.1206-7 T | 0.8785 |
| **(b) Electric current** | | |
| Conductor 1 | 0.5087 A | 0.4135 |
| Conductor 2 | 0.6846 A | 2.5162 |
| Conductor 3 | 0.5621 A | 2.0923 |
| **Load imbalanced state** | | |
| **(c) Magnetic field** | | |
| Conductor 1 | 7.5661-8 T | 0.9656 |
| Conductor 2 | 1.1150-7 T | 0.5572 |
| Conductor 3 | 1.1119-7 T | 1.0982 |
| **(b) Electric current** | | |
| Conductor 1 | 0.7571 A | 0.8129 |
| Conductor 2 | 0.6114 A | 0.7312 |
| Conductor 3 | 0.4941 A | 0.3099 |

*Source:* Reprinted with permission from A.H. Khawaja et al., Estimation of current and sag in overhead power transmission lines with optimized magnetic field sensor array placement. *IEEE Transactions on Magnetics*, , vol. 53, no. 5, pp. 1–10, May 2017.

### Results and Discussion

We have successfully proposed and demonstrated a cost-effective and robust solution for diagnosis of HVTLs. The method estimates current and sag with a dual-axis sensor array with sensors equal to the number of conductors placed at the support tower. The developed algorithm identifies the sag in conductors and then estimates the electric current accurately. We tested the proposed array design and algorithm by means of numerical simulations and laboratory experiment. The method was tested on a number of scenarios encompassing symmetrical and unsymmetrical sag in adjacent conductors. The error for sag estimation and current reconstruction remained within 1% during numerical simulations. For laboratory experiments, the error in calculated sag is less than 1% whereas the percentage difference between measured and reconstructed electric current remains less than 2.6% for load current different states. Under experimental conditions in comparison with numerical simulation, the error for sag remains the same due to the small incremental step size used in the algorithm, whereas a slight increase in error for phase reconstruction arises from noise sources in the real sensing system. Nevertheless, as the designed system remains well conditioned the error stays well within the measurement standard.

## 3.3 Monitoring of Operation Parameters of Power Transmission Lines

### 3.3.1 Conductor Elongation and Motion

A typical three-phase one-circuit tower configuration is presented in Figure 3.16. For any such tower configuration, sag can be defined as the conductor elongation over a period of time. Conductor elongation reduces ground clearance and is the result of insufficient transfer of heat between the conductor and the surrounding air. The main reasons for heating the conductor over a prolonged period of time is extreme hot weather conditions and overloading of the transmission line during peak hours. Some other reasons may include reduction of mechanical tension at support towers where conductors are attached. With the increase in sag, conductors become exposed to wind-induced vibrations of larger amplitude. In regions with snowfall, accumulation of snow creates an airfoil that increases conductor elongation and results in wind-induced oscillations of higher amplitudes. As evident from Figure 3.16, sag and conductor motion are related to one another. The characteristics of wind-induced conductor motion include the maximum angle of rotation $\theta$, the maximum amplitude from mean position $x^r(t_0 : t_1)$, and frequency of rotation $(1/x^r(t_0 : t_1))$.

Apart from the application of efficient sensing technology, another aspect of magnetic field measurement in estimation of electric and spatial parameters is knowledge of the field strength at a sensing point near HVTL lines. For the same effect, one can use the Biot–Savart law to determine the magnetic field from each of the phase conductors at the sensing point. As in our scenario of extremely low frequency (ELF) (50 or 60 Hz), there is no mutual coupling between the magnetic and electric fields, and their near-field effect can be analyzed separately.

To evaluate the impact of the magnetic field on the catenary suspended between two transmission towers, as shown in Figure 3.16, consider a small cross-sectional length $d\vec{l}_p$ located at distance $\vec{r}$ generating magnetic field $d\vec{B}_p$ when current $I$ flows through a phase conductor $p$. Its effect at sensing point $(x_s, y_s, z_s)$ on the support tower can be calculated

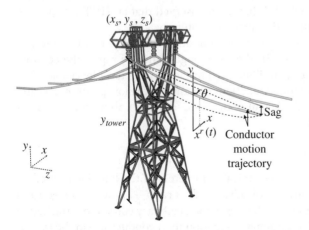

**Figure 3.16** Profile of sag and conductor motion for HVTLs.

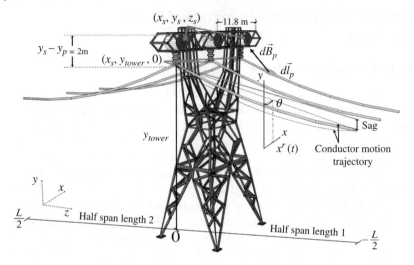

**Figure 3.17** Detailed profile of sag and conductor motion for HVTLs.

by the Biot–Savart law. According to the Biot–Savart principle, the magnetic field $d\vec{B}_p$ is generated by the small cross-sectional length and evaluated as

$$d\vec{B}_p = \frac{\mu_0 I_i}{4\pi} \frac{d\vec{l}_p \times \vec{r}}{|\vec{r}|^3} \tag{3.21}$$

As is evident from Figure 3.17, the small cross-section $d\vec{l}_p$ varies along the $y$ and $z$ axes. The mathematical relationship can be expressed as

$$d\vec{l}_p = 0\hat{i}_x + dy_p\hat{i}_y + dz_p\hat{i}_z \tag{3.22}$$

$$d\vec{l}_p = 0\hat{i}_x + \frac{dy_p}{dz_p}dz_p\hat{i}_y + dz_p\hat{i}_z \tag{3.23}$$

Here, variable $y_p$ represents the coordinates of the sagged conductor and requires further resolution. For the same effect, it has been observed that an HVTL suspended between two adjacent towers follows the shape of a catenary.

The shape of a catenary is a function of the conductor weight per unit length, $w$, the horizontal component of tension, $H$, the span length, and the sag of the conductor. Conductor sag and span length are illustrated in Figure 3.17 for a level span. The exact catenary equation uses hyperbolic functions. Relative to the low point of the catenary curve shown in Figure 3.18, the height of the conductor, $y(z)$, above this low point is given by the following equation

$$y(z) = \frac{H}{w} \cosh\left(\left(\frac{w}{H}x\right) - 1\right) \tag{3.24}$$

Sag can be directly related to the mechanical parameters of the line determined by a constant $\alpha$, where $\alpha = \frac{w}{H}$. The inverse of this, which is ratio $\frac{H}{w}$, is also referred to as the catenary constant by some authors. Therefore, for a catenary conductor attached to tower at height $y_{tower}$, the underlying catenary equation for a conductor $p$ in the $(x, y, z)$ coordinate system is:

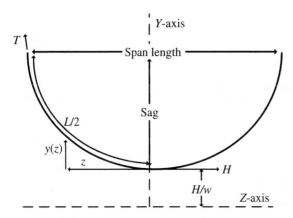

**Figure 3.18** Catenary curve for a level span.

$$y_p = y_{tower} + \frac{1}{\alpha}(\cosh(\alpha_p z_p) - 1), -\frac{L}{2} \leq z_p \leq \frac{L}{2} \tag{3.25}$$

Here, $L$ denotes the length of span between adjacent towers.

Now, $\frac{dy_p}{dz_p} = \sinh(\alpha_p z_p)$.

By substituting $\frac{dy_p}{dz_p}$ in $d\vec{l}_p$, this becomes

$$d\vec{l}_p = \sinh(\alpha_p z_p)dz_p\hat{i}_y + dz_p\hat{i}_z \tag{3.26}$$

Distance vector $\vec{r} = (x_s - x_p)\hat{i}_x + (y_s - y_p)\hat{i}_y + (z_s - z_p)\hat{i}_z$.

Now substituting $d\vec{l}_p$ and $\vec{r}$ in $d\vec{B}_p$, this becomes

$$d\vec{B}_p = \frac{\mu_o I_i}{4\pi}\left[\frac{(\sinh(\alpha_p z_p)dz_p\hat{i}_y + dz_p\hat{i}_z) \times ((x_s - x_p)\hat{i}_x + (y_s - y_p)\hat{i}_y + (z_s - z_p)\hat{i}_z)}{|\vec{r}|^3}\right] \tag{3.27}$$

The cross-product becomes

$$d\vec{B}_p = \frac{\mu_o I_i}{4\pi}\left[\frac{-\sinh(\alpha_p z_p)dz_p\hat{i}_z + (z_s - z_p)\sinh(\alpha_p z_p)\hat{i}_x dz_p + (x_s - x_p)\hat{i}_y dz_p - (y_s - y_p)\hat{i}_x dz_p}{|\vec{r}|^3}\right] \tag{3.28}$$

Rearranging,

$$d\vec{B}_p = \frac{\mu_o I_i}{4\pi}\left[\frac{(((z_s - z_p)\sinh(\alpha_p z_p)) - (y_s - y_p))\hat{i}_x + (x_s - x_p)\hat{i}_y - \sinh(\alpha_p z_p)\hat{i}_z}{|\vec{r}|^3}\right] dz_p \tag{3.29}$$

Now, integrating $d\vec{B}_p$ to evaluate the magnetic field vector $\vec{B}_s$ from two adjacent half-span lengths of transmission line, this becomes

$$\vec{B}_s = \frac{\mu_o I_p}{4\pi}\int_{-\frac{L}{2}}^{\frac{L}{2}}\left[\frac{(z_s - z_p)\sinh(\alpha_p z_i) - (y_s - y_p)}{|\vec{r}|^3}\hat{i}_x + \frac{(x_s - x_p)}{|\vec{r}|^3}\hat{i}_y + \frac{(-\sinh(\alpha_p z_p)(x_s - x_p))}{|\vec{r}|^3}\hat{i}_z\right] dz_i \tag{3.30}$$

where $\mu_o$ is permeability of free space.

Then,

$$\vec{B}_s = \vec{B}_{X_s} + \vec{B}_{Y_s} + \vec{B}_{Z_s} \tag{3.31}$$

where

$$\vec{B}_{X_s} = \hat{i}_x \frac{\mu_o I_p}{4\pi} \int_{-\frac{L}{2}}^{\frac{L}{2}} \frac{((z_s - z_p)\sinh(\alpha_p z_p)) - (y_s - y_p)}{|\vec{r}|^3} dz_p \tag{3.32}$$

$$\vec{B}_{Y_s} = \hat{i}_y \frac{\mu_o I_p}{4\pi} \int_{-\frac{L}{2}}^{\frac{L}{2}} \frac{x_s - x_p}{|\vec{r}|^3} dz_p \tag{3.33}$$

$$\vec{B}_{Z_s} = \hat{i}_z \frac{\mu_o I_p}{4\pi} \int_{-\frac{L}{2}}^{\frac{L}{2}} \frac{-((x_s - x_p)\sinh(\alpha_p z_p))}{|\vec{r}|^3} dz_p \tag{3.34}$$

We have considered the horizontal magnetic field component for detection since it contains the maximum flux density for the given geometry. Here, the field radiated only from the neighboring span lengths is considered. Moreover, only half-span lengths on both sides of the tower are considered based on the fact that the magnetic field decays rapidly by a factor of $|\vec{r}|^3$. Furthermore, the sensing point $(x_s, y_s, z_s)$ is located on the tower where the magnetic fields from both span lengths superimpose at the sensing point. Therefore, the influence can be calculated by adding the effect from both span lengths, where the half-span length 1 corresponds to the flux density from $-L/2$ to $0$, and the half-span length 2 corresponds to the flux density from $0$ to $L/2$. For the same effect both half-span lengths are integrated separately, and $\vec{B}_{X_s}$ becomes

$$\vec{B}_{X_s} = \hat{i}_x \frac{\mu_o I_p}{4\pi} \left[ \begin{array}{l} \int_{-\frac{L}{2}}^{0} \left[ \frac{((z_s - z_p)\sinh(\alpha_p z_p)) - (y_s - y_p)}{[(x_s - x_p)^2 + (y_s - y_p)^2 + (z_s - z_p)^2]^{3/2}} \right] dz_p \\ + \int_{0}^{\frac{L}{2}} \left[ \frac{((z_s - z_p)\sinh(\alpha_p z_p)) - (y_s - y_p)}{[(x_s - x_p)^2 + (y_s - y_p)^2 + (z_s - z_p)^2]^{3/2}} \right] dz_p \end{array} \right] \tag{3.35}$$

Another source of the magnetic field is the image current induced in the conducting earth, which is superimposed on the one generated by the phase conductors. For calculations at a larger distance from the source conductors, the effect of the magnetic field generated by the image current should be considered. Contrary to that, our application calls for the measurements to be performed in close vicinity to current-carrying conductors, where the influence of the magnetic field generated by image currents is negligible and therefore has not been considered in the calculations.

### 3.3.2 Detection and Estimation

#### 3.3.2.1 Overall System Design

Spatial monitoring of HVTLs refer to gain some real-time knowledge of physical positioning of conductors, particularly the conductor elongation parameters known as the sag and wind-induced conductor motion parameters. For the same effect in this chapter, we propose a simple estimation technique using a single-axis magnetic field sensor mounted at the transmission tower. Our system detects the change in the transmission line induced by sag and wind-induced conductor motion, and estimates its characteristics through robust algorithms, which is a significant development over methods that require a number of sensors. These algorithms are optimized by means of an empirical formula that reduces the computation time for each iteration. Towards the tower, our method requires magnetic field measurements from only a single axis, as any change in sag and wind-induced conductor motion results in a detectable variation of magnetic field along each axis for the same current amplitude. At the substation, electric current measurements are observed that are readily available from the regular current transformer (CT). Overall, the sensor provides non-contact magnetic field measurements that are time-stamped and transmitted to the substation. The data are then simultaneously processed along with electric current readings from the CT. The sensing and measurement unit installed at the tower can be powered by solar panels or electromagnetic power harvesting.

A key factor in uninterrupted power delivery is the physical safety of HVTLs. The physical health of power transmission lines is chiefly threatened by two environmental conditions:

- Increased power flow during hot weather increases the line ampacity, which raises the conductor temperature. This results in increased sag as the heat is not transferred efficiently to the air around the line conductors. With the passage of time, the sag exceeds safe ground clearance.
- In regions with snowfall, ice accumulates on sagged conductors and creates an airfoil. In the presence of strong winds, this swings the weighted conductors. The swinging effect is referred to as wind-induced conductor motion. Beyond the safe operational range, sag not only increases the chances of a short circuit from nearby vegetation in presence of wind but also increases the motion amplitude. Conductor motion results in phase-to-phase short circuits and permanent structural damage to the support towers.

We propose the use of a single sensing unit to be mounted on transmission tower. Since a transmission network forms a closed circuit, the current measurements at the substation are directly proportional to the electric current at the suspension tower where the sensor is installed. By using the line impedance parameters (resistance, inductance, and capacitance), we can accurately calculate the current at the tower. The current measurements for this system can be performed by the CT readily available in substations. For magnetic field measurements, a tower-mounted sensor requires conditioning circuitry with appropriate gain, micro-controller handling instrumentation chain, and data communication link between the tower and the substation. Each sensor measurement is required to be synchronized with the current measurements at the substation to estimate the physical parameters, i.e. sag and wind-induced conductor motion in the line. The synchronization can be achieved by using global positioning system (GPS) signals, a method referred to as time-stamping. In our system, the time-stamped magnetic field and current data can be processed to estimate the sag and conductor motion parameters at the data center in the substation. The overview of the complete conceived system is shown in Figure 3.19.

### 3.3.2.2 Cost-effective Alternative to Time Stamping by GPS Receivers

Synchronization between the measurements at the tower and substation ensures precise data processing. The time stamps generated with a GPS have a precision of up to a tenth of a microsecond [16]. Nevertheless, such a solution may incur installation and operational costs. Considering a single measurement cycle of magnetic field and electric

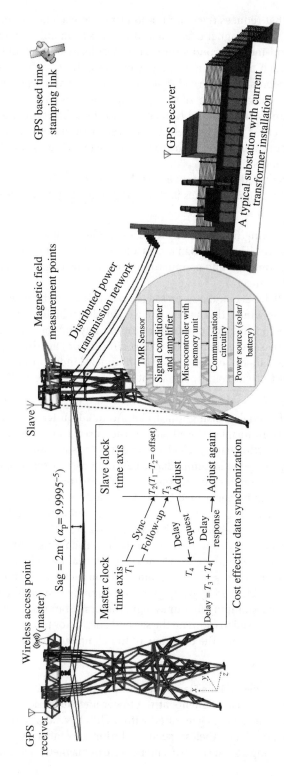

**Figure 3.19** Overview of the complete system outlook for a typical three-phase tower configuration 3L1. Inset: Time synchronization between Master and Slave towers by IEEE 1588. (*Source:* Reprinted with permission from A.H. Khawaja et al., Estimating sag and wind-induced motion of overhead power lines with current and magnetic-flux density measurements. *IEEE Transactions on Instrumentation and Measurement*, April 2017.)

current, our system requires synchronization at 50/60 Hz. Any less precise time synchronization mechanism is feasible for our system. Alternativly, we can use fewer GPS receivers while grouping neighboring towers to form a cluster. The GPS time stamp can be utilized only for longer distance synchronization, i.e. substation to the Master towers. For shorter distances IEEE Standard 1588 can be implemented in a Master–Slave wireless access topology, which is shown in the inset of Figure 3.19, whereby the data from neighboring towers is synchronized with wireless communication based on IEEE Standard 1588. The Standard uses a four-step method to adjust the clock timing of devices configured as Master and Slave. In this context, we have also implemented the IEEE Standard 1588 for wireless synchronization in substation [17]. Based on rapid adaptation of IEEE Standard 1588 by industry, we have conceived a similar approach for this system. In our scheme, one tower with a wireless access point (AP) and GPS receiver becomes a Master and the other towers in the wireless coverage of the Master operate in Slave mode. Adjacent towers to the Master are configured in Slave mode and communicate with the Master as wireless nodes (Figure 3.19). The Master employs a GPS time stamp to adjust its clock, and then adjusts the Slave devices accordingly with a radio link. The radio link or the coverage of the Master depends on the wireless signal strength of the AP. A typical inexpensive AP has a coverage of 15 km [18] in the line of sight. We anticipate that this approach will significantly trim down the equipment cost within the uncertainty limits acceptable for our system.

### 3.3.2.3 Calculation of Magnetic Field From Sagged Conductor by the Simple Approximation Method

One can use the Biot–Savart law to determine the magnetic field from each of the phase conductors at the sensing point. As in our scenario of ELF (50 or 60 Hz), there is no mutual coupling between the magnetic and electric fields, and their near-field effect can be analyzed separately. To evaluate the impact of the magnetic field on the catenary suspended between two transmission towers, consider the magnetic field $(d\vec{B}_i)$ generated by a small cross-sectional length $(d\vec{l}_i)$ at distance $(\vec{r})$ when current $I$ flows through the phase conductor $(i)$. Its effect at sensing point $(\vec{r})$ on the support tower (Figure 3.20) can be calculated by the Biot–Savart law as follows:

$$d
\vec{B}_i = \frac{\mu_o I_i}{4\pi} \frac{d\vec{l}_i \times \vec{r}}{|\vec{r}|^3} \tag{3.36}$$

where

$$d\vec{l}_i = 0\hat{i}_x + dy_i\hat{i}_y + dz_i\hat{i}_z \tag{3.37}$$

$$d\vec{l}_i = 0\hat{i}_x + \frac{dy_i}{dz_i}dz_i\hat{i}_y + dz_i\hat{i}_z \tag{3.38}$$

An HVTL suspended between two adjacent towers at same height follows the shape of a catenary and sag can be directly related to the mechanical parameters of the line determined by a constant $\alpha$. The underlying catenary equation in the $(x, y, z)$ coordinate system is:

$$y_i = \frac{1}{\alpha}(\cosh(\alpha z_i) - 1), -\frac{L}{2} \leq z_i \leq \frac{L}{2} \tag{3.39}$$

where $L$ denotes the length of span between adjacent towers.

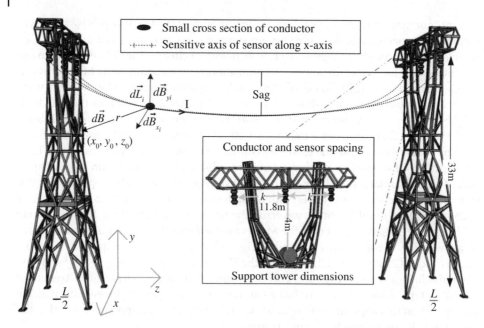

**Figure 3.20** Application of Biot–Savart law for sagged conductors for a typical three-phase tower. (*Source*: Reprinted with permission from A.H. Khawaja et al., Characteristic estimation of high-voltage transmission line conductors with simultaneous magnetic field and current measurements. *2016 IEEE International Instrumentation and Measurement Technology Conference Proceedings*, Nov. 2016.)

Now, $\frac{dy_i}{dz_i} = sinh(\alpha z_i)$ and distance vector $\vec{r} = (x_0 - k)\hat{i} + (y_0 - y_i)\hat{i}_y + (z_0 - z_i)\hat{i}_z$.

Substituting $d\vec{l}_i$ and $\vec{r}$ into $d\vec{B}_i$ and evaluating the magnetic field for one span of transmission length:

$$\vec{B}_i = \frac{\mu_o I_i}{4\pi} \int_{-\frac{L}{2}}^{\frac{L}{2}} \left[ \frac{(z_0 - z_i)\sinh(\alpha z_i) - (y_0 - y_i)}{|\vec{r}|^3}\hat{i}_x + \frac{(x_0 - k)}{|\vec{r}|^3}\hat{i}_y + \frac{(-\sinh(\alpha z_i)(x_0 - k))}{|\vec{r}|^3}\hat{i}_z \right] dz_i$$

$$(3.40)$$

where $\mu_0$ is permeability of free space and $k$ is the spacing between adjacent transmission lines.

To detect magnetic field variations in a practical system, sensors are fabricated to detect the field along one to three axis dimensions depending on usage. From $\vec{B}_i$, it is clear that the magnetic field along the $x$ axis is affected by change in sag, hence it is suitable to employ a sensor with one sensitive axis. Using only one axis data will reduce the complexity and power requirements of the system. Hence, the magnetic field at a sensing point $(x_0, y_0, z_0)$ along the $x$ axis from a sagged phase conductor is

$$\vec{B}_{X_i} = \hat{i}_x \frac{\mu_o I_i}{4\pi} \int_{-\frac{L}{2}}^{\frac{L}{2}} \frac{((z_0 - z_i)\sinh(\alpha z_i)) - (y_0 - y_i)}{|\vec{r}|^3} dz_i \tag{3.41}$$

The expression in (3.41) requires integration for very small regions of the catenary over the complete span of the line. This demands many computations based on the integration technique used. For implementation in practical monitoring systems, a robust

**Table 3.4** Magnetic field calculation for a typical three-phase tower configuration: method, computation time, and error.

| Working condition | | Magnetic field coefficients radiated from each of the conductors (C) along the x axis | | | | | | Computation time, SAM/IST (seconds) |
| | | Integration by symbolic substitution (IST) | | | Simple approximation method (SAM) | | | |
| Span | Sag | C1 | C2 | C3 | C1 | C2 | C3 | |
| --- | --- | --- | --- | --- | --- | --- | --- | --- |
| 300 | 2 | 0.0012 | 0.0013 | 0.0012 | 0.0013 | 0.0013 | 0.0013 | 0.000177/33.09 |
| 300 | 3 | 0.0015 | 0.0016 | 0.0015 | 0.0015 | 0.0016 | 0.0015 | 0.000199/37.82 |
| 350 | 3 | 0.005 | 0.0058 | 0.005 | 0.0052 | 0.006 | 0.0052 | 0.000202/38.99 |
| 350 | 4 | 0.0056 | 0.0065 | 0.0056 | 0.0058 | 0.0068 | 0.0058 | 0.000217/41.23 |
| 400 | 4 | 0.04 | 0.1304 | 0.04 | 0.0413 | 0.1374 | 0.0413 | 0.000245/46.30 |
| 400 | 5 | 0.0418 | 0.1168 | 0.0418 | 0.043 | 0.1224 | 0.043 | 0.000243/46.29 |

*Source:* Reprinted with permission from A.H. Khawaja et al., Characteristic estimation of high-voltage transmission line conductors with simultaneous magnetic field and current measurements. *2016 IEEE International Instrumentation and Measurement Technology Conference Proceedings*, Nov. 2016.

solution is desired. For this, a classical method known as a simple approximation method can provide adequate results well within practical error limits. Using this approach, it is appropriate to compute the magnetic field only at points separated by a fixed distance and take advantage of the closed-loop form. Here the line is assumed to be straight, with different heights at a number of points along the span of the line. The empirical formula derived from (3.41) using this approach becomes:

$$B_{x_i} = \frac{\mu_0 I_i}{4\pi} \sum_{z=-\frac{L}{2}}^{\frac{L}{2}} \frac{(z_0 - z_i)\sinh \alpha z_i - (y_0 - y_i)}{|\vec{r}|^3} \tag{3.42}$$

The combined magnetic field channeled from $n$ phase conductors is summed according to superposition principle as follows:

$$B_{x_{total}} = \sum_{i=1}^{n} B_{x_i} \tag{3.43}$$

We consider a typical one-circuit configuration (inset in Figure 3.20). Tests with sag ranging between 2 and 5 m and with span length 300–400 m between support towers were conducted to evaluate integration (3.36) with our simple approximation method (3.42). Integration for (3.36) is performed by a symbolic substitution technique considering a small cross-sectional length of phase conductors $dL_i$ of 0.1 m (Figure 3.20). The error between the calculations obtained from (3.36) and (3.42) is linear and complies with the measurement uncertainty error of 10% suggested in IEEE Standard 644. The error between both methods under different operating conditions remains within 5%. Table 3.4 summarizes the results for various working conditions.

In addition to calculation error, which remains within 5% using the empirical formula for different working conditions, the error from the measurement system with the magnetoresistive (MR) sensor arises from amplifier noise and intrinsic noise sources such

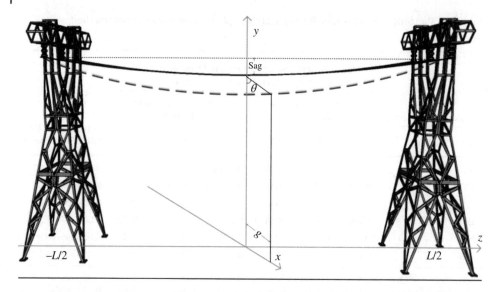

**Figure 3.21** Conductor in motion phenomena annotated as *g*. (*Source:* Reprinted with permission from A.H. Khawaja et al., Characteristic estimation of high-voltage transmission line conductors with simultaneous magnetic field and current measurements. *2016 IEEE International Instrumentation and Measurement Technology Conference Proceedings*, Nov. 2016.)

as thermal shot noise, thermal magnetic noise, electronic 1/f noise, and random telegraphic noise [69]. As sampling frequency for this method is very low, this noise can be significantly reduced with an isolated power source such as a battery to eliminate noise from AC power and selecting an ultra-low noise instrumentation amplifier. In fact, uncertainty from hardware error for magnetic field measurements can be limited to less than 1% using a 12-bit ADC.

### 3.3.2.4 MF of Conductor in Motion

During wind-induced oscillations, wind-induced conductor motion can be viewed as the motion of a sagged phase conductor suspended between two towers. A common shape adopted by HVTLs during conductor motion is single-loop motion proportional to the wind force on the conductors, as shown in Figure 3.21. To compute the emanated magnetic field for such HVTLs, the motion trajectory can be viewed as a conductor in a rotated coordinate system. For a conductor rotation in time $t$, the coordinates of the rotated conductor in the $x$–$y$ plane becomes

$$x'(t) = x(t_0 + \delta t)\cos\theta + y(t_0 + \delta t)\sin\theta \tag{3.44}$$

$$y'(t) = -x(t_0 + \delta t)\sin\theta + y(t_0 + \delta t)\cos\theta \tag{3.45}$$

By using (3.44) and (3.45), one can simulate single-loop conductor motion and also calculate its amplitude $x'(t)$ for $\theta$ degree rotation. In Figure 3.22 the coordinates for such a motion are generated. The magnetic field from such a conductor in motion is calculated by substituting $x'(t)$ and $y'(t)$ in (3.41), which becomes

$$B_{x_i}(t) = \frac{\mu_0 I_i}{4\pi} \sum_{z=-L/2}^{L/2} \frac{(z_0 - z_i)\sinh\alpha z_i - (y_0 - y'_i)}{[((x_0 + x'_i) - k)^2 + (y_0 - y'_i)^2 + (z_0 - z_i)^2]^{3/2}} \tag{3.46}$$

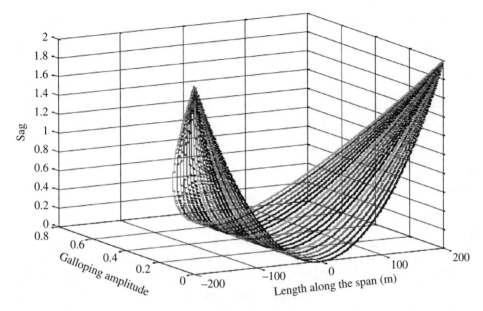

**Figure 3.22** Three-dimensional view of motion trajectory. (*Source:* Reprinted with permission from A.H. Khawaja et al., Characteristic estimation of high-voltage transmission line conductors with simultaneous magnetic field and current measurements. *2016 IEEE International Instrumentation and Measurement Technology Conference Proceedings,* Nov. 2016.)

where $B_{x_i}(t)$ is magnetic field of the $i$ th phase conductor with $\theta$ degrees during a conductor motion.

Under the influence of equal wind force on each phase conductor, the motion characteristics (frequency of oscillation, amplitude, direction, and degree of rotation) are the same for each phase conductor. By applying the superposition principle for (3.46) at regular intervals of time, the change in magnetic field strength can be related to motion characteristics. Figure 3.23 demonstrates the impact of the magnetic field on the sensor at the sensing unit installed under phase conductor B at a distance of 4 m on the support tower. Here the variation in field strength is the result of moving away and then towards the sensing point. The maxima point of field strength represents the mean position whereas two minima points are indicative of maximum conductor motion amplitude away from the sensor.

## 3.4 Spatial Monitoring of HVTLs in Real-world Scenarios

### 3.4.1 Mathematic Model of HVTLs in Real-world Scenarios

The physical safety of HVTLs is a prime contributor to uninterrupted power delivery. To this effect, we foresee two chief environmental factors, as described below.

1) Hot weather conditions lead to an increase in the line ampacity, which in turn raises the conductor temperature. This results in an increased sag when the heat is not transferred to the air surrounding the conductors due to inefficient convective cooling. Increased sag can result in the distance between the bottom part of the conductor

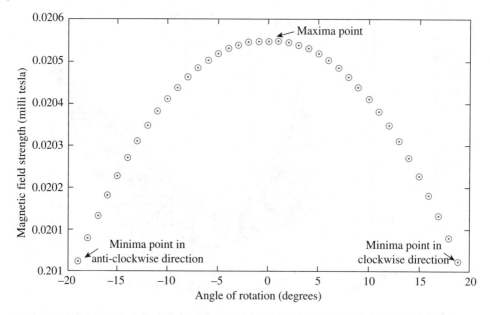

**Figure 3.23** Pattern of magnetic field variation of a conductor in motion. (*Source:* Reprinted with permission from A.H. Khawaja et al., Characteristic estimation of high-voltage transmission line conductors with simultaneous magnetic field and current measurements. *2016 IEEE International Instrumentation and Measurement Technology Conference Proceedings*, Nov. 2016.)

and the ground becoming less than the ground clearance. Another contributor to increased sag is ice accumulation on conductors with snowfall. The ice increases the tension in the bundled conductor, which results in conductor elongation.

2) Airfoil occurs due to ice accumulation on sagged conductors. In this case, strong winds induce swinging of these weighted conductors. This effect is classified as aeo-line vibrations (AV) and conductor galloping (CG), depending upon the frequency and displacement amplitude.

Once sag exceeds the safe operational limit, it becomes correlated to the conductor motion. Conductor sag not only enhances the probability of short circuit with nearby vegetation but also increases the amplitude of the conductor motion in the presence of wind. Large amplitude motion, referred to as CG, cause short circuits with adjacent conductors and collapse of the support towers. To account for these factors, we formulated an efficient method to detect and estimate a change in sag and conductor swinging motion by means of magnetic field and electric current measurements at the support tower and substation, respectively. For the magnetic field, uniaxial magnetic field sensors equal to the number of conductors are installed on the support tower. Since a transmission network forms a closed circuit from the generator to the substation, for electric current measurement the sensed values at the substation are proportional to the electric current flowing in the conductors near the suspension tower. We calculate this using the known line impedance parameters. Next, as the measured magnetic field at the tower varies with a change in conductor position, such variations can be interpreted into corresponding sag and conductor motion by processing them alongside the measured electric current at the substation. These measurements require to be precisely synchronized at all times to ensure accurate estimation of the parameters.

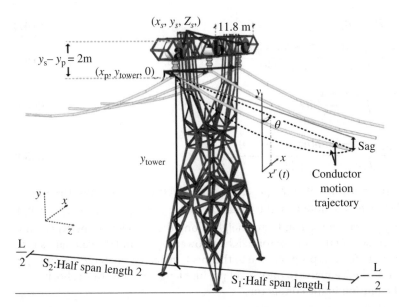

**Figure 3.24** Profile of sag and conductor motion for HVTLs. (*Source:* Reprinted with permission from A.H. Khawaja and Q. Huang, Estimating sag and wind-induced motion of overhead power lines with current and magnetic-flux density measurements. *IEEE Transactions on Instrumentation and Measurement*, vol. 66, no. 5, pp. 897–909, May 2017.)

The overall view of the conceived system is presented in Figure 3.19. The electric current is recorded by CT and relevant supervisory control and data acquisition (SCADA) terminals readily available in substations. For magnetic field measurements, tower-mounted sensors are installed with conditioning and sampling circuitry, with a micro-controller which handles sampled magnetic field data and establishes a communication link between the tower and the substation. Magnetic field data are required to be synchronized with the electric current measurement at the substation to estimate the aforementioned spatial parameters. This synchronization can be achieved by employing a GPS-based time-stamping mechanism. However, installation of a GPS-based time-stamping module may incur high installation and maintenance costs, therefore we found a relatively less expensive solution for distributed time synchronization. Once a synchronized link is established, the data transmitted from magnetic field sensors are processed alongside CT measurements for estimation of sag and motion parameters at a data center in the substation. Such an arrangement is also emphasized in the road map for implementation of a self-aware smart grid, where a synchronized communication link exists between all the components of a distributed power system.

To develop a mathematical model for magnetic field distribution from a number of sagged conductors attached to a support tower, one needs to start with the influence of the magnetic field from a single conductor at one sensing point. We can recall that the magnetic field impact from a current-carrying conductor at a point on the support tower can be calculated using the Biot–Savart law.

For the configuration presented in Figure 3.24, the vector projections $(B_{X_s}, B_{Y_s}, B_{Z_s})$ from the magnetic field vector $(\vec{B}_s)$ measured at any of the sensing points $s$ located at coordinates $(x_s, y_s, z_s)$ are given by

$$\vec{B}_s = \vec{B}_{X_s} + \vec{B}_{Y_s} + \vec{B}_{Z_s} \tag{3.47}$$

$$\vec{B}_{X_s} = \hat{i}_x \frac{\mu_o I_p}{4\pi} \int_{-\frac{L}{2}}^{\frac{L}{2}} \frac{((z_s - z_p)\sinh(\alpha z_p)) - (y_s - y_p)}{|\vec{r}|^3} dz_p \tag{3.48}$$

$$\vec{B}_{Y_s} = \hat{i}_y \frac{\mu_o I_p}{4\pi} \int_{-\frac{L}{2}}^{\frac{L}{2}} \frac{x_s - x_p}{|\vec{r}|^3} dz_p \tag{3.49}$$

$$\vec{B}_{Z_s} = \hat{i}_z \frac{\mu_o I_p}{4\pi} \int_{-\frac{L}{2}}^{\frac{L}{2}} \frac{-(\sinh(\alpha z_p)(x_s - x_p))}{|\vec{r}|^3} dz_p \tag{3.50}$$

where $\mu_o$ is the permeability of free space, $I_p$ is the electric current, and the sensor to conductor distance is denoted by the distance vector $\vec{r} = (x_s - x_p)\hat{i}_x + (y_s - y_p)\hat{i}_y + (z_s - z_p)\hat{i}_z$. Here, $y_p$ refers to the conductor height along the span length between support towers. It is the sum of the conductor height at tower ($y_{tower}$) and the catenary shape adopted by the conductor. It is proportional to the mechanical parameter $\alpha_p$, where $\alpha_p$ is a catenary constant which equates to the weight per unit length divided by the horizontal component of the tension in the conductor:

$$y_p = y_{tower} + \frac{1}{\alpha_p}(\cosh(\alpha_p z_p) - 1), -\frac{L}{2} \leq z_p \leq \frac{L}{2} \tag{3.51}$$

where $L$ denotes the length of span between adjacent towers.

The detailed discussion of the derivation of vector $\vec{B}$ can be found in the literature. It is evident that the field strength for each of these components is affected by a change in sag or the position of the conductors. The magnitude of the magnetic field along all three components can be used to find an optimal sensing direction that receives the maximum flux density, and therefore a sensor sensitive in the same direction can be utilized for measurements. The optimal sensing locations can be determined based on conductor geometry and tower configuration. For instance, in a typical one-circuit tower arrangement for overhead HVTLs, a secure location of sensor placement can be inside the empty space located 2 m above the conductor. The magnetic field has been calculated for a sensing point 2 m above each conductor in the same vertical plane. It has been observed that the magnetic field coefficient along the horizontal component contributes to up to 89.61% of the total field strength per unit electric current (Table 3.5). Since the horizontal magnetic field component varies sufficiently for a change in the spatial position of the conductor, uniaxial magnetic field sensors arranged to detect this component are appropriate for our method. To achieve the same effect, we propose using the same number of sensors as conductors and placing them in the same vertical plane. The sensor–conductor distance should be adjusted according to the magnetic field strength, the resolution of the sensor, and the amplification factor of the conditioning circuitry.

The fact that the sensing points for our proposed measurement system are located on the support tower implies that the magnetic field strength from current-carrying conductors from both span lengths, towards and away from sensors, contribute to the magnitude of the magnetic field at sensing points. In order to ease the computation load from the integral calculation for two full span lengths, we observe that the term $|\vec{r}|^3$ in (3.48) indicates that the magnetic field decays significantly with distance. Due to the same effect, we have considered the magnetic field radiated only from half-spans,

**Table 3.5** Comparison of magnetic field strength along each of the vector projections.

| | Magnetic field strength*$(10_T^{-6})$ contributions from each conductor (C) | | | | | | | | | |
|---|---|---|---|---|---|---|---|---|---|---|
| | Sensing point (a) | | | Sensing point (b) | | | Sensing point (c) | | | Contribution to total strength (%) |
| Vector projections | C1 | C2 | C3 | C1 | C2 | C3 | C1 | C2 | C3 | |
| $x$ axis | 0.450 | 0.002 | 0.000 | 0.002 | 0.450 | 0.002 | 0.000 | 0.002 | 0.450 | 89.61 |
| $y$ axis | 0 | 0.033 | 0.016 | 0.033 | 0 | 0.033 | 0.016 | 0.033 | 0 | 9.93 |
| $z$ axis | 0 | 0.001 | 0.000 | 0.001 | 0 | 0.001 | 0.000 | 0.001 | 0 | 0.458 |

Magnetic field strength per unit electric current for sensing points located 2 m above each conductor

*Source:* Reprinted with permission from A.H. Khawaja and Q. Huang, Estimating sag and wind-induced motion of overhead power lines with current and magnetic-flux density measurements. *IEEE Transactions on Instrumentation and Measurement*, vol. 66, no. 5, pp. 897–909, May 2017.

contrary to the full span lengths found in the literature [4]. Based on this premise, the measured magnetic field complies with the calculated one. Grounded on this important finding, now we only require to integrate the two half-span lengths for the magnetic field calculation. This shortens the integral length to half and save computational cost, in terms of time and energy, for the algorithms. Since the sensing point at $x_s, y_s, z_s$ is sandwiched between two span lengths, it is required to integrate the span lengths separately. The first integral for the half-span $(s_1)$ starts from the maximum sag point $(-L/2)$ and ends at the tower. Similarly, the second integral for the half-span $(s_2)$ begins from the tower and ends at the maximum sag point $(L/2)$ of the other span away from tower. For the same effect, the horizontal magnetic field component becomes

$$\vec{B}_{X_s} = \hat{i}_x \frac{\mu_o I_p}{4\pi} \left[ \int_{-\frac{L}{2}}^{0} \left[ \frac{((z_s - z_p)\sinh(\alpha_p z_p)) - (y_s - y_p)}{[(x_s - x_p)^2 + (y_s - y_p)^2 + (z_s - z_p)^2]^{3/2}} \right] dz_p \right.$$
$$\left. + \int_{0}^{\frac{L}{2}} \left[ \frac{((z_s - z_p)\sinh(\alpha_p z_p)) - (y_s - y_p)}{[(x_s - x_p)^2 + (y_s - y_p)^2 + (z_s - z_p)^2]^{3/2}} \right] dz_p \right] \tag{3.52}$$

where $-\frac{L}{2}$ and $\frac{L}{2}$ refer to half-span lengths ending towards and away from the sensor, respectively. By integrating the variables in $\vec{B}_{X_s}$, we can obtain the magnetic field coefficient $A_X$ for each of the conductors based upon the sensor to conductor distance, span length, and conductor sag. The horizontal magnetic field component $\vec{B}_{X_s}$ can be represented as a product of $\frac{\mu_o I_p}{4\pi}$ and coefficient $A_X$ as

$$\vec{B}_{X_s} = \hat{i}_x \frac{\mu_o I_p}{4\pi} A_X \tag{3.53}$$

A number of HVTLs are arranged on a support tower, which implies that the other conductors affect the magnetic field distribution at the sensing point. The combined magnetic field channeled from $j$ conductors is summed according to superposition principle. The same can be applied to generalize the equation, and compute the magnetic field strength at a point $s$ from $j$ conductors with electric current ($I_j$):

$$B_{X_s} = \sum_{p=1}^{j} \frac{\mu_o I_j}{4\pi} A_j \tag{3.54}$$

As discussed above, the effects of image current are not considered as we perform the measurements in close vicinity to current-carrying conductors. The equation can be adapted for any tower configuration consisting of $i$ sensors and $j$ conductors to obtain the coefficient matrix. It will take the shape of a square matrix as we propose to use sensing points equal to the number of conductors ($i = j$). The resultant horizontal magnetic field component at $i$ sensor locations is summed as

$$B_{X_i} = \left[ \frac{\mu_o}{4\pi} \begin{pmatrix} A_{11} & \cdots & A_{1m} \\ \vdots & \ddots & \vdots \\ A_{m1} & \cdots & A_{mm} \end{pmatrix} \right] * I_j \tag{3.55}$$

The product of magnetic field coefficients and constant $\left( \frac{\mu_o}{4\pi} \right)$ can be denoted as the sensitivity matrix $S_x$ for the horizontal axis and can be written as

$$B_{X_i} = S_x I_j \tag{3.56}$$

Overhead HVTLs are arranged in various circuit and tower configurations depending on the capacity of the line. To verify the suitability of the sensor location on different tower configurations, one can use the described equations to find sensing points with good resolution. As discussed earlier, we found that the horizontal field strength contribution from conductor to sensing point is a maximum when both are located in the same vertical plane. Therefore, the proposed design consists of a similar arrangement where three uniaxial sensors are placed on the support tower above each of the conductors to obtain the measurements. We calculated the maximum magnetic field strength at sensing points located 2 m above each conductor during a typical cycle of the electric current for a three-phase 500 kVA line. The sag is varied between 2 and 4 m, symmetrical and unsymmetrical, with a span length of 300 m considered on both sides of the support towers. Table 3.6 summarizes the results for these test conditions. It can be deduced that a change in sag can be detected by calculating the difference between the measured and calculated magnetic fields, where the calculated magnetic field equates to the product of the sensitivity matrix $S_x$ and the electric current, obtained simultaneously along

**Table 3.6** Magnetic field strength for symmetrical and unsymmetrical sag along the $x$ axis and comparison of magnetic field strength along each of the vector projections.

| Maximum magnetic field strength/cycle for electric current: 1 kA peak–peak | | | | | |
|---|---|---|---|---|---|
| Sag (m) in conductor (C) | | | Strength at sensing point ($10^{-3}$T) | | |
| C1 | C2 | C3 | a | b | c |
| 2 | 2 | 2 | 0.3795 | 0.3799 | 0.3792 |
| 2 | 3 | 2 | 0.3801 | 1.0597 | 0.3799 |
| 2 | 3 | 4 | 0.3804 | 1.0606 | 1.3909 |
| 4 | 4 | 4 | 1.3922 | 1.3966 | 1.3913 |

*Source:* Reprinted with permission from A.H. Khawaja and Q. Huang, Estimating sag and wind-induced motion of overhead power lines with current and magnetic-flux density measurements. *IEEE Transactions on Instrumentation and Measurement*, vol. 66, no. 5, pp. 897–909, May 2017.

with the magnetic field measurement. This observation can be translated to formulate an algorithm to accurately identify a change in sag.

### 3.4.2 MF of HVTLs in Motion for Real-world Scenarios

Overhead HVTLs oscillate primarily because of excitations induced by wind. We can recall from the introduction that wind-induced conductor motion is categorized into AV and CG, where AV refer to conductor oscillation of small amplitude and high frequency (10–100 Hz) and CG refers to large amplitude motion with very low frequency (0–10 Hz). AV is frequently associated with winds blowing the conductor past its critical frequency, whereas CG is linked with wind-induced vibration of ice-accredited conductors [19; 20]. Moreover, researchers have extensively studied and observed the conductor motion, devised methods for prevention, and characterized this motion into various degrees of freedom (DoF). The DoF sums up as torsional motion (conductor rotation around its own axis), horizontal and vertical displacement as 3-DoF, torsional and vertical displacement as 2-DoF, and only vertical displacement as 1-DoF [21; 22]. Additionally, the conductors are observed to oscillate as single-loop, two-loop, and three-loop waves during CG of large amplitude [19]. During a 3-DoF motion the conductor is displaced along the horizontal and vertical axis. Such a displacement induces damage to the clamps holding the conductors on the tower and to the support tower itself. Also, during large amplitude oscillations, there exists a possibility of short-circuit fault with nearby conductors and vegetation. For the same reason, first we devised a mathematical model of oscillating HVTLs to analyze the behavior of the magnetic field of HVTLs in motion. Then unsymmetrical oscillations similar to horizontal and vertical motion for HVTLs are generated, and their impact on magnetic field variation in time is analyzed to develop an efficient algorithm.

To develop a mathematical framework of wind-induced conductor oscillations, the effect can be viewed as motion of a sagged conductor suspended between two towers during a time interval. Conductor motion frequency refers to the time to complete one complete rotation around the mean position. With a conductor rotation $\theta$ in time $t$, the coordinates of catenary conductor $y_p$ in (3.51) in the rotated $x$–$y$ plane will become

$$x^r_p(t) = x_p(t_o) \cos \theta_p + y_p(t_o) \sin \theta_p \tag{3.57}$$

$$y^r_p(t) = -x_p(t_o) \sin \theta_p + y_p(t_o) \cos \theta_p \tag{3.58}$$

For a conductor $p$ in motion whether the type of motion is CG or AV, the magnetic field varies with time. The corresponding variation can be calculated by

$$\vec{B}_{X_s}(t) = \hat{i}_x \frac{\mu_o I_p}{4\pi} \left[ \begin{array}{l} \int_{-\frac{L}{2}}^{0} \left[ \frac{((z_s-z_p)\sinh(\alpha_p z_p))-(y_s-y^r_p(t))}{[(x_s-x^r_p(t))^2+(y_s-y^r_p(t))^2+(z_s-z_p)^2]^{3/2}} \right] dz_p \\ + \int_0^{\frac{L}{2}} \left[ \frac{((z_s-z_p)\sinh(\alpha_p z_p))-(y_s-y^r_p(t))}{[(x_s-x^r_p(t))^2+(y_s-y^r_p(t))^2+(z_s-z_p)^2]^{3/2}} \right] dz_p \end{array} \right] \tag{3.59}$$

where $B_{X_s}(t)$ is the magnetic field radiated from conductor $p$ at time $t$, at an angle rotated $\theta$ degrees and adopting the shape based on $\alpha_p$.

Under the effect of unequal wind force on each of the conductors, motion characteristics (frequency of oscillation, displacement amplitude, direction and degree of rotation) are unsymmetrical for all the individual conductors. Change in magnetic field strength can be related to these characteristics by applying the superposition principle for $j$ conductors at regular intervals of time

$$B_{X_s}(t) = \sum_{p=1}^{j} \frac{\mu_o I_j}{4\pi} A_j(t) \tag{3.60}$$

The magnetic field of such conductors primarily increases when the conductor moves close to the sensing point and reduces when it moves away. As shown in Figure 3.25, the

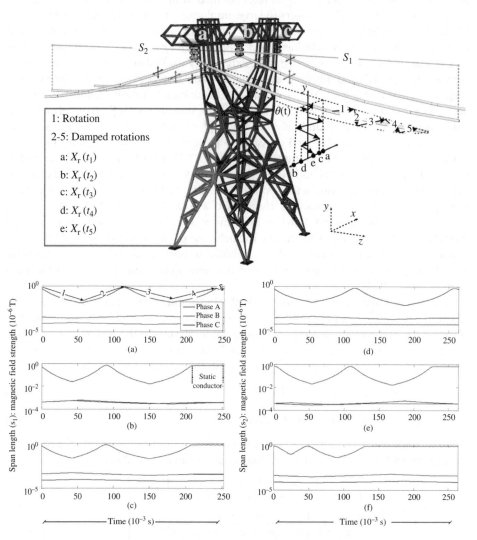

**Figure 3.25** Complex conductor motion outlook. Magnetic field contribution from unsymmetrical conductor motion on both sides of the span length at each sensing point: (a) and (d) first sensor, (b) and (e) second sensor, and (c) and (f) third sensor. (*Source:* Reprinted with permission from A.H. Khawaja and Q. Huang, Estimating sag and wind-induced motion of overhead power lines with current and magnetic-flux density measurements, *IEEE Transactions on Instrumentation and Measurement*, vol. 66, no. 5, pp. 897–909, May 2017.)

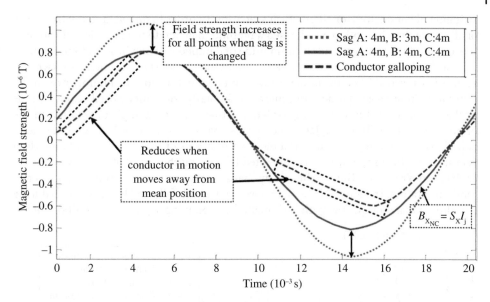

**Figure 3.26** Magnetic field profile comparison for change in sag and during the conductor motion. (*Source:* Reprinted with permission from A.H. Khawaja and Q. Huang, Estimating sag and wind-induced motion of overhead power lines with current and magnetic-flux density measurements, *IEEE Transactions on Instrumentation and Measurement*, vol. 66, no. 5, pp. 897–909, May 2017.)

trajectory can be viewed as a conductor in a rotated coordinate system to compute the emanated magnetic field for such HVTLs. It is contrary to the magnetic field of a conductor when sag has increased, where the field strength remains uniformly constant (low or high) for same electric current amplitude. As shown in Figure 3.26, for a sensing point located 2 m above conductor 1, magnetic field strength is increased when the sag in conductor 2 is reduced from 4 m to 3 m. On the other hand, the strength of the magnetic field reduces when the conductor moves away from the center position, and is the same as that of a static conductor when it is close to its mean position. This shows that to detect conductor motion, the magnetic field measured at the tower can be referenced with the product of sensitivity matrix ($S_x$) and electric current ($I_j$) measured at the substation during the observation window. The magnetic field of a conductor in motion shows a pattern of increasing and decreasing field strength. However, motion is not symmetrical in all conductors which means first we must understand this unsymmetrical motion of conductors for both the span lengths, and then calculate the magnetic field strength for these conductors. To cater for this, we have generated a bi-directional unsymmetrical oscillation in all conductors, and then observed the magnetic field at the proposed sensing points.

### 3.4.3 MF of Conductors for Random Bi-directional Motion

Under the influence of AV or CG, the conductor motion adopts a complex trajectory. It is based on factors such as wind speed, angle between direction of wind and cross-sectional area of conductor, geographical location of HVTLs, and proximity to water reservoirs, river crossings, and the sea [20]. Thus, it is difficult to generalize this swinging of conductors for all practical scenarios. In order to understand the trajectory of all conductors in motion, we devised a method to generate the direction of rotation, the amplitude of rotation, and the amplitude of damped rotation in a random manner.

First the direction of rotation for each conductor, clockwise or counter-clockwise, is selected arbitrarily. Then to simulate a random conductor motion, a number of angles of rotation in increasing order are generated. Similarly, for damped oscillations various angles are generated in decreasing order, less than a random angle of rotation in the opposite direction. The complete trajectory is a vector of rotation $V_{rot}$ along the horizontal axis, containing rotations and damped rotations of each conductor. Similarly, a vector for vertical motion $V_{vert}$ is generated by varying the shape of the conductor for each random rotation. This is achieved by changing the mechanical parameter $\alpha_p$ at each random rotation. In Figure 3.27 we can observe the path followed by conductors during a bi-directional motion. The path is simulated by obtaining $V_{rot}$ and $V_{vert}$ for both of the

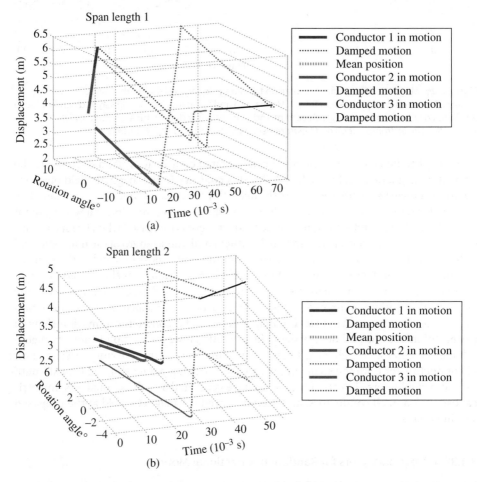

**Figure 3.27** Random bi-directional unsymmetrical path of each of the conductors on the support tower along (a) span length 1 and (b) span length 2. (*Source:* Reprinted with permission from A.H. Khawaja and Q. Huang, Estimating sag and wind-induced motion of overhead power lines with current and magnetic-flux density measurements, *IEEE Transactions on Instrumentation and Measurement*, vol. 66, no. 5, pp. 897–909, May 2017.)

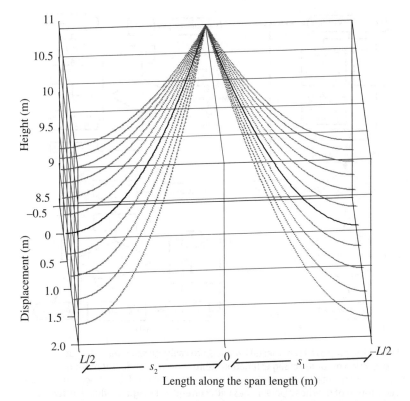

**Figure 3.28** Three-dimensional view of conductor motion at both sides of span lengthes $s_1$ and $s_2$. The black line at the center shows the mean position of the conductor during oscillations. (*Source:* Reprinted with permission from A.H. Khawaja and Q. Huang, Estimating sag and wind-induced motion of overhead power lines with current and magnetic-flux density measurements, *IEEE Transactions on Instrumentation and Measurement*, vol. 66, no. 5, pp. 897–909, May 2017.)

span lengths, $s_1$ and $s_2$, for each of the conductors. Using $V_{rot}$ and $V_{vert}$ simultaneously, one can simulate a bi-direction motion of the conductors. In Figure 3.28, an oscillating conductor is demonstrated, where the trajectory adopted by conductors for $s_1$ and $s_2$ will be different from each other and comprises a magnetic field superimposed from conductors at both sides. Hence, an oscillating conductor at time $t$, for a rotation of $\theta$ degrees, adopts the shape based on $\alpha_p$. The amplitude of such a rotation can be obtained by replacing $V_{rot}$ with $\theta$ and $V_{vert}$ with $\alpha_p$ in the respective equations.

The magnetic field is generated when electric current flows through the conductors. Its strength is proportional to the magnitude of the electric current when the conductors are at rest. However, under the influence of AV or CG, the magnetic field strength generated during each cycle of electric current also depends on displacement amplitude. We can visualize the strength of the magnetic field of each conductor for non-symmetrical bi-directional motion in both span lengths. Here, the dominant magnetic field coefficients are from the conductor closest to the sensor. Motion induced in other conductors has a negligible influence on the magnetic field at the sensing point and therefore can be ignored for motion estimation. It can be further deduced that the magnetic field of the conductor closest to the sensor is only affected by the motion of the conductor.

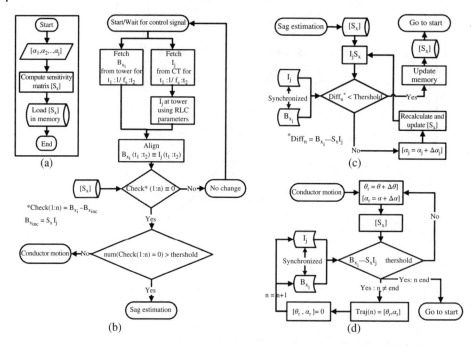

**Figure 3.29** Flowchart of the proposed scheme: (a) execution to calibrate sag parameters at installation, (b) main detection module, (c) sag estimation subroutine, (d) conductor motion estimation module. (*Source:* Reprinted with permission from A.H. Khawaja and Q. Huang, Estimating sag and wind-induced motion of overhead power lines with current and magnetic-flux density measurements, *IEEE Transactions on Instrumentation and Measurement*, vol. 66, no. 5, pp. 897–909, May 2017.)

### 3.4.4 A Unified Algorithm for Sag and Conductor Motion Detection

An algorithm has been developed based upon the mathematical model and the signal characteristics of the measured magnetic field for variation in the sag or wind-induced motion of conductors. It detects when the conductors experience a change in sag or undergo any motion and then estimates the corresponding factors. The algorithm is based on two observations:

1) The magnetic field strength at sensing points varies with a change in sag in one or more conductors.
2) A wind-induced oscillation in one or more conductors varies the magnetic field strength. The magnetic field of such conductors approximates the magnetic field of static conductors when the conductor is close to the mean position.

The developed algorithm is summarized in Figure 3.29. It can be executed at the data center, once the time-stamped data from the magnetic field sensing units from the tower and CT are acquired. At other times, magnetic field sensing units may wait for an acquisition signal from the data center in sleep mode. The initial sensitivity matrix $S_x$ is calculated on sensing unit installation and stored in memory. At this stage $S_x$ is based upon the initial mechanical parameter $\alpha_p$ for each of the conductors, and conductor to sensor distances at the support tower. The same memory recognizes changes in the sensitivity matrix $S_x$ for the other steps in execution. The process is summarized in Figure 3.29a.

The algorithm comprises a parent module that is responsible for distinguishing the change in sag from the conductor motion. Once triggered by data center, it fetches the time-stamped magnetic field sensor data from the tower and CT measurements at the substation. It then calculates the electric current for the tower at distance $d$ transmitting sensor data, by adjusting the CT measurement based on the impedance parameter (resistance (R), inductance (I), capacitance (C)) for the section of transmission line. Finally, the electric current ($I_j$) and magnetic field ($B_{X_i}$) at the tower are aligned according to the time stamps. The next step is to analyze the time synchronized samples spaced by sampling frequency ($F_s$) recorded during the observation window of $t$ seconds. Now, to detect a change in sag or conductor motion, $S_x$ is loaded from memory and multiplied with $I_j$. The product of $S_x$ and $I_j$ is denoted by $B_{X_{iNC}}$. It equates to magnetic field sensor data $B_{X_i}$ when there is no change in system parameters. On the other hand, the magnetic field coefficients in $S_x$ are changed in a pattern for a variation in sag or during the conductor motion which consequently varies the measured magnetic field strength. In particular, $B_{X_i}$ differs from $B_{X_{iNC}}$ throughout the observation window, implying that the sag has been increased. On the other hand, if $B_{X_i}$ is same as $B_{X_{iNC}}$ at some points, and reduces at others, this indicates conductor motion. Thus, sag and conductor motion can be distinguished by a sample-to-sample difference between $B_{X_i}$ and $B_{X_{iNC}}$, termed *Check*. If there are non-zero elements in *Check* for a number greater than a minimum threshold, this indicates an increase in sag, otherwise conductor motion is indicated. These steps are illustrated in Figure 3.29b.

Once any of these two states is flagged, corresponding subroutines handle the estimation process separately. The algorithm for sag estimation computes a difference $Diff_n$ between samples of $B_{X_i}$ and $B_{X_{iNC}}$. If $Diff_n$ exceeds the threshold, the comparison is carried out iteratively for updated $S_x$ obtained by incrementing $\alpha_p$ in (3.51). Results are iteratively compared to find out $\alpha_p$ where the difference between $B_{X_i}$ and $B_{X_{iNC}}$ returns a small $Diff_n$. Once the difference is less than the threshold, the subroutine returns corresponding sag $\alpha_p$ in all conductors, updating the memory value of $S_x$. The execution jumps to the starting point of parent module on completion of the estimation process. The sag estimation subroutine is summed up in Figure 3.29c.

On the other hand, for conductor motion flagged by the parent module, the subroutine retrieves the bi-directional trajectory for the conductor in motion. The motion in conductors other than the one closest to the sensing point has a negligible influence on magnetic field strength and can be ignored. Therefore, the conductor motion is estimated by recompiling the magnetic field coefficient for the conductor nearest to the sensor in the sensitivity matrix $S_x$. In particular, it is achieved by incrementing the rotation angle $\theta$ and the mechanical sag parameter $\alpha_p$ for the closest conductor in (3.57) and (3.58). The algorithm updates the sensitivity matrix, multiplies it with $I_j$, and then compares it with $B_{X_i}$. Complete motion trajectory $Traj(n)$ can be obtained by adjusting $S_x$ for all sampled points, such that the difference between $B_{X_i}$ and $S_x * I_j$ is minimized. Here $i$ and $j$ refer to the sensor and conductor on the same vertical plane, closest to the conductor in motion. The execution moves back to the starting point of the parent module on completion of the estimation process. The conductor motion estimation subroutine is summed up in Figure 3.29d.

### 3.4.5 Validation of the Proposed Approach

We have simulated a distributed transmission network considering a load of 700 MW being delivered from the generation site to the consumer by a substation located at 350 km. We assume that synchronized magnetic field and electric current measurements are recorded and processed with our proposed algorithms at the substation. We tested the designed algorithm to perform conductor sag and motion detection, and then estimation for a three-phase typical support tower configuration 3L1.

The method is validated with numerical simulations processed in MATLAB and verified with a scaled laboratory setup. For numerical simulations, typically one circuit configuration is utilized, where three sensors are considered placed at a height of 2 m above each of the conductors. The sensor-to-sensor distance is kept the same as the conductor-to-conductor distance, i.e. 11.8 m.

The algorithm proposed has been validated by two test scenarios. The first test case resembles an event where an unsymmetrical sag change occurs in all phase conductors. We assume that this sag differs from the one used in the calculation of $S_x$, which is stored in the memory unit. This value is assumed to be recorded at the installation phase or during the last update by the algorithm. Test parameters along with iteration history and results are presented in Table 3.7. The primary module detects this variation in sag where $S_x * I_j$ is different from $B_{Xi}$ measured.

Once detected by the main algorithm, the subroutine estimates the unsymmetrical sag for each of the conductors. With every iteration, sag is increment by 0.01 m and used in the algorithm. This ensures that the algorithm can resolve a variation of up to 0.01 m in sag in any of the phase conductors. The subroutine for sag estimation calculates the sag in all phase conductors with error less than 1%. However, for implementation in a measurement system the resolution is limited by the sensitivity of the sensor and the magnetic field strength of the conductors.

The other test case involves an event of conductor motion, where a large amplitude unsymmetrical CG is generated with the method detailed. The magnetic field of conductor 1 under CG referenced with the magnetic field of a static conductor ($S_x * I_j$) is shown in Figure 3.30. Such a motion is detected by signal variations during the observation window when compared with $S_x * I_j$. Once detected, it is then processed by the estimation subroutine. Averaged unsymmetrical mean rotation from both span lengths and displacement of conductor 1 is retrieved from magnetic field of CG during an observation period of 200 ms. The root mean square errors for rotation and displacement retrieval were 0.3623 and 0.0833, respectively. Table 3.8 summarizes the results for all the conductors. The results validate the efficacy of the developed algorithm for an unknown change in sag or conductor motion.

For evaluation of sag and wind-induced conductor motion, we have established a scaled setup in the laboratory similar to standard one circuit configuration, 3L1, to demonstrate compliance of our method for practical scenarios. Three conductors were placed in flat formation with the help of two support towers. The complete experimental setup is shown in Figure 3.31. A position on the vertical plane is selected for sensor placement in compliance with the proposed method. Consequently, each sensor is placed below each of the conductors at a distance of 27 cm. The conductors were adjusted at an unsymmetrical sag from each other, measured with a calibrated Vernier caliper. This sag was later estimated by the algorithm. Three resistive loads, each of 7.25 ohms with power

**Table 3.7** Estimation of unsymmetrical sag in conductors.

| Test conditions for each conductor (C) | Scenario of unsymmetrical sag change | | |
| --- | --- | --- | --- |
| | C1 | C2 | C3 |
| Sag in memory (m) | 1.2000 | 1.4000 | 1.6000 |
| New sag (m) | 2.0000 | 3.0000 | 2.0000 |
| Sag calculated (m) | 2.0001 | 3.0002 | 2.0001 |
| Test current (A) | 845.3216 | −389.5919 | −454.1749 |
| **Iteration history for sag estimation ($Diff_n = S_x * I_j - B_{Xi}$)** | | | |
| Iteration | $Diff_1$ | $Diff_2$ | $Diff_3$ |
| 1 | 1.28 | 3.59 | 0.41 |
| 30 | 0.93 | 3.40 | 0.13 |
| 60 | 0.46 | 3.14 | 0.26 |
| 90 | 0.17 | 2.75 | 0.87 |
| 120 | 1.12 | 2.14 | 1.93 |
| 150 | 2.69 | 1.03 | 4.20 |
| 180 | 5.72 | 1.63 | 12.60 |
| 210 | 14.17 | 16.63 | 42.48 |
| 240 | 161.27 | 12.39 | 7.08 |
| 270 | 20.39 | 1.39 | 1.49 |
| **min@Iteration** | 81 | 161 | 41 |
| **min@$Diff_n$** | 0.0043 | 0.0007 | 0.0002 |

Initial sag: 1 m (symmetrical), increment per iteration = 0.01 m

*Source:* Reprinted with permission from A.H. Khawaja and Q. Huang, Estimating sag and wind-induced motion of overhead power lines with current and magnetic-flux density measurements, *IEEE Transactions on Instrumentation and Measurement*, vol. 66, no. 5, pp. 897–909, May 2017.

rating of 6 kW, emulated a three-phase load state. The magnetic field around HVTLs ranges from a few to hundreds of microTesla based on the mutual distance between the sensing point and the conductors. Nevertheless, it is detectable with commercially available sensing devices. With the aforementioned configuration, a magnetic field strength of a few microTesla is emulated at sensing points during the laboratory experiment, which is identical to the magnetic field strength around real power lines.

For validation of the method, selection of a sensitive magnetic field sensor which is compatible with simple conditioning circuitry is deemed important. Among other types of magnetic field sensors available, MR magnetometers are widely in use and are considered to be a common type of magnetic field sensors. Their resistance varies proportionally to magnetic field strength along the sensitive axis, therefore these sensors can measure any of the field components in a given geometry [13]. MR sensors based on tunnel (T) technology are reported to be more efficient than those based on anisotropic (A) fabrication technology. In contrast to AMR sensors, TMR sensors require low power to operate and a straightforward conditioning circuitry without any set/reset pulse. There is no disorientation in TMR sensors on exposure to higher magnetic field

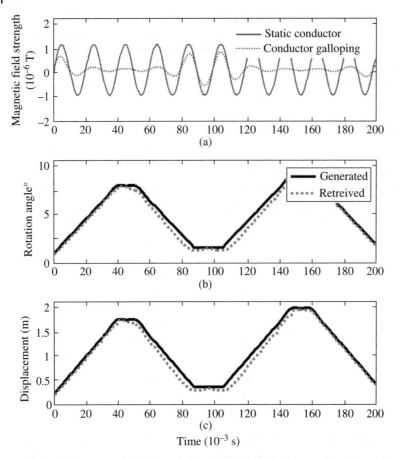

**Figure 3.30** Simulated conductor galloping: (a) magnetic field strength for a conductor in motion varies during each cycle, (b) the angle of rotation during vibration is retrieved, and (c) conductor displacement is retrieved. (*Source:* Reprinted with permission from A.H. Khawaja and Q. Huang, Estimating sag and wind-induced motion of overhead power lines with current and magnetic-flux density measurements, *IEEE Transactions on Instrumentation and Measurement*, vol. 66, no. 5, pp. 897–909, May 2017.)

strengths. Also, TMR sensors show a better time response at exposure to frequency varying uniform magnetic field and high strength transients as demonstrated. It has been verified that TMR sensors are low cost, have a large measurement range, can operate in a range of temperatures and consume far less power relative to AMR and Hall effect magnetic field sensors [14].

We measured the horizontal field component as it contains the maximum contribution of flux density for flat tower configuration. We therefore utilized three uniaxial TMR sensors MMLP57F, arranged along the horizontal axis at the sensing points. These sensors are configured in a full Wheatstone bridge arrangement, with a rated sensitivity of 4.9 mV/0.1 mT and 10 nT resolution. For a voltage bias of 5 V, and an amplification factor of 100, the sensor output is 24.5 mV/10 $\mu$T. The sensed signal is filtered for noise and ambient interference with a dedicated circuitry designed for each of the

**Table 3.8** Estimation results for conductor motion.

| Test case: Conductor (C) galloping | | | |
|---|---|---|---|
| **Test condition** | **C1** | **C2** | **C3** |
| **Sag in memory (m)** | 4 | 4 | 4 |
| **Frequency of complete period (Hz)** | 4.63 | 5.68 | 4.93 |
| **Original displacement (m)** | 1.98 | 1.60 | 1.81 |
| **Maximum mean rotation angle (°):** $max(V_{rot_{s1}} + V_{rot_{s2}})/2$ | 8.56 | 7.10 | 7.95 |
| **Displacement (m) retrieved** | 1.944 | 1.57 | 1.77 |
| **Rotation angle (°) retrieved** | 8.40 | 6.90 | 7.70 |

*Source:* Reprinted with permission from A.H. Khawaja and Q. Huang, Estimating sag and wind-induced motion of overhead power lines with current and magnetic-flux density measurements, *IEEE Transactions on Instrumentation and Measurement*, vol. 66, no. 5, pp. 897-909, May 2017.

three sensors. A detailed discussion on amplification and noise considerations is presented in Section 5.5.3. The electric current measurements are performed with three Hantek CC-65 clamp ammeters with rated error of ±2%. All the measured entities are sampled with a six-channel simultaneous sampling ADC model AD7656 configured on evaluation board model number EVAL-AD7656-1SDZ. Another important factor is the synchronization between magnetic field and electric current measurement. Since, the TMR sensors and clamp ammeters were in close proximity to each other, synchronization was achieved by means of concurrent sampling from all six channels of the ADC, i.e. three channels connected to magnetic field measurement units, and three channels to clamp ammeters. The data is then acquired for data processing with a System Demonstration Platform model number SDP-B from Analog Devices which interfaced the evaluation board to acquisition software on a computer. The process is summarized in a functional block diagram of Figure 3.31(b).

Once the conductors had been attached to the resistive load, electric current was allowed to flow from the three-phase AC mains to the load. To achieve the same effect, magnetic field distribution along horizontal axis and electric current measurement samples were recorded. The experiment includes the sag estimation and conductor vibration measurement in two stages. During the first stage, $S_x$ is computed using a symmetrical sag value of 10 cm stored in memory. The real sag in conductors is kept to 13 cm in conductor 1, 15 cm in conductor 2, and 20.5 cm in conductor 3, which resembles an unsymmetrical sag scenario. The recorded magnetic field and electric current data are then processed in MATLAB. Figure 3.32 shows the waveform of the magnetic field measured for unsymmetrical sag and symmetrical sag represented by $S_x * I_j$ for each of the three conductors. When processed by the detection algorithm, the data is classified as a case of sag estimation, which is processed by the sag estimation subroutine. Incremental step size is kept at 0.1 cm to cater for small changes in sag. The subroutine for sag estimation retrieves the sag in all conductors with an error of less than 1%.

During the second stage of the experiment, a scenario of conductor motion is emulated, where a table fan induces air flow by blowing the air at the angle shown in Figure 3.31. The wind from the fan emulates a small amplitude oscillation with high frequency, similar to AV in conductors. Once motion is physically observed, the data are captured for a period of 1 s at sampling frequency $F_s$ of 5 kHz and then processed

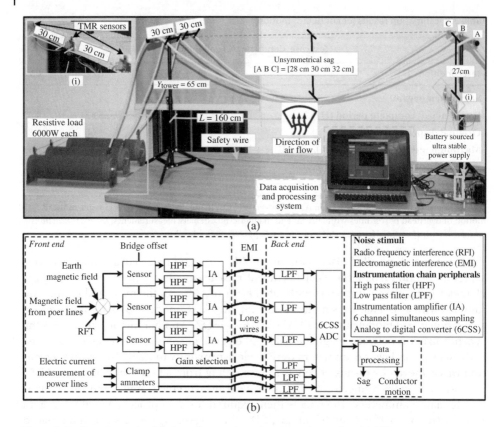

(a)

(b)

**Figure 3.31** Experimental validation of sag and conductor motion retrieval. (a) Measurement arrangement for a scaled test bed of typical one circuit configuration 3L1. (b) Generic block diagram illustrating the signal processing stages under the influence of various noise stimuli. (*Source:* Reprinted with permission from A.H. Khawaja and Q. Huang, Estimating sag and wind-induced motion of overhead power lines with current and magnetic-flux density measurements, *IEEE Transactions on Instrumentation and Measurement*, vol. 66, no. 5, pp. 897–909, May 2017.)

in MATLAB with the proposed algorithm. The detection algorithm classifies the data as an event of conductor motion. The small difference between the measured magnetic field and $S_x * I_j$ can be observed in Figure 3.33. Since the vibration amplitude is small with high frequency, it is only visible at peaks and indicates AV. The subroutine for conductor motion estimation processes the signals and retrieves the rotation and amplitude. As expected, the frequency for this vibration was relatively high, i.e. 100 Hz with a maximum vibration amplitude of 0.15 m. The retrieved conductor trajectory affirms the efficacy of our method.

### 3.4.6 Noise Tolerance and Uncertainty Analysis

Sources of noise and error can originate from the measurement environment, the sensor itself, and analog front-end electronics. The sources of noise and their remedial measures are illustrated in the schematic of the experimental setup shown in Figure 3.31.

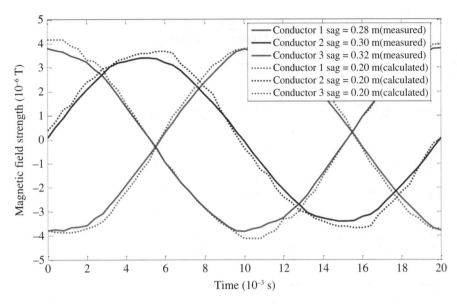

**Figure 3.32** Magnetic field strength versus time for the experimental setup. Waveforms show a comparison between the recorded magnetic field of unsymmetrical sagged conductors, calculated by $S_x * Ij$ (sag value in $S_x = 20$ cm). (*Source:* Reprinted with permission from A.H. Khawaja and Q. Huang, Estimating sag and wind-induced motion of overhead power lines with current and magnetic-flux density measurements, *IEEE Transactions on Instrumentation and Measurement*, vol. 66, no. 5, pp. 897–909, May 2017.)

Earth magnetic field acts as a main source of environmental disturbance. It remains constant at a point in space, and appears as an offset voltage at the sensor output. The presence of the Earth's magnetic field might saturate the high gain amplifier for our system. Any offset in bridge output from the difference in the MR element can introduce a voltage bias. This is prevented by connecting the differential output of the sensors to high pass filters before amplification, which blocks any DC component added as an offset from the Earth's magnetic field or due to resistance variation in the Wheatstone bridge. For the actual system, the sensors require to be mounted on steel towers, preferably close to conductors. The steel structure has a high permeability compared to free air, which may attenuate the magnetic field strength. However, the attenuation due to the steel structure is negligible, which is evident from the FEA presented in [6]. The TMR sensor used in our arrangement exhibits a low intrinsic noise level of 10 nT/Hz $^{1/2}$. The noise is dominated by resistance fluctuation inside the active MR elements and is independent of the external magnetic field [23]. In the case of extrinsic noise, the probable reason is the differential output of the Wheatstone bridge, which appears as common mode noise. Radio-frequency interference (RFI) is another source of disturbance and superimposes a varying DC offset proportional to the radio frequency signal. The extrinsic noise is first suppressed by a low noise instrumentation amplifier (INA333) with a high common mode rejection ratio (CMMR) of 100 dB and built in RFI filters close to the sensor. Each sensor is attached to a nearby installed instrumentation amplifier, which is then fed to the sampling unit located at a distance from the sensing units. The amplified signal is prone to low-frequency electromagnetic interference (EMI). The operational

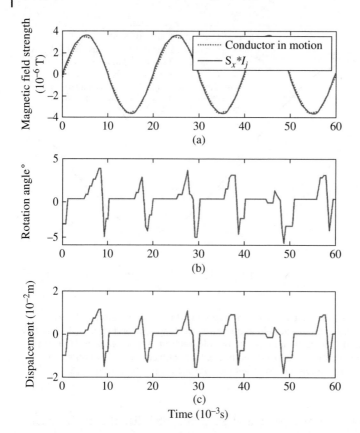

**Figure 3.33** Aeolian vibration generated in conductor 1: (a) the magnetic field varies slightly due to the small amplitude of vibration, (b) the retrieved angle of rotation, oscillation frequency ≈100 Hz, and (c) the retrieved conductor displacement. (*Source:* Reprinted with permission from A.H. Khawaja and Q. Huang, Estimating sag and wind-induced motion of overhead power lines with current and magnetic-flux density measurements, *IEEE Transactions on Instrumentation and Measurement*, vol. 66, no. 5, pp. 897–909, May 2017.)

amplifier (AD8597) filters this noisy signal, which is subsequently sampled with a 16-bit simultaneous ADC. All these signal processing steps are shown in the block diagram in Figure 3.31b.

Likewise, the data processing algorithm is also vulnerable to computational errors. The algorithm is based on an iterative comparison between the measured and calculated magnetic fields. Hence, errors in the data processing algorithm are minimized with small iterative steps. This method increases the precision while incurring large computation overhead. Additionally, the computational model utilized in this chapter considers single conductors mounted on a tower, which are utilized in most power transmission configurations. Some other configurations may also contain bundled conductors. In our work we demonstrated that the computational error in the presence of such conductors is 0.01% and can be neglected.

Lastly, the uncertainty for the proposed system may arise from several other sources. Since we have opted for a time-stamping mechanism for synchronization, this implies that uncertainty may arise from the difference in time in stamps at the substation and the tower. A similar source of uncertainty is synchronization between neighboring towers when implemented with the protocol based on IEEE Standard 1588. Such jitters can be handled by implementing a data alignment approach at a zero crossing of the measured magnetic field for each electric current cycle if synchronization mismatch occurs within this range. This step is part of the proposed algorithm and is shown as the Align block in the flowchart (Figure 3.29). A stable power source is the backbone of the system, but when the power is generated with solar panels or an electric field scavenging mechanism, the efficacy of such devices adds a degree of uncertainty. The final source of uncertainty is the sensitivity deviation for each sensor from the one mentioned by the manufacturer. Such a deviation results in an error that propagates throughout the instrumentation chain. Nevertheless, this systematic error can be avoided by defining a sensitivity scale factor for each of the sensors under the uniform magnetic field generated in a Helmholtz coil. Some sources of uncertainty are unpredictable, e.g. corona discharge from nearby conductors can appear as magnetic field noise when picked up by the sensing elements [24]. It may result in uncertainty in cases when the frequency of corona discharge is not attenuated by filters and sampled along the magnetic field data from power lines. A thunder strike event on a tower will generate a strong transient magnetic field pulse which adds uncertainty and can even damage the measurement setup.

We have successfully demonstrated an accurate sag and wind-induced motion detection method for overhead HVTLs. It requires time-synchronized magnetic field measurement at the support tower and electric current at substation. We have proved that having the number of magnetic field sensors equal to that of conductors, sensitive along the horizontal axis, can provide precise results. Based on the mathematical framework and characterization of magnetic field signals, a novel unified algorithm has been developed to detect and estimate conductor sag and motion. The algorithm has been tested on numerical simulation of a typical one-circuit tower configuration and on a scaled laboratory setup. The test cases include events such as unsymmetrical change in sag and conductors undergoing random bi-direction motion. During the experiments, the error for sag estimation remained less than 1%. In addition, for conductor motion retrieval the root mean square error for angle of rotation was 0.3623 and for displacement amplitude it was 0.0833. The low error validates the efficacy of the developed algorithm.

## 3.5 Unified Current Reconstruction and Operation Parameters of HVTLs

This section is from one of the authors' pending patent [25].

In the previous sections, we learnt that overhead HVTLs are a vital mode of transportation of power for widely distributed transmission networks. In this section, we explain various components of this method, which aims to overcome the deficiencies of prior methods and provide a way to monitor the electric current and spatial parameters of overhead transmission lines with a non-contact magnetic field sensor array. It can provide real-time load monitoring as well as obtaining spatial information for phase conductor positions in rest and motion. This is achieved by utilizing measurements from magnetic field sensors, which can be mounted on support towers or arranged in a portable device close to each of the conductors. The method also describes a way of determining the optimal positioning of sensing units.

A novel method for monitoring the electric current and spatial parameters of overhead transmission lines with non-contact magnetic field sensor array is provided. The objectives of the present approach are summarized in the following steps.

*Step 1: Sensitivity matrix calculation based on the sensor to conductor distance*

For each of the towers where monitoring is desired, magnetic field sensors are mounted. Different versions of magnetic field sensors are available to detect the field strength from one to three field components, i.e. $x$ axis, $y$ axis, and $z$ axis. The field strength along all these components varies with different tower configurations. For our method, we require double the number of sensed components at various locations. The sensed components can be the same component measured at different locations or different components measured at the same location. In all the examples provided here the sensed components are horizontal field components measured at six different locations for three phase conductors of a flat tower configuration.

The sensing points can be divided into two groups each containing an equal number of sensors and phase conductors. For horizontal magnetic field components at double the number of points, the sensitivity coefficient $A'_{ij}$ of each horizontal component of each sensing point $i$ with respect to position of the phase conductor is calculated to constitute a sensitivity matrix $S'_x = (A'_{ij})_{NN}$. Among them, $i = 1, 2, ..., N$ and $j = 1, 2, ..., N$, where $N$ is the number of phase conductors. Similarly, the sensitivity coefficient $A''_{ij}$ of each sensing point in the second group is calculated to form the sensitivity matrix $S''_x = (A''_{ij})_{NN}$. For horizontal component measurement, the number of uniaxial magnetic field sensors in each group is the same as the number of phase conductors. For cases where two axis measurements are performed using one sensor, one group may comprise horizontal field components and the other group the vertical field components. Once the sensitivity matrix for each group is obtained, they are stored in memory.

*Step 2: Conductor motion detection*

During a specific time interval of $t$ seconds where $t = t_1 : t_2$, the magnetic field measured at each sensing point is processed to detect if the conductors are in motion commonly induced by wind forces. The motion is detected if the magnetic field at sensors closest to the conductors shows a strong variation at adjacent cycle peak points. For accurate detection, the sum of such variations for a number of cycles is compared with a threshold value. Once motion in one or more conductors is detected, the motion characteristics, such as frequency, displacement amplitude, and rotation trajectory, are calculated. If no motion in conductors is established, i.e. the sum of variations at peak points in each cycle is less than a threshold value, the estimation program moves to step 3.

*Step 3: Electric current and sag reconstruction*

1. Read the sensitivity matrices $(S'_x, S''_x)$ from the cache.
2. Reconstruct the electric current calculation of the one-dimensional electric current matrix I:

$$\mathbf{I} = \mathbf{S}'_x \mathbf{B}'_x \qquad (3.61)$$

Here, $\mathbf{I} = [I_1, \ldots, I_N]^T$ and $B_x' = [B_{x_1}', \ldots, B_{x_N}']^T$, where the MF matrix $B_x'$ is a one-dimensional matrix that contains the magnetic field strength obtained from the first to the $N$ th sensors of the first group.

3. Verify the electric current reconstruction. Since the electric current is the same for both sensor groups, the reconstructed electric current obtained using the first group can be verified:

$$\mathbf{B}_x'' = \mathbf{S}_x''I \qquad (3.62)$$

where

$$\mathbf{B}_x'' = \begin{bmatrix} B_{x_1}'' \\ \vdots \\ B_{x_N}'' \end{bmatrix} \qquad (3.63)$$

where MF matrix $B_x''$ is the magnetic field strength obtained from the second group of $N$ magnetic field sensors.

If the equality does not hold true, this implies that the spatial position of the conductors has changed due to an increase in conductor elongation for one or more conductors. In this case, go to step 4. If the equality holds, the electric current and sag parameters are correctly identified. In this case, go to step 5.

4. Jump to the next value in the look-up table and update the sensitivity matrix. The algorithm will now utilize a new combination from the look-up table of sag values to update the sensitivity matrix coefficients for both of the sensor groups. At this point, go back to step 2. The look-up table is part of the embedded system that consists of different sag combinations for all phase conductors on both sides of the span lengths. The look-up table can also take the form of an iterative algorithm to generate such combinations, set according to the concrete implementation situation.

5. Update the cache and return. Update both the sensitivity matrices $S_x'$ and $S_x''$ , communicate the sag and electric current reconstruction results, and return to step 2. The present method contains a monitoring solution with magnetic sensing to estimate the electric current in the phase conductors as well as the spatial position of the phase conductors. In particular, the conductor motion can be detected and estimated on one side while electric current and sag in conductors is reconstructed on the other.

It should be noted that the similar modules are designed by similar reference numerals although they are illustrated in different figures. Also, in the following description, a detailed description of the known functions and configurations incorporated herein will be omitted as it may obscure the subject matter of the present method.

Figure 3.34 is the implementation flow chart of the proposed approach. Here, a sensitivity matrix is computed for the set of two sensor groups for computation of the overhead transmission line electric current and spatial position monitoring of the conductors. Each group constitutes a number of sensors or sensing directions equal to the number of phase conductors. The method for monitoring the electric current and spatial state of an overhead transmission line of the proposed method, as described in Figure 3.34, includes the following steps.

In this scheme, as shown in Figure 3.35, magnetic field sensors are arranged on top of support tower. The magnetic field sensors can be considered as being divided into two groups for algorithm implementation on the measured signals, where each group

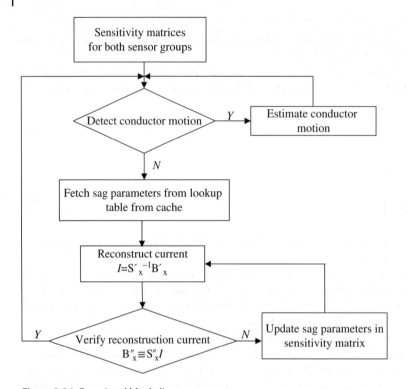

**Figure 3.34** Functional block diagram.

consists of sensors along the same horizontal axis. In this example, the overhead transmission line consists of three phase conductors, i.e. the number of phase conductors, denoted by $N$, is 3. In Figure 3.35 the first group on the left-hand side comprises the magnetic field sensors, marked $L_{L1}$, $L_{L2}$, $L_{L3}$ from the outside to the inside. Similarly, on the right-hand side in Figure 3.35 the second group of magnetic field sensors are marked as conductors $L_{R1}$, $L_{R2}$ and $L_{R3}$ from the outside to the inside. In this example, two magnetic field sensors are placed above each side of the phase conductors, extending along the neighboring span lengths. These sensors can be considered as part of separate groups, with $SR_{L1}$, $SR_{L2}$, $SR_{L3}$ as the first group, and the magnetic field sensors $SR_{R1}$, $SR_{R2}$, $SR_{R3}$ as the second group. In one embodiment, the method consists of hardware with on-board processing, as shown in Figure 3.36.

Here, the magnetic field sensors based on MR technology are utilized. The differential sensor output is amplified close to the sensors and then sent to the computing unit located on the tower some distance away. At this point, the signals from sensors are filtered, amplified, and sampled digitally. A simultaneous sampling analog to digital converter digitizes all the sensor inputs and sends a data-ready signal to the processor, where a computerized estimation method analyzes the data for extraction of relevant parameters. In one embodiment of electronic circuitry, the embedded computing unit consists of a secure communication circuitry that transmits retrieved operational parameters and receives control signals. All the units and sensing equipment are powered by a stable power supply connected with a long-life battery consisting of an array of cells. The

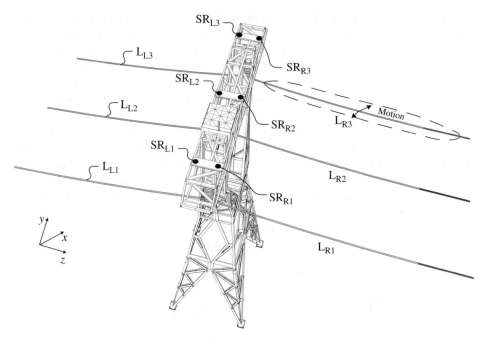

**Figure 3.35** Configuration of magnetic field sensors.

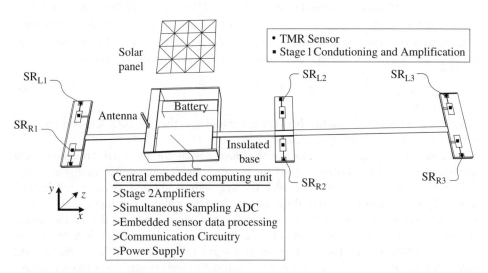

**Figure 3.36** Hardware implementation structure.

battery unit can be charged with a solar panel or use an electromagnetic induction mechanism.

In a preferred embodiment the uniaxial magnetic field sensor based on TMR effect is used, in which the sensors are sensitive only along the horizontal component of the magnetic field vector. The number of sensors is twice the number of phase conductors. The relationship between magnetic field, electric current, and spatial position of catenary phase conductors is estimated using the Biot–Savart law. The magnetic field sensing points require to be positioned at points where the error in the reconstruction process is minimized. This ensures the convergence of the error within an acceptable range of the industrial standard set by IEEE Standard 1588, i.e. less than 5%. In order to determine the optimal magnetic field sensing position, it is necessary to fully understand the source current, as well as the geometric relationship between the sensing point and the phase conductor.

The magnetic field at sensor point $s$, which satisfies the Biot–Savart law, is denoted as $\vec{B} = \vec{B}_{X_s} + \vec{B}_{Y_s} + \vec{B}_{Z_s}$. The majority of the total magnetic flux density generated by the electric current in phase conductors is concentrated along the horizontal component. It varies to a detectable level when the electric current or spatial position of the conductors is modified. In the proposed method, for the same reason, a magnetic field sensor is a linear one-axis magnetic field sensor mounted to detect the horizontal component. According to the Biot–Savart law, the electric current in a phase conductor positioned at $(x_p, y_p, z_p)$ generates the magnetic field $\vec{B}_{X_s}$ at sensing point $s(x_s, y_s, z_s)$ and satisfies

$$\vec{B}_{X_{sp}} = \hat{i}_x \frac{\mu_0 I_p}{4\pi} \left[ \begin{array}{c} \int_{-\frac{L}{2}}^{0} \left[ \frac{((z_s - z_p)\sinh(\alpha_p z_p)) - (y_s - y_p)}{[(x_s - x_p)^2 + (y_s - y_p)^2 + (z_s - z_p)^2]^{3/2}} \right] dz_p \\ + \int_{0}^{\frac{L}{2}} \left[ \frac{((z_s - z_p)\sinh(\alpha_p z_p)) - (y_s - y_p)}{[(x_s - x_p)^2 + (y_s - y_p)^2 + (z_s - z_p)^2]^{3/2}} \right] dz_p \end{array} \right] \tag{3.64}$$

where $\hat{i}$ is the unit vector for the horizontal direction, $\mu_0$ is the permeability of free space, $L$ is the span length, and the distance vector between the sensor and the phase conductor is represented as $\vec{r} = (x_s - x_p)\vec{i}_x + (y_x - y_p)\vec{i}_y + (z_s z_p)\vec{i}_z$, where $\hat{i}_y$ is the vertical unit vector, $\hat{i}_z$ is the unit vector of longitudinal (i.e. electric current transmission) direction, and $y_p$ is the conductor height of a sagged conductor, which can be represented as

$$y_p = y_{tower} + \frac{1}{\alpha_p}(\cosh(\alpha_p z_p) - 1), -\frac{L}{2} \le z_p \le \frac{L}{2} \tag{3.65}$$

where $y_{tower}$ is the conductor height at the tower and $\alpha_p$ is the mechanical tension parameter between the adjacent span lengths.

Because the sensing point is close to the phase conductor, the Earth's magnetic field can be ignored in the calculation. Since the magnetic fields generated by each phase conductor are superimposed on each other, the magnetic field intensity measured at each magnetic field sensor is a sum of the magnetic field from all the conductors. The field is scaled by a sensitivity coefficient. The sensitivity coefficient depends solely on the sensor to conductor distance. In this embodiment, the position of the magnetic field sensors is shown in Figure 3.35. For the first set of magnetic field sensors, the magnetic field strength generated by the first group with $i$ magnetic field sensors and $j$ phase conductors is:

$$B'_{X_{ij}} = \frac{\mu_0 I_j}{4\pi} \int_{-L/2}^{L/2} \frac{(z_i - z_j)\sinh \alpha_j z_j - (y_i - y_j)}{|\vec{r}|^3} dz_j \tag{3.66}$$

To simplify, the sensitivity coefficient can be separated in above equation as follows:

$$A'_{ij} = \frac{\mu_0}{4\pi} \int_{-L/2}^{L/2} \frac{(z_i - z_j)\sinh \alpha_j z_j - (y_i - y_j)}{|\vec{r}|^3} dz_j \tag{3.67}$$

The sensitivity coefficient of $N$ conductors for $M$ magnetic field sensors in the first set of magnetic field sensors will become $N = M$ as the number of conductors is the same as the number of sensors in the first group, so, $S'_x = (A'_{ij})_{N \times N}$. In the same way, the magnetic field intensity measured by $N$ magnetic field sensors of the first group from $B'_{x_1}$ to $B'_{x_N}$ is generated by the electric current from $I_1$ to $I_N$, where $N$ is the number of current-carrying conductors. This relationship takes form:

$$
\begin{bmatrix} B'_{x_1} \\ \vdots \\ B'_{x_N} \end{bmatrix} = \begin{pmatrix} A_{11} & \cdots & A_{1N} \\ \vdots & \ddots & \vdots \\ A_{N1} & \cdots & A_{NN} \end{pmatrix} * \begin{bmatrix} I_1 \\ \vdots \\ I_N \end{bmatrix}
\tag{3.68}
$$

Inverse electric current reconstruction is prone to errors. For the method to provide adequate results, it is vital to place the sensors at points where the error in reconstruction with equation (3.68) is minimized. In the present embodiment, therefore, a magnetic field sensor placement position optimization method is also provided. According to this method, the horizontal and vertical coordinates of $N$ magnetic field sensors $(x_s, y_s)$ for both groups of sensors are adjusted $(x_s + \delta x_s, y_s + \delta y_s)$ to find a condition number of the sensitivity matrix for positions where this number is minimized, preferable close to unity, using $|S'_x|_2|S'^{-1}_x|_2 \approx 1$ and $|S''_x|_2|S''^{-1}_x|_2 \approx 1$.

Overhead HVTLs are commonly encountered with wind-induced conductor oscillations. This effect can be viewed as rotation of sagged phase conductors suspended between end points attached to a tower during a certain time period. Conductor motion frequency refers to the time taken to complete one rotation around the mean position. Figure 3.37 shows the field strength of one phase conductor during a time interval of 500 ms. Figure 3.37a shows the comparison of the magnetic field intensity of a static conductor with a conductor in motion. Here, a displacement of the phase conductor away from the mean position varies the magnetic field strength during each cycle and vice versa. On the other hand, a change in sag changes the magnetic field strength for all the samples retrieved during an observation window.

In this approach, the conductor displacement is first distinguished from the variation of sag scenario. This is achieved by processing the magnetic field data during a time interval of $t$ seconds. This is illustrated in Figure 3.37a, where the measured magnetic field at peak points $k_p$ from each of the sensors during adjacent cycles $(p = 1,...,p)$ exceeds a threshold that implies conductor motion. For accuracy, the results from a number of cycles can be summed up as follows:

$$
\sum_{p=1}^{P} k_p > \text{threshold}, p = 1, 2, \ldots, P
\tag{3.69}
$$

where $P$ is the number of adjacent cycles within the interval of $t$ seconds. In Figure 3.37a, to simplify the graph surface we draw out the variation rates of the first and second adjacent cycles ($k_1$ and $k_2$).

Often wind and other factors lead to oscillation of overhead transmission lines, which displaces the sagged conductor periodically, forming motion trajectory as shown in Figure 3.35 (for simplicity, the diagram only indicates motion in conductor $L_{R3}$). The oscillation frequencies of each of the phase conductors can be related to the time taken to complete one period by the phase conductor from the mean position to the maximum displacement point. The oscillation path of each phase conductor is different.

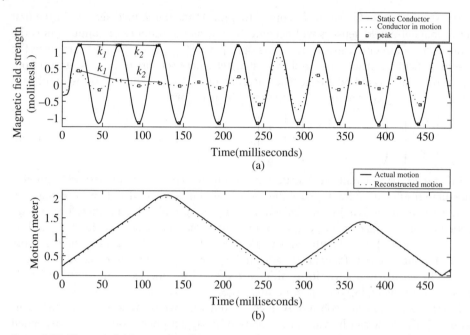

**Figure 3.37** Magnetic field strength and reconstructed motion curves: (a) magnetic field strength of static and conductor in motion, and (b) reconstructed motion and actual motion.

Therefore, for a phase conductor in motion during a time interval of $t$ seconds, the coordinates are rotated in the $x$–$y$ plane as

$$x_p^r = x_p(t_0) \cos \theta_p + y_p(t_0) \sin \theta_p \tag{3.70}$$

$$y_p^r = x_p(t_0) \sin \theta_p + y_p(t_0) \cos \theta_p \tag{3.71}$$

For the phase conductors in motion, the magnetic field strength changes with time, increasing when the phase conductor is at a point near the magnetic field sensor and vice versa. To calculate this field strength, equations (3.70) and (3.71) are replaced in (3.64) as

$$\vec{B}_{X_s p}(t) = \hat{i} \frac{\mu_0 I_p}{4\pi} \int_{-\frac{L}{2}}^{\frac{L}{2}} [\frac{(z_s - z_p) \sinh \alpha_p z_p - (y_s - y_p^r(t))}{|(x_s - x_p^r(t)) + (y_s - y_p^r(t)) + (z_s - z_p)|^3}] dz_p \tag{3.72}$$

Here, the magnetic field intensity $B_{X_s p}(t)$ at sensing point $s$ depends on the rotation angle and mechanical tension parameter $\alpha_p$. The influence of the wind force on each phase conductor is different in the real environment. The oscillation characteristics (oscillation frequency, amplitude, rotation direction, and angle) of each phase conductor can be extracted from the measured magnetic field strength at the sensing points.

In this embodiment, the monitoring system triggers the calculation program of all these parameters once a conductor motion is detected. The oscillation frequency of the $j$ phase conductor is calculated from the magnetic field signal obtained from the magnetic field sensor $B_{x_j}'$ closest to the conductor. Once the phase conductor displacement reaches the peak value, the magnetic field strength is lowest. In contrast, when the displacement

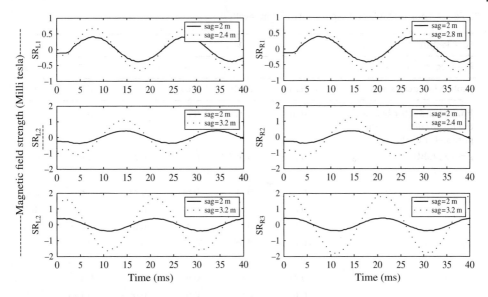

**Figure 3.38** Magnetic field strength in different sag conditions.

of the phase conductor reaches 0, the magnetic field magnitude is largest. Therefore, it equates to the reciprocal of the interval between adjacent maximum and minimum energy.

Similarly, the trajectory can be retrieved by calculation of the magnetic field by an iterative increase in (3.70) and (3.71). The coordinates of the phase conductors in the rotated $x$–$y$ plane are obtained by updating $x_j^r \leftarrow x_j^r(\theta_j + \delta\theta)$, $y_j^r \leftarrow y_j^r(\theta_j + \delta\theta)$. Finally, at time $t$, according to (3.72), where $s = i, p = j$, the algorithm gets the magnetic field strength generated by the phase conductor $i$ at each magnetic field sensor. Measured magnetic field strength $B'_{x_i}$, $B''_{x_i}$ is compared with calculated field $B'_{x_i,calculated}(x_j^r, y_j^r), B''_{x_i}B'_{x_i,calculated}(x_j^r, y_j^r)$. Once the comparison is successful, the displacement of each phase conductor is reconstructed according to the rotation angle of each phase conductor. It should be noted that the comparison here means that the difference between the two magnetic field strengths is less than the set threshold. In this example, as shown in Figure 3.37b, when the phase conductor is in motion, the displacement obtained according to the reconstruction is almost equal to the real displacement, which validates the reconstruction method. The same process continues for the next time interval until the conductor comes to rest.

With the passage of time and heating of phase conductors, the mechanical tension parameter of the phase conductors has a tendency to increase, which in turn increases the conductor elongation (sag), therefore the magnetic field intensity calculated from the sensitivity matrix of the one in memory is different from the measured values, as shown in Figure 3.38. Such a change can be detected by a comparison between calculated and measured magnetic field values only when the conductors are not in motion.

First, according to (3.61), the magnetic field strength from the first group of magnetic field sensors is obtained for sensor 1 $B'_{x_1}$ to sensor $N$ $B'_{x_N}$. Then, the last stored sensitivity matrix for the same group is obtained from memory $S'_x$ to reconstruct the electric current $I_1$ to $I_N$. To verify, the reconstructed electric current values are multiplied with

**Table 3.9** Look-up table and sag reconstruction results.

| Span length | | 1 | | | 2 | | |
|---|---|---|---|---|---|---|---|
| Phase conductor | | 1 | 2 | 3 | 1 | 2 | 3 |
| Caption in Figure 3.35 | | (LL1) | (LL2) | (LL3) | (LR1) | (LR2) | (LR3) |
| **Phase conductor based on the mechanical tension parameter in the cache** | | 0 m | 0 m | 0 m | 0 m | 0 m | 0 m |
| Arc sag table | | 0 m | 0 m | 0 m | 0 m | 0 m | 0.1 m |
| | | . | . | . | . | . | . |
| | | 0 m | 0 m | 0 m | 0 m | 0 m | 3.9 m |
| | | 0 m | 0 m | 0 m | 0 m | 0 m | 4.0 m |
| | | 0 m | 0 m | 0 m | 0 m | 0.1 m | 0 m |
| | | 0 m | 0 m | 0 m | 0 m | 0.1 m | 0.1 m |
| | | . | . | . | . | . | . |
| | | 0 m | 0 m | 0 m | 0 m | 0.1 m | 3.9 m |
| | | 0 m | 0 m | 0 m | 0 m | 0.1 m | 4.0 m |
| | | ... | ... | ... | ... | ... | ... |
| | | 4.0 m | 4.0 m | 4.0 m | 4.0 m | 3.9 m | 0 m |
| | | 4.0 m | 4.0 m | 4.0 m | 4.0 m | 3.9 m | 0.1 m |
| | | . | . | . | . | . | . |
| | | 4.0 m | 4.0 m | 4.0 m | 4.0 m | 3.9 m | 3.9 m |
| | | 4.0 m | 4.0 m | 4.0 m | 4.0 m | 3.9 m | 4.0 m |
| | | 4.0 m | 4.0 m | 4.0 m | 4.0 m | 4.0 m | 0 m |
| | | 4.0 m | 4.0 m | 4.0 m | 4.0 m | 4.0 m | 0.1 m |
| | | . | . | . | . | . | . |
| | | 4.0 m | 4.0 m | 4.0 m | 4.0 m | 4.0 m | 3.9 m |
| | | 4.0 m | 4.0 m | 4.0 m | 4.0 m | 4.0 m | 4.0 m |
| **Actual sag** | | 2.4 m | 3.2 m | 3.2 m | 2.8 m | 2.4 m | 3.2 m |
| **Reconstructed arc** | | 2.4 m | 3.2 m | 3.2 m | 2.8 m | 2.4 m | 3.2 m |

$S_x''$ of the other sensor group. If the comparison returns a false result, new sag parameters are obtained from a look-up table, as shown in Table 3.9, to update the sensitivity matrix. The process is repeated and, if correct, the new sensitivity matrices and positions of the new sag parameters in look-up table are stored in memory. The resultant electric current and sag parameters represent the corresponding line conditions.

The sag table and reconstruction contrast for this embodiment are shown in Table 3.9. We can see that as long as sag data are found in the sag table, the reconstructed electric current satisfies $B''_x \equiv S_x'' I$. The obtained sag data form the reconstruction of the sag. From Table 3.9, the reconstructed sag is consistent with the actual sag.

From Table 3.10, we can see that the reconstructed electric current is less than 0.3% of the actual electric current. After electric current and sag reconstruction have been completed, the execution returns to step 2 and monitoring of the next time interval $t$ continues.

**Table 3.10** Comparison between the reconstructed electric current and the actual electric current.

| Signal to noise ratio of the magnetic field sensor | $SR_{L1}$ | $SR_{L2}$ | $SR_{L3}$ | $SR_{R1}$ | $SR_{R1}$ | $SR_{R3}$ |
|---|---|---|---|---|---|---|
| | 21.575 | 20.7282 | 20.8983 | 20.7599 | 20.7987 | 20.7154 |
| 2.5 s magnetic field sensor Average magentic field strength (Tesla) | $B_{p1}$ | $B_{p2}$ | $B_{p3}$ | $B_{p4}$ | $B_{p5}$ | $B_{p6}$ |
| | 0.0004 | 0.0007 | 0.001 | 0.0004 | 0.0008 | 0.0012 |
| | Conductor1 | | Conductor2 | | Conductor2 | |
| Actual current (A) | 845.32 | | −389.82 | | −455.5 | |
| Reconstructed current (A) | 847.12 | | −390.65 | | −456.48 | |

## 3.6 Fault Location in Overhead HVTLs

With the rapid development of smart grids over the past few years, HVTLs, which form one significant part for delivering electric current, have been further developed and constructed. As a consequence, the number of HVTLs and their total length have increased, binging about more challenges for the safe and reliable operation of power grids. With the structure of the transmission lines becoming very large, the possibility of faults occurring in the power system, due to lightning, short circuits, faulty equipment, mis-operation, human errors, overload, growing vegetation (or swaying trees), and aging, etc. has increased [26; 27; 28]. Among all the possible faults, short-circuit faults are the most common [29]. The reliability of the transmission system is threatened when short-circuit faults occur.

In addition, encouraged by deregulation and liberalization, those who are trying to benefit from the power and energy markets cause further problems for the operation and reliability levels in transmission systems. This phenomenon makes it necessary to guarantee the reliability of the power grid [30].

In short, it is important to detect the faults and locate their position in transmission systems in order to take timely remedial action and restore power delivery if faults happen. Against this background, novel methods for detecting short-circuit fault locations in high-voltage overhead power transmission lines have been proposed.

### 3.6.1 Types of Short-circuit Faults

When a power system is in its normal operation state, except for the neutral point in a three-phase four-wire system, each two phases or the phase and the ground are insulated from each other. However, when the insulation distance is removed, a short-circuit occurs between the phases or the phase and the ground.

#### 3.6.1.1 Definition of Short-circuit Faults
In a three-phase current system, short-circuit faults are categorized into four types [31]: (i) single-phase earth faults, (ii) two-phase earth short-circuits, (iii) two-phase short-circuits, and (iv) three-phase short-circuits (see Figure 3.39). A single-phase earth fault, called a single-phase fault for short, occurs when one phase is short-circuited

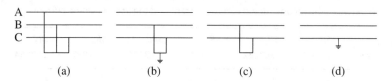

**Figure 3.39** Types of short-circuit faults.

to the ground directly. A single-phase short-circuit is the most common short-circuit fault in a three-phase network, accounting for almost 65% of all short-circuit faults.

In a two-phase earth short-circuit the three-phase alternating power supply, two-phase conductors, and the ground work are in an equipotential state. In other words, the two phases and the ground are connected without impedance. Two-phase earth short-circuits make up 20% of all short-circuit faults.

When two phase conductors are directly connected by metal or low impedance, a two-phase short circuit occurs. The occurrence probability of a two-phase short circuit is 10%.

A three-phase short-circuit happens when three phases are short-circuited. It is also called a symmetrical short circuit because when under these conditions the three phase currents and voltages are balanced due to equation of the three-phase impedance. Although of the four types of short-circuit, the possibility of occurrence of a three-phase short circuit is the lowest, accounting for 5% of all short-circuit faults, it is the most harmful to the stability of the power system because the fault current magnitude is the largest, so the power deviation of the system is the greatest.

### 3.6.1.2 Fault Reasons

The main reason for short-circuit faults is the destruction of the phase-to-phase or phase-earth insulation, which is usually, the faults are caused by one of the following [32]:

- The deterioration of the insulation as a result of aging electrical equipment.
- Iinappropriate design, manufacturing, installation, and maintenance.
- Environmental reasons, such as deterioration of the insulation as a result of extremely high temperature or thunder lightning and thunder.
- Mis-operation by operators.
- Animal disturbance, such as birds perching on bare current-carrying lines or mice biting the lines in high-voltage switchgears.

### 3.6.1.3 Damage Caused by Short-circuit Faults

When short-circuit faults happen, the short-circuit current will be much larger than the rated current magnitude, which results the equipment heating up. If the short-circuit lasts a long time, the thermal effect will cause equipment burnout and casue the conductors fuse and break, which is how electric fires start.

Short-circuit current can also cause electrodynamic force between two conductors. When the electrodynamic force is large enough, the bus lines will be distorted and the switches will be destroyed. The voltage in the power system will therefore decrease due to the short-circuit fault, which threatens the stability of the power system. In particular, an electric power plant in parallel operation will lose synchronization when short-circuit faults happen, which can induce power failure in large regions.

In addition, the strong magnetic field generated by the fault current can interfere with adjacent communication networks and control systems.

The safe and stable operation of a power system therefore must include systems that reduce the occurrence of the short-circuit faults.

### 3.6.1.4  Prevention of Short-circuit Faults

Actions to prevent short-circuit faults from happening can be classified into four types:

- Regular inspection of the insulation in circuits to avoid problems caused by aging equipment.
- The short-circuit fault currents should be calculated correctly to ensure the appropriate electrical equipment, which has the correct current-carrying capacity to meet the requirements of the circuit currents, is chosen.
- Different kinds of conductors and equipment should be used according to the environment they are applied in. Equipment with good sealing performance is employed in harsh environments to ensure it is dust-proof and waterproof.
- Operators should strictly obey safe principles in power systems.
- Management and monitoring should be in place to preventing animals from damaging equipment.

## 3.6.2  Fault Detection with Magnetic Sensors

### 3.6.2.1  Introduction

The traveling-wave technique and impedance measurement are two currently available methods of detecting faults [33]. In the traveling-wave technique, either the transient created by a fault is captured or impulses are injected into the line and the reflected traveling wave is detected with time-domain reflectometry (TDR). The fault location is then determined by timing analysis of the traveling wave. As the faulted signal obtained at the end of the transmission line is mingled with noise, a modern signal-processing technique, such as a wavelet [34], is used for fault location. In the impedance measurement based technique, the voltage and current during the pre-fault and post-fault periods are acquired and analyzed. The line parameters can then be calculated using the transmission line model and the fault can be located. These methods may be one-ended, two-ended, or even multi-ended, depending on how the receiving devices are deployed. As the development of relay protection is fast, it is now possible to collect data from the synchronized phasor measurement unit embedded in a digital relay or a fault recorder in order to achieve the goal of fault location [35]. It has been shown that synchronized measurement can improve the performance of fault location.

All these methods produce reasonable fault location results. However, they all need to connect a device to the HVTL and the required devices are generally costly. For example, in the traveling wave based approach, the accuracy of the location is highly dependent on the performance of a costly high-speed data acquisition system. Furthermore, these methods fully depend on an assumption that the parameters of the transmission line are uniform. Considering the fact that the transmission lines are distributed over a large geographical area, this assumption is generally not entirely true. For example, the non-uniform spacing of the phase conductors may affect the inductance (although phase conductors are transposed to reduce the effect), temperature may affect the resistance, and sag may affect the capacitance in different transmission line segments [36].

Operation experience shows that most systems can limit the fault location error within 1–2% of the monitored line length. In certain cases, the error may reach 5% or even more [37]. The accuracy factor becomes even more important for long transmission lines because even a relatively small error would result in an uncertainty of a few kilometers, causing significant delay in the maintenance crew finding the fault location [37]. Particulary in areas over rough terrain (e.g. mountainous area, such as Liangshan in China), the installed system may cause an error of 3 km in 500 kV transmission lines. The maintenance crew may have to walk 6 km in a mountainous area to determine the exact location of the fault point. For non-permanent faults such as flashover caused by sag of stressed lines, because of their temporary nature it may take even more effort to locate the fault point as the fault is not permanent and the system may have already resumed its normal condition by the time the maintenance crew arrive.

Most of the faults in power systems are short-circuit related. Hence, fault detection may be accomplished by the comparison of the current under normal conditions and fault conditions, and current measurement can be achieved by magnetic field measurement. Traditional current measurement, using Amperes' law, involves placing a coil wound around the conductor. In high-voltage engineering applications, this is not an ideal solution since the coil needs to be close to the high-voltage conductor. Current measurement can in fact be carried out by magnetic field measurement. Once there is current flowing in a conductor, there will be a magnetic field generated around it. The magnetic field strength increases with its line ampacity. The strength, direction, and distribution of the magnetic field emanating from the conductors contain information on the electric power parameters, such as teh amplitude, frequency, and phase of the electric current. It should be noted that measurements based on modern magnetic field sensors can provide accurate and reliable data without physical contact.

A low-cost high-precision solution has been introduced that does not need to connect the device to the HVTL and is highly sensitive to fault location [38]. The new method proposed here uses a novel type of sensitive MR sensor to measure the transient current. The sensors are placed on the tower and the data are transmitted to a data-processing center where analysis software can determine which span holds the fault location. The collected magnetic field data can also be used to identify the fault type and even locate it within the fault span.

### 3.6.2.2 Design of the Location Scheme

For fault location application purposes, it is enough to accurately locate the fault in a certain span (whose length may be between 400 and 1000 m). Once the location of the faulty span is known, it is relatively easy for the line inspector to locate the specific point of the fault within the span. In this scheme, we propose installing sensitive MR sensors at every tower (for long spans, additional monitoring terminals can be added in the middle of the span as terminal devices are not expensive and can work at a distance from the current-carrying conductors). The overall scheme of the proposed solution is shown in Figure 3.40. The data collected from the remote monitoring terminals can be visualized in client software with the aid of a geographical information system (GIS) to help the operating crew to locate the fault position promptly

The monitoring terminal is a small device integrated on a PCB. The whole system comprises a microprocessor (CPU) and its peripheral devices (sensor module, data acquisition module, storage module, communication module, signal pre-conditioning module,

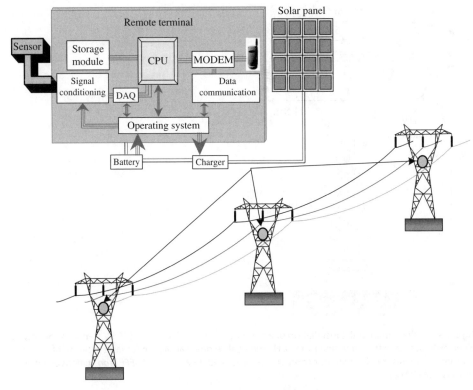

**Figure 3.40** Overall scheme of the proposed solution for fault location in HVTLs (the circles indicate the positions of the sensors). (*Source:* Reprinted with permission from Q. Huang et al., A novel approach for fault location of overhead transmission line with noncontact magnetic-field measurement, *IEEE Transactions on Power Delivery*, July 2012.)

and power supply). The microprocessor controls the whole system and performs data acquisition of the magnetic signal continuously.

The information in the signal can be extracted by simple analysis (e.g. amplitude calculation). Once a certain change in the signal is detected, the data are stored and sent to the center station through the communication channel. Since the system aims to serve mountainous areas, a radio station communication solution (i.e. not dependent on commercial service) is preferable. The sensor module comprises a magnetic sensor chip capable of measurement in the $x$, $y$, and $z$ axes, and its amplifying and filtering circuits. The sensor is designed to be independent of the disturbance coming from the monitoring unit power supply by separating it from the main board. The whole system can be powered by a solar power module composed of a solar panel, a charger, and a battery, or it can be powered through coupling with the transmission lines.

The magnetic field of current-carrying transmission line conductors can be calculated by Maxwell's equations. Under certain assumptions, the analytical expression for the relative position between the measured point and the conductors can be formed. Since there are many types of towers and the tower configuration affects the distribution of the magnetic field, the following calculation and simulation are based on a typical 500 kV transmission tower, shown in Figure 3.41, in which the length of the insulator string

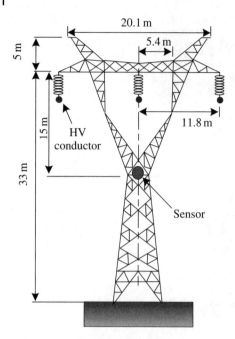

**Figure 3.41** Illustration of the installation of a magnetic sensor on a typical 500 kV transmission tower. (*Source:* Reprinted with permission from Q. Huang et al., A novel approach for fault location of overhead transmission line with noncontact magnetic-field measurement, *IEEE Transactions on Power Delivery*, July 2012.)

is assumed to be 4 m and the sensor is installed at the middle of the tower (18 m above the ground level). Figure 3.42 shows the decomposition of the resultant magnetic field at the measurement point, which can be written as (assuming the conductors have no sag and the transmission lines are infinitely long with respect to the distance between the measurement point and the conductors):

$$\vec{B} = \hat{i}_x B_x + \hat{i}_y B_y + \hat{i}_z B_z = \hat{i}_x (B_b + (B_a + B_c) \cos \theta) B_x + \hat{i}_y (B_a + B_c) \sin \theta + \hat{i}_z 0 \qquad (3.73)$$

where $B_a, B_b$ and $B_c$ are the magnetic field components for $i_A, i_B$ and $i_C$, respectively, $\hat{i}_x, \hat{i}_y$ and $\hat{i}_z$ are the unit vectors along the $x$, $y$, and $z$ axes, respectively, and $\theta$ is the angle between $\vec{B}_a$ and $\vec{B}_b$ or $\vec{B}_c$ and $\vec{B}_b$.

If the system is not symmetric, the magnetic field should be calculated according to the Biot–Savart law. However, for simple estimation, the effect of conductor sag is neglected, the line is assumed to be of infinite length, and then $B_a$, $B_b$ and $B_c$ can be calculated as:

$$B_a = \frac{\mu_0 i_A}{2\pi r_A}, B_b = \frac{\mu_0 i_b}{2\pi r_B}, B_c = \frac{\mu_0 i_C}{2\pi r_C}, \qquad (3.74)$$

where $\mu_0$ is the magnetic constant and $r_A, r_B$ and $r_C$ are defined in Figure 3.42.

### 3.6.2.3 Method for Location of the Fault

Since the three components of the magnetic field in three-dimensioanl space can be separately measured by MR sensors, they can be used to derive the change in the three-phase current. For a radial transmission system, in which only the sending end

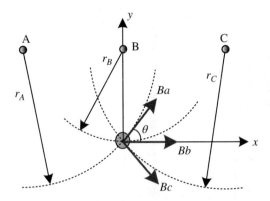

**Figure 3.42** Calculation of the magnetic field at the point of the sensor head. The direction of current is assumed to be along the z axis, which is pointing toward the observer. (*Source:* Reprinted with permission from Q. Huang et al., A novel approach for fault location of overhead transmission line with noncontact magnetic-field measurement, *IEEE Transactions on Power Delivery*, July 2012.)

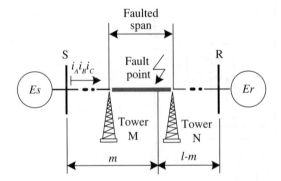

**Figure 3.43** Calculation of the magnetic field at the point of the sensor head. The direction of current is assumed to be along the z axis, which is pointing toward the observer. (*Source:* Reprinted with permission from Q. Huang et al., A novel approach for fault location of overhead transmission line with noncontact magnetic-field measurement, *IEEE Transactions on Power Delivery*, July 2012.)

has power sources, it is simple to locate the fault point by simply checking where the short-circuit current exists in the transmission lines. The output of the terminal closest to the fault point can only be used for qualitative analysis because its output may be affected by the fault (noting that the infinite length assumption may not be true in this case). However, the other outputs can be used to identify the type of fault. For a two-ended system, i.e. both sides have power sources (e.g. tie line), the fault can be located and analyzed by simply noting that the short-circuit level at the two sides is different. Since the location is limited to a span, it is relatively easy to find the exact fault location, even for non-permanent faults.

### 3.6.2.4 Numerical Simulations

In this section, different cases are studied. All the case studies are based on the system shown in Figure 3.43. The implemented model in Matlab/Simulink® is shown in Figure 3.44. The parameters of the system are listed in Table 3.11. Assume that a load

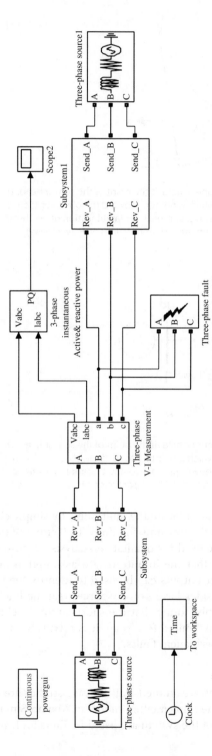

**Figure 3.44** Matlab implementation of the model. (*Source:* Reprinted with permission from Q. Huang et al., *Innovative Testing and Measurement Solutions for Smart Grids*, Wiley-IEEE Press, April 2015.)

**Table 3.11** The parameters of the transmission line under test.

| Voltage (kV) | | 500 | Line length (km) | | 200 |
|---|---|---|---|---|---|
| Sending end short-circuit level (MVA) | | 54560 | Receiving end short-circuit level (MVA) | | 43300 |
| Positive sequence parameters | $R_{pos}(\Omega/km)$ | 0.022 | Zero sequence parameters | $R_0(\omega/km)$ | 0.170 |
| | $L_{pos}(mH/km)$ | 0.940 | | $L_0(mH/km)$ | 2.37 |
| | $C_{pos}(\mu F/km)$ | 0.0122 | | $C_0(\mu F/km)$ | 0.0065 |

$f = 50$ Hz

*Source:* Reprinted with permission from Q. Huang et al., A novel approach for fault location of overhead transmission line with noncontact magnetic-field measurement, *IEEE Transactions on Power Delivery*, vol. 27, no. 3, pp. 1186–1195, July 2012.

of 630 MW is being delivered from the sending end to the receiving end of the power system. In the following numerical simulations, it is assumed that the fault occurs at 0.1 s and disappears at 0.2 s. In all cases, the tower configuration shown in Figure 3.41 is used. With these parameters and assumptions, a distributed model was built and various simulations performed.

### A. Location of the Fault Span

Typical faults on transmission lines include single-phase short circuits, double-phase short circuits, and three-phase short circuits. In the impedance-based approach, the effects of the fault resistance and ground resistance must be considered. In the proposed scheme, these effects do not affect the accuracy of the location.

Figure 3.45 shows the waveform of the current measured at the tower right ahead of the fault point, under the condition where a three-phase short circuit occurs at the middle of the transmission line. When the three-phase short-circuit occurs, the measured magnetic field increases. The ratio of the magnetic field during the fault to that during the normal situation is same as the ratio of the short-circuit current to the current at normal conditions (as discussed in the previous section), provided that the transient process is not considered. The transient generally further increases the value of the magnetic field.

In order to obtain a full view of the transient process along the whole transmission line to help find the fault point, the magnetic field measured at every tower along the transmission line, which connects the two systems, is plotted in Figure 3.46. The x axis represents the evolution of time (during the life cycle of the fault, i.e. pre-fault, fault, and post-fault), the y axis represents the distribution along the transmission line (0 to the length of the line), and the z axis represents the measured magnetic field. Note that the high magnitude at the beginning of the fault is caused by the decaying DC components in the three-phase currents. The magnitude of the magnetic field at steady state under fault conditions is determined by the short-circuit levels of the two systems connected at the two sides. The difference in the short-circuit level can help to locate the fault to a span by simply comparing the values of the measured magnetic field. It is noted that at one side, although there is difference in the measured magnetic field along the

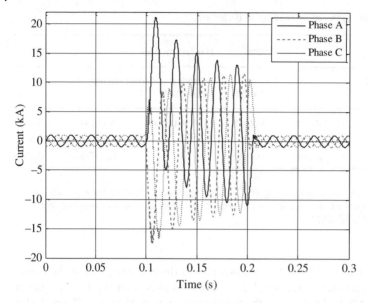

**Figure 3.45** Typical waveform of the current flowing in three phases during the occurrence of a three-phase short-circuit fault ($m = 100$ km). (*Source:* Reprinted with permission from Q. Huang et al., A novel approach for fault location of overhead transmission line with noncontact magnetic-field measurement, *IEEE Transactions on Power Delivery*, vol. 27, no. 3, pp. 1186–1195, July 2012.)

transmission line, the difference is very small. This may help a lot considering that the data at certain monitoring terminal may not be collected back in time due to component failure. For this reason, any data from the monitoring terminal can be used to identify the fault type, and this is discussed in the next section.

If the fault happens to occur at a point where the short-circuit level at two sides is almost the same, making the magnitude of the magnetic field at the two sides very close, then the direction of the magnetic field may be a good indicator to determine the fault span. The magnitude and direction of the magnetic field measured at the two sides (towers M and N in Figure 3.43) of the fault point are shown in Figure 3.47. The small circle at the center describes the magnetic field under normal conditions. Since the radial component (the $z$ component in Figure 3.42) is generally zero, only $B_x$ and $B_y$ are plotted. During the life cycle of a fault, both $B_x$ and $B_y$ evolve with time. It is shown in the figure that, when the system is under normal state, the measured magnetic fields at M and N have the same direction and magnitude, but once a fault is initiated, they have almost opposite directions. If the fault disappears or is cleared by reclosure, the two measured magnetic fields move back to the same point again (the center small circle).

### B. Identification of the Fault Type

The measured magnetic field can be used not only for locating the fault but also for identifying the fault type. According to (3.73) and (3.74), when a fault occurs the type of fault can be identified according to the magnitude and direction of the measured magnetic field. Since the short-circuit current is generally much larger than the normal current, the magnetic field caused by current flowing in an unfaulted phase can be neglected.

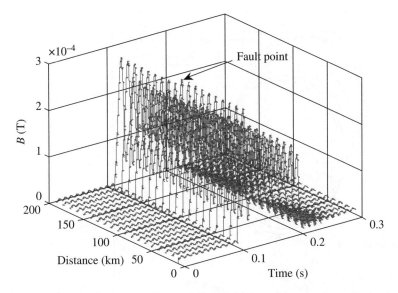

**Figure 3.46** The distribution of the magnetic field along the transmission line under three-phase short-circuit conditions ($m = 100$ km). (*Source:* Reprinted with permission from Q. Huang et al., A novel approach for fault location of overhead transmission line with noncontact magnetic-field measurement, *IEEE Transactions on Power Delivery*, vol. 27, no. 3, pp. 1186–1195, July 2012.)

Therefore, when a single-phase fault occurs, the measured magnetic field satisfies the following conditions (refer to Figure 3.42 for the meaning of the parameters):

$$\begin{cases} \frac{B_y}{B_x} \approx \frac{\sin\theta}{\cos\theta} = tg\theta, & \texttt{fault at phase A} \\ B_y \approx 0, & \texttt{fault at phase B} \\ \frac{B_y}{B_x} \approx \frac{-\sin\theta}{\cos\theta} = -tg\theta, & \texttt{fault at phase C} \end{cases} \tag{3.75}$$

Figure 3.48 displays the measured magnetic fields under a single-phase short circuit and Figure 3.49 displays the calculated output of the sensor along different directions. It is observed that equation (3.75) is true.

When a double-phase fault occurs, one should distinguish the difference between a line-to-line fault and a double line-to-ground fault. When a line-to-line fault is considered (transient neglected), the relation of the measured magnetic field components according to (3.73) is (refer to Figure 3.42 for the meaning of the parameters):

$$\begin{cases} \frac{B_y}{B_x} \approx \frac{\frac{1}{r_A}\sin\theta}{\frac{1}{r_B}+\frac{1}{r_A}\cos\theta}, & \texttt{AB line-to-line fault} \\ B_x \approx 0, & \texttt{AC line-to-line fault} \\ \frac{B_y}{B_x} \approx \frac{-\frac{1}{r_C}\sin\theta}{\frac{1}{r_B}+\frac{1}{r_C}\cos\theta}, & \texttt{BC line-to-line fault} \end{cases} \tag{3.76}$$

When a double line-to-ground fault occurs, there is no such simple relation among the components. However, by incorporating the estimation of the short-circuit current (since the fault point is known), the fault type can be identified. Table 3.12 lists the characteristics under different fault conditions to hep in identification. The identification procedure consists of the following steps:

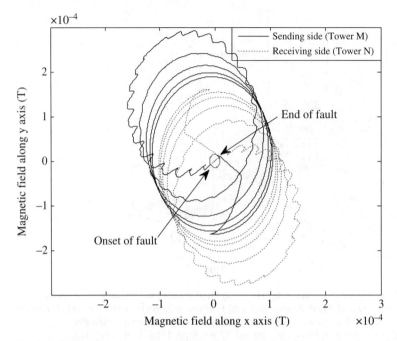

**Figure 3.47** The direction and magnitude of the magnetic field under the three-phase short-circuit conditions (fault occurs at 0.2 s and is cleared at 0.3 s, $m$ = 100 km). (*Source:* Reprinted with permission from Q. Huang et al., A novel approach for fault location of overhead transmission line with noncontact magnetic-field measurement, *IEEE Transactions on Power Delivery*, vol. 27, no. 3, pp. 1186–1195, July 2012.)

*Step 1*: Locate the fault according to the data sent back from the terminal installed along the transmission line.

*Step 2*: Approximately estimate the short-circuit current level using the fault distance determined in step 1.

*Step 3*: Compare the measured magnetic field and the calculated value to determine if the fault is single-phase, double-phase or three-phase (third column of Table 3.12).

*Step 4*: Further determine the type of the fault according to equations (3.75) and (3.76), and the rules in Table 3.12.

*Step 5*: Locate within the fault span according to the result of previous location and identification.

### C. Location Within the Fault Span

The previously described method can locate the fault span. It is observed that the measured magnetic fields have no significant difference along the faulted transmission line. However, the measured magnetic fields at the two sides (e.g. M and N in Figure 3.43) of the fault span are different. This difference can be used to estimate the distance between the fault point and the tower. Figure 3.50 shows the model for estimation of the distance to a tower in a faulted span for a single conductor. Assume that the sensor is installed under a faulted conductor on which a current $i(t)$ is flowing through at a distance of $a$. If $a$ is far less than the length of a span, then the left side of the sensor point can be

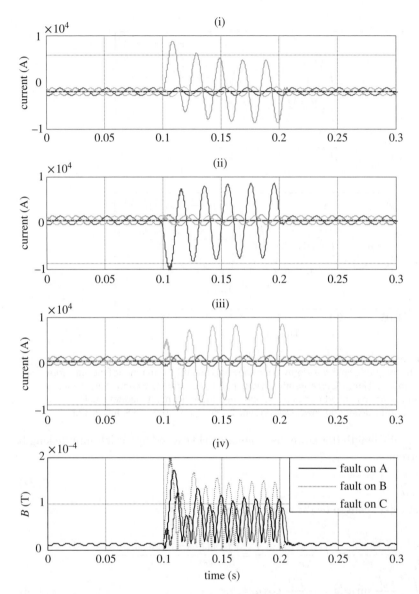

**Figure 3.48** Magnetic field under single-phase fault conditions: (a) current waveform under fault on phase A, (b) current waveform under fault on phase B, (c) current waveform under fault on phase C, and (d) the resulted magnetic field waveform under single-phase short-circuit conditions. (*Source:* Reprinted with permission from Q. Huang et al., A novel approach for fault location of overhead transmission line with noncontact magnetic-field measurement, *IEEE Transactions on Power Delivery*, vol. 27, no. 3, pp. 1186–1195, July 2012.)

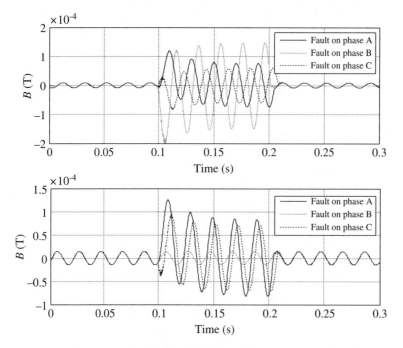

**Figure 3.49** The measured magnetic fields along different axes under single-phase short-circuit conditions: (a) the magnetic field waveform along the $x$ axis and (b) the magnetic field waveform along the $y$ axis ($m = 100$ km). (*Source:* Reprinted with permission from Q. Huang et al., A novel approach for fault location of overhead transmission line with noncontact magnetic-field measurement, *IEEE Transactions on Power Delivery*, vol. 27, no. 3, pp. 1186–1195, July 2012.)

regarded as infinite length, therefore the magnetic field caused by the left infinite length section of current is:

$$B = \frac{\mu_0 i}{4\pi a} \tag{3.77}$$

The magnetic field caused by the right section of current is:

$$\begin{aligned} B &= \int_0^l \frac{\mu_0 i}{4\pi} \frac{\vec{dl} \times \vec{r}}{r^3} = \int_0^l \frac{\mu_0 i}{4\pi} \frac{dl \sin \alpha}{r^2} = \int_0^l \frac{\mu_0 i}{4\pi} \frac{d(-actg\alpha)\sin^3 \alpha}{a^2} \\ &= \int_{\frac{\pi}{2}}^{\alpha_f} \frac{\mu_0 i}{4\pi a} \sin \alpha d\alpha = -\frac{\mu_0 i}{4\pi a} \cos \alpha_f \end{aligned} \tag{3.78}$$

Therefore, the ratio of the measured magnetic field beside the fault point to the other measured magnetic field is:

$$\frac{B_{xf}}{B_x} = \frac{B_{yf}}{B_y} = \frac{1 - \cos \alpha_f}{2} \tag{3.79}$$

where $B_x$ and $B_y$ are assumed to be the measured magnetic field components far away from the fault span.

Double-phase and three-phase faults are more complex because a simple analytical expression cannot be found due to the difference in phase angle in the three phases.

**Table 3.12** Identification of the fault type by magnetic field measurement.

| Fault type | | Field strength | Supplementary rules |
|---|---|---|---|
| | | **Characteristics for identification** | |
| **Three-phase fault** | | Largest | All components are the largest |
| Double-phase line-to-line fault | AB | Medium magnetic field | Satisfies the first equation of (3.76) |
| | AC | | Satisfies the second equation of (3.76) |
| | BC | | Satisfies the third equation of (3.76) |
| Double-phase line-to-ground fault | AB | | $y$ component is $\frac{1}{\sqrt{3}}$ of that in a three-phase fault and $\frac{B_y}{B_x} > 0$ |
| | AC | | $y$ component is as large as in a three-phase fault |
| | BC | | $y$ component is $\frac{1}{\sqrt{3}}$ of that in a three-phase fault and $\frac{B_y}{B_x} < 0$ |
| Single-phase fault | A | Smallest magnetic field | Satisfies the first equation of (3.75) |
| | B | | Satisfies the second equation of (3.75) |
| | C | | Satisfies the third equation of (3.75) |

*Source:* Reprinted with permission from Q. Huang et al., A novel approach for fault location of overhead transmission line with noncontact magnetic-field measurement, *IEEE Transactions on Power Delivery*, vol. 27, no. 3, pp. 1186–1195, July 2012.

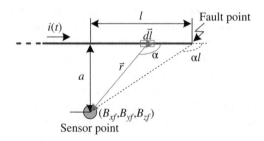

**Figure 3.50** Estimation of the fault distance in a fault span. (*Source:* Reprinted with permission from Q. Huang et al., A novel approach for fault location of overhead transmission line with noncontact magnetic-field measurement, *IEEE Transactions on Power Delivery*, vol. 27, no. 3, pp. 1186–1195, July 2012.)

However, once the type of the fault is determined, the location within the fault span can be accomplished using $B_x$ or $B_y$ components. For example, in single-phase fault cases, the location can be estimated using either $B_x$ or $B_y$; under double-phase and three-phase line-to-line faults, one can choose the $B_y$ component to compare. Figure 3.51 shows the ratio of the magnetic field measured at the tower next to the fault point to that measured at other towers as a function of the fault location. When the fault distance is less than 100 m, a significant difference can be observed. If there is no difference observed, it can be concluded that the fault point is out of the range of 100 m, which can also help the maintenance crew to expedite the location procedure.

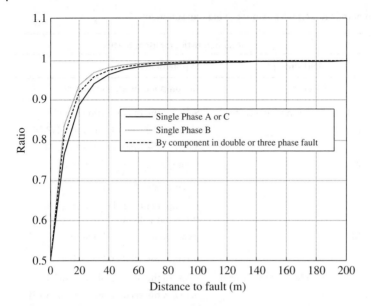

**Figure 3.51** Ratio of the magnetic field measured at the tower next to the fault point to that measured at the other towers under different fault types. (*Source:* Reprinted with permission from Q. Huang et al., A novel approach for fault location of overhead transmission line with noncontact magnetic-field measurement, *IEEE Transactions on Power Delivery*, vol. 27, no. 3, pp. 1186–1195, July 2012.)

### 3.6.2.5 Error Estimation

The error of this proposed scheme comes from following aspects: (1) the measurement error of the device, which is highly dependent on the characteristics of the MR sensor and the signal conditioning circuit, (2) the error caused by the sag of the transmission lines, and (3) when the approach is used for a multiple conductor transmission system (e.g. the typically used quad-bundled conductor), the non-uniform distribution of the current will cause a certain amount of error or, if twisting of conductors occurs, it may bring errors to the measurement of magnetic field. Other factors, such as the skin effect, can be neglected since the sensor is placed far enough away from the conductors.

*A. Measurement Error*

One of the advantages of MR sensors is that solid-state sensors have typical bandwidths in the megaHertz range and resolutions of tens of microGauss [39]. Modern circuit design technology can easily facilitate a signal conditioning circuit of megaHertz bandwidth. However, if only for fault location purposes, it is reasonable to limit the sampling frequency to under 2000 Hz. In the laboratory experiment, a test 12-bit A/D converter can easily limit the measurement error in a range of 1%.

*B. Sag Effect*

When sag of the conductors is considered, the magnetic field at the measurement point will be different from that generated by the current-carrying power line conductors, which are assumed to be straight horizontal wires. The evaluation of the error caused by sag can be accomplished using a combination of the Biot–Savart law and the catenary equation[9].

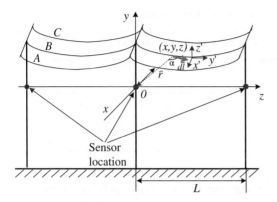

**Figure 3.52** Magnetic field generated by the transmission system considering the sag effect. (*Source:* Reprinted with permission from Q. Huang et al., A novel approach for fault location of overhead transmission line with noncontact magnetic-field measurement, *IEEE Transactions on Power Delivery*, vol. 27, no. 3, pp. 1186–1195, July 2012.)

Consider the transmission system shown in Figure 3.52. For evaluation purposes, assume that the sags of two adjacent leveled spans are symmetric and all three phase conductors have the same amount of sag (different sags can also be similarly handled). To simplify the calculation, two coordinate systems, $(x, y, z)$ and $(x', y', z')$, are used. The $(x, y, z)$ coordinate system is centered at the location of the sensor, in order to be congruent with the coordinate system in Figure 3.41, while $(x', y', z')$ is centered at the maximum sag point of the middle phase conductor (phase B in Figure 3.52), in order to be congruent with the popular use of the catenary equation. The calculation of the magnetic field at an arbitrary point in the space can then be implemented as follows.

Considering phase conductor B in Figure 3.52, the governing catenary equation in the $(x', y', z')$ coordinate system is:

$$z' = \frac{1}{\alpha}(\cosh(\alpha y') - 1), -\frac{L}{2} \le y' \le \frac{L}{2} \tag{3.80}$$

where $L$ is the length of the span and $\alpha$ is a constant determined by the mechanical parameters of the transmission lines.

Consider a small section of any one of the transmission lines $d\vec{l}$. According to the Biot–Savart law, it generates a magnetic field at a point $(x'_0, y'_0, z'_0)$:

$$d\vec{B} = \frac{\mu_0 I}{4\pi} \frac{d\vec{l} \times \vec{r}}{\vec{r}^3} \tag{3.81}$$

According to (3.80), $d\vec{l}$ can be expressed as

$$d\vec{l} = dy\hat{i}_y + dz'\hat{i}_z = dx\hat{i}_x + dy\hat{i}_y + \frac{dz'}{dy'}dy'\hat{i}_z \tag{3.82}$$

where $\frac{dz'}{dy'} = \sinh\alpha y'$ and

$$\vec{r} = (x'_0 - x')\hat{i}_x + (y'_0 - y')\hat{i}_y + (z'_0 - \frac{1}{\alpha}(\cosh\alpha y' - 1))\hat{i}_z \tag{3.83}$$

**Table 3.13** Evaluation of the sag effect on the calculation of the magnetic field.

| | Coefficient along three spatial axes | | | | | |
| | Straight horizontal wire | | | Sag effect considered | | |
| Phase conductors | x | y | z | x | y | z |
| --- | --- | --- | --- | --- | --- | --- |
| A | 0.0423 | 0.0453 | 0 | 0.0434 | 0.0469 | 0 |
| B | 0.0909 | 0 | 0 | 0.0954 | 0 | 0 |
| C | 0.0423 | 0.0453 | 0 | 0.0434 | 0.0469 | 0 |

Span, 400 m; phase space, 11.8 m; sag, 5 m.
*Source:* Reprinted with permission from Q. Huang et al., A novel approach for fault location of overhead transmission line with noncontact magnetic-field measurement, *IEEE Transactions on Power Delivery*, vol. 27, no. 3, pp. 1186–1195, July 2012).

Substituting (3.82) and (3.83) into (3.81), and if only one span is considered, the magnetic field generated at point $(x_0, y_0, z_0)$ is

$$\vec{B} = \frac{\mu_0 i}{4\pi} \int_{-L/2}^{L/2} \left[ \frac{z_0' - \frac{1}{a}(\cosh \alpha y' - 1) + (y' - y_0')\sinh \alpha y'}{|\vec{r}|^3} \hat{i}_x \right. \\ \left. + \frac{(x_0' - x')\sinh \alpha y'}{|\vec{r}|^3} \hat{i}_y + \frac{-(x_0' - x')}{|\vec{r}|^3} \hat{i}_z \right] dy' \tag{3.84}$$

When only one of the three conductors is considered (the other two phase currents are assumed to be zero), since the two adjacent spans are symmetric, integration in two spans will double the magnetic field along the $\hat{i}_x$ and $\hat{i}_z$ directions, and cancel the magnetic field along the $\hat{i}_y$ direction.

Then, with the symmetry assumption, the magnetic field generated by phases A, B and C at the middle sensor in Figure 3.52 can be obtained by the following equations:

$$\vec{B}_a = \frac{\mu_0 i_A}{4\pi} \int_{-L/2}^{L/2} \left[ \frac{z_0' - \frac{1}{a}(\cosh \alpha y' - 1) + (y' - y_0')\sinh \alpha y'}{M_a} \hat{i}_x \right. \\ \left. + \frac{(x_0' - k)\sinh \alpha y'}{M_a} \hat{i}_y + 0\hat{i}_z \right] dy' \tag{3.85}$$

$$\vec{B}_b = \frac{\mu_0 i_B}{4\pi} \int_{-L/2}^{L/2} \left[ \frac{z_0' - \frac{1}{a}(\cosh \alpha y' - 1) + (y' - y_0')\sinh \alpha y'}{M_b} \hat{i}_x \right. \\ \left. + \frac{x_0' \sinh \alpha y'}{M_b} \hat{i}_y + 0\hat{i}_z \right] dy' \tag{3.86}$$

$$\vec{B}_c = \frac{\mu_0 i_C}{4\pi} \int_{-L/2}^{L/2} \left[ \frac{z_0' - \frac{1}{a}(\cosh \alpha y' - 1) + (y' - y_0')\sinh \alpha y'}{M_c} \hat{i}_x \right. \\ \left. + \frac{(x_0' + k)\sinh \alpha y'}{M_c} \hat{i}_y + 0\hat{i}_z \right] dy' \tag{3.87}$$

where $\hat{i}_x$, $\hat{i}_y$, and $\hat{i}_z$ are the unit vectors along the $x'$, $y'$, and $z'$ axes, respectively, $(x_0, y_0, z_0)$ is the coordinate of the sensor point under the $(x', y', z')$ coordinate system (origin in the $(x, y, z)$ coordinate system), and $k$ is the space between the phase conductors (11.8 m, as shown in Figure 3.41).

$$M_a = \left[ (x_0' - k)^2 + (y_{0'} - y')^2 + (z_{0'} - \frac{1}{\alpha}(\cosh \alpha y' - 1))^2 \right]^{\frac{3}{2}} \tag{3.88}$$

$$M_b = \left[ x_{0'}^2 + (y_{0'} - y')^2 + (z_{0'} - \frac{1}{\alpha}(\cosh \alpha y' - 1))^2 \right]^{\frac{3}{2}} \tag{3.89}$$

$$M_c = \left[ (x_0' + k)^2 + (y_{0'} - y')^2 + (z_{0'} - \frac{1}{\alpha}(\cosh \alpha y' - 1))^2 \right]^{\frac{3}{2}} \tag{3.90}$$

Comparing these equations with (3.73), it is clear that one only needs to compare the definite integrals with $\frac{1}{r_A}$, $\frac{1}{r_B}$, and $\frac{1}{r_C}$ in (3.73) and their components along the three spatial axes, respectively. For a transmission system configured as shown in Figure 3.41, assuming the span length $L$ at 400 m and further assuming a sag of 5 m on all three phase conductors, the comparison is shown in Table 3.13. It is shown that the error is always below 5%.

To evaluate the error caused by bundled conductors, a typical quad-bundle configuration with 0.3 m spacing is considered. For example, if phase B is considered, and further assuming the current is uniformly flowing in four conductors, the following formula can be used to evaluate the error by assuming that the magnetic field is generated by a single conductor:

$$\left( \frac{1}{2} \left( \frac{1}{sqrt(r_B - 0.15)^2 + 0.15^2} + \frac{1}{sqrt(r_B + 0.15)^2 + 0.15^2} \right) \right) / \left( \frac{1}{r_B} \right) \tag{3.91}$$

Numerical results show that, for the configuration discussed in this paper, this may only cause 0.01% error.

### 3.6.2.6 Comparison of the approaches

According to error estimate, one can safely conclude that the error brought by the sag and bundled conductor will not affect the effectiveness of the approaches discussed above, hence the proposed approach is feasible. Table 3.14 summarizes the general requirements, advantages, and disadvantages of the available approaches for fault location of the power transmission line. One of the great advantages of the proposed approach is that its maximum location error is one span and it does not assume a homogeneous line. Traditional approaches have to rely on the assumption of a homogeneous line and the evaluation of the performance is obtained by the fault location error defined in IEEE PC37.114 [33], i.e. a percentage error in fault location estimate based on the total line length:

$$error = \frac{\text{instrument reading} - \text{exact distance to the fault}}{\text{total line length}} \tag{3.92}$$

This means that even a small error may cause a large physical distance for a long transmission line. Furthermore, unlike the impedance measurement based approach, the ground resistance does not affect the performance of the proposed approach. The only disadvantages of the proposed approach are that it has to rely on distributed measurement and, because it depends on current measurement, it cannot be used for open-circuit fault detection. However, considering the fact that the most common and dangerous fault that occurs in a power system is a short-circuit, not working for open-circuit faults would not limit the use of the proposed approach. With modern communications, information, and integrated circuit technology, it is very easy to implement such a distributed measurement system.

### 3.6.2.7 Monitoring Software Development

The design of the fault location software system includes the design of the database, each functional module, the fault location module, and the fault type identification module [40]. A sequential query language (SQL) server is used to build the system database and

**Table 3.14** Comparison of the available fault location approaches.

| Approaches | Descriptions | |
|---|---|---|
| Impedance measurement approach | Requirements | Measurement of voltage and current |
| | | Parameters to know: line length and transmission impedance |
| | Advantages | No additional hardware cost |
| | Disadvantages | Have to assume homogeneity in line parameter |
| | | Relative location error |
| | | Does not work well for faults with fault resistance |
| | | Performance is highly dependent on signal processing techniques |
| | | Affected by CT saturation |
| | | Does not work for open-circuit faults |
| | | Hard to map the electrical distance to actual geographical location |
| Traveling wave approach | Requirements | Measurement of current or voltage traveling wave |
| | | Parameters to know: line length and wave speed |
| | Advantages | Works for open-circuit faults |
| | Disadvantages | Has to assume uniform wave speed |
| | | High-speed data acquisition, high cost |
| | | Relative location error |
| | | Performance is highly dependent on signal processing techniques |
| | | Hard to map the electrical distance to actual geographical location |
| Magnetic field measurement approach | Requirements | Measurement of current |
| | | Parameters to know: none |
| | Advantages | Non-contact |
| | | Absolute location error: one span |
| | | Independent of distributed line parameters |
| | | Not affected by saturation |
| | | Location of fault is mapped to actual geographical location and can be easily visualized |
| | Disadvantages | Distributed measurement, requiring reliable data communication |
| | | Does not work for open-circuit faults |

*Source:* Reprinted with permission from Q. Huang et al., A novel approach for fault location of overhead transmission line with noncontact magnetic-field measurement, *IEEE Transactions on Power Delivery*, vol. 27, no. 3, pp. 1186–1195, July 2012.

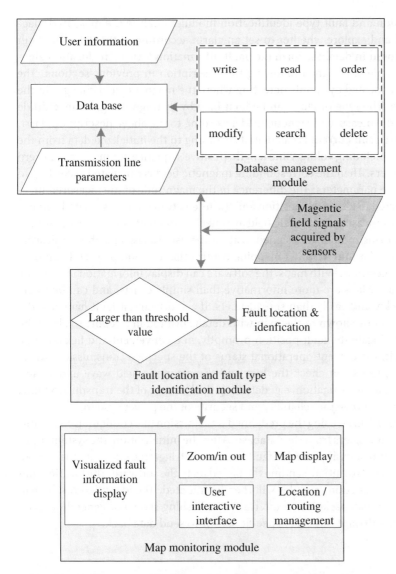

**Figure 3.53** Architecture and flowchart of the software system. (*Source:* Reprinted with permission from Q. Huang et al., Design and implementation of a non-contact magnetic field measurement based fault location system for overhead transmission line, *IEEE International Conference on Smart Instrumentation, Measurement and Applications (ICSIMA)*, Nov. 2013.)

there are three main databases to manage the data, including the measured magnetic field signal, the transmission tower parameters (such as the location of the tower, the voltage of the line, the configuration of the transmission lines at each tower, etc.), and the fault information. The software was developed by VC++ 6.0 and the flow chart showing it is given in Figure 3.53. In the database management module, the database classes are defined, and properties and methods are encapsulated, such as record reading, modifying, ordering etc., enabling flexible access and operation of database forms at any time.

The fault location and fault type identification modules analyze the measured magnetic field signal and explore whether to set an alarm according to the data. The fault information is stored in database form (i.e. fault information form). The location algorithm of this module is designed following the description in previous sections. The fault point can be located by simply detecting where the magnetic field changes in the transmission line. The type of fault can be identified by distinguishing the magnitude and direction of the measured magnetic field according to the above description. First, the level of short-circuit current is calculated according to the initialized data from the transmission line, including voltage, line length, positive sequence parameters, and zero sequence parameters. Then the corresponding magnetic field is calculated according to the tower structure parameters. The difference in the magnitude of the measured magnetic field is detected to locate the position and the intensity of fault is estimated. Finally, the components of measured magnetic field are checked to identify the type of fault.

The map monitoring module is designed to allow the user to visualize the monitoring interface based on MapInfo MapX, a mapping control that can add powerful mapping capabilities to applications. With maps, the software can display information in an easily understood format. Maps are more informative than simple charts, and can be interpreted more quickly and easily than spreadsheets. If a short-circuit fault happens, the position of fault will be shown in the map with details printed in a report to allow the operating crew to locate the fault position promptly, in a convenient and flexible way. If there is no fault, the current operational status of the specified transmission line is shown. Moreover, users can check the logs and the magnetic field wave under fault conditions for further investigation, e.g. detecting the hot spot of the transmission line, which is helpful to improve the reliability and security of the power system.

The software is initialized using the predefined transmission line configuration parameters and system parameters from the database. After the initialization, the system monitors the specified transmission line. If a fault occurs, the triggering system in the remote terminal will be started to initiate sending the fault data to the software. The location and identification of the function modules will then be initiated. The identified results will be displayed on the map background interface in a flashing style. For general purpose measurement, the software can order specific sensor to send data back.

## Bibliography

1 "IEC Smart Grid Standardization Roadmap," SMB Smart Grid Strategic Group, Report 1.0 - 2009-12, June 2010.

2 CPWR. (2018) The Construction Chart Book, 4th edition. [Online]. Available: http://www.elcosh.org/document/1059/269/d000038/sect38.html

3 W. Wang and S. Pinter, *Dynamic line rating systems for transmission lines*. U.S. Department of Energy, 2014.

4 X. Sun, Q. Huang, Y. Hou, L. Jiang, and P. W. T. Pong, "Noncontact operation-state monitoring technology based on magnetic-field sensing for overhead high-voltage transmission lines," *IEEE Transactions on Power Delivery*, vol. 28, no. 4, pp. 2145–2153, 2013. [Online]. Available: 10.1109/TPWRD.2013.2264102

5 A. Phillips, "Evaluation of instrumentation and dynamic thermal ratings for overhead lines," Office of Scientific and Technical Information, US Department of Energy, Report, 2013.

6 X. Sun, Q. Huang, L. Jiang, and P. Pong, "Overhead high-voltage transmission-line current monitoring by magnetoresistive sensors and current source reconstruction at transmission tower," *IEEE Transactions on Magnetics*, vol. 50, no. 1, pp. 1–5, 2014.

7 MicroMagnetics. (2018) Low-noise TMR magnetic field sensor. [Online]. Available: http://www.micromagnetics.com/product_page_stj300.html

8 E. R. Council, *Transmission line reference book: 345kV and above*. Palo Alto, CA: Electric Power Research Institute, 1975.

9 A. Mamishev, R. Nevels, and B. Russell, "Effects of conductor sag on spatial distribution of power line magnetic fields," *IEEE Transactions on Power Delivery*, vol. 11, no. 3, pp. 1571–1576, 1996.

10 X. Sun, K. Lui, K. Wong, W. Lee, Y. Hou, Q. Huang, and P. Pong, "Novel application of magnetoresistive sensors for high-voltage transmission-line monitoring," *IEEE Transactions on Magnetics*, vol. 47, no. 10, pp. 2608–2611, 2011.

11 Z. Lei, G. Li, W. Egelhoff, P. Lai, and P. Pong, "Review of noise sources in magnetic tunnel junction sensors," *IEEE Transactions on Magnetics*, vol. 47, no. 3, pp. 602–612, 2011. [Online]. Available: 10.1109/TMAG.2010.2100814

12 W. Shen. Lecture notes for introduction to numerical computation. [Online]. Available: https://www.math.psu.edu/shen_w/451/NoteWeb

13 J. Lenz and S. Edelstein, "Magnetic sensors and their applications," *IEEE Sensors Journal*, vol. 6, no. 3, pp. 631–649, 2006.

14 C. Duret and U. Shintarou, "TMR: A new frontier for magnetic sensing," NTN-SNR Roulements Research and Innovation Mechatronics, Report, 2012.

15 MultiDimension. 3 axis TMR linear sensor. Accessed: 2016-04-30. [Online]. Available: http://www.dowaytech.com/en/1879.html

16 J.-A. Jiang, Y.-H. Lin, J.-Z. Yang, T.-M. Too, and C.-W. Liu, "An adaptive PMU based fault detection/location technique for transmission lines. II. PMU implementation and performance evaluation," *IEEE Transactions on Power Delivery*, vol. 15, no. 4, pp. 1136–1146, 2000. [Online]. Available: 10.1109/61.891494

17 S. Jing, Q. Huang, J. Wu, and W. Zhen, "A novel whole-view test approach for onsite commissioning in smart substation," *IEEE Transactions on Power Delivery*, vol. 28, no. 3, pp. 1715–1722, 2013.

18 TP-Link. 2.4 GHz high power wireless outdoor CPE. Accessed: 2016- 01-11. [Online]. Available: https://www.tp-link.com/us/products/details/cat-5039_TL-WA5210G.html

19 M. Farzaneh, *Atmospheric Icing of Power Networks*. Springer, 2008.

20 H. M. Ryan, *High Voltage Engineering and Testing*. IET Digital Library, 2013.

21 B. Liu, K. Zhu, X. Sun, B. Huo, and X. Liu, "A contrast on conductor galloping amplitude calculated by three mathematical models with different DOFs," *Shock and Vibration*, 2014.

22 Y. K. Liu, Z. C. Fu, and X. H. Yang, "Galloping amplitude analysis and observation on full-scale test overhead transmission lines." IEEE Power & Energy Society General Meeting, July 2015, pp. 1–5.

23 N. A. Stutzke, S. E. Russek, D. P. Pappas, and M. Tondra, "Low-frequency noise measurements on commercial magnetoresistive magnetic field sensors," *Journal of Applied Physics*, vol. 97, no. 10, 2005.

**24** G. Zhao, J. Hu, Y. Ouyang, Z. Wang, S. X. Wang, and J. He, "Tunneling magnetoresistive sensors for high-frequency Corona discharge location," *IEEE Transactions on Magnetics*, vol. 52, no. 7, pp. 1–4, 2016.

**25** Q. Huang, A. H. Khawaja, A. Zheng, S. J. J. L. Z. Zhang, and J. Yi, "Method for simultaneous monitoring of phase current and spatial parameters of overhead transmission lines with non-contact magnetic field sensor array," Patent, Dec., 2017, US Patent 20180143224A1, https://patents.google.com/patent/US20180143224A1/en.

**26** T. Adu, "A new transmission line fault locating system," *IEEE Transactions on Power Delivery*, vol. 16, no. 4, pp. 498–503, 2001.

**27** Y. Liao and M. Kezunovic, "Optimal estimate of transmission line fault location considering measurement errors," *IEEE Transactions on Power Delivery*, vol. 22, no. 3, pp. 1335–1341, 2007.

**28** H. Ha, B. Zhang, and Z. Lv, "A novel principle of single-ended fault location technique for EHV transmission lines," *IEEE Transactions on Power Delivery*, vol. 18, no. 4, pp. 1147–1151, 2003.

**29** T. Bouthiba, "Fault location in EHV transmission lines using artificial neural networks," *International Journal of Applied Mathematics and Computer Science*, vol. 14, no. 1, pp. 69–78, 2004.

**30** M. Bockarjova, A. Sauhats, and G. Andersson, "Statistical algorithms for fault location on power transmission lines," in *IEEE Russia Power Tech*, 2005, pp. 1–7.

**31** P. Anderson and A. Fouad, *Power System Control and Stability*. Iowa State University Press, 1977.

**32** IEEE/CIGRE Joint Force On Stability Terms definitions, "Definition and classification of power system stability," *IEEE Transactions on Power Systems*, 2004.

**33** "IEEE guide for determining fault location on ac transmission and distribution lines," *IEEE Std C37.114-2004*, pp. 1–36, 2005.

**34** A. Abur and F. H. Magnago, "Use of time delays between modal components in wavelet based fault location," *International Journal of Electrical Power and Energy Systems*, vol. 22, no. 6, pp. 397–403, 2000.

**35** S. M. Brahma, "Iterative fault location scheme for a transmission line using synchronized phasor measurements," *International Journal of Emerging Electric Power Systems*, vol. 8, no. 6, pp. 1–14, 2007.

**36** *Electrical Transmission and Distribution Reference Books*. ABB Power Transmission and Distribution Company Inc, 1997.

**37** A. Sauhats and M. Danilova, "Fault location algorithms for super high voltage power transmission lines," in *IEEE Bologna Power Tech Conference Proceedings*, vol. 3, 2003, pp. 1–6.

**38** Q. Huang, W. Zhen, and P. W. T. Pong, "A novel approach for fault location of overhead transmission line with noncontact magnetic-field measurement," *IEEE Transactions on Power Delivery*, vol. 27, no. 3, pp. 1186–1195, 2012.

**39** B. Wincheski and J. Simpson, "Development and application of wide bandwidth magneto-resistive sensor based eddy current probe," ser. American Institute of Physics Conference Series, D. O. Thompson and D. E. Chimenti, Eds., vol. 1335, 2011, pp. 388–395.

**40** Q. Huang, R. Yao, F. Li, and W. Zhen, "Design and implementation of a non-contact magnetic field measurement based fault location system for overhead transmission line," in *IEEE International Conference on Smart Instrumentation, Measurement and Applications (ICSIMA)*, 2013, pp. 1–4.

# 4

# Magnetic Field Measurement for Modern Substations

## 4.1  Introduction to GIS-based Substations

### 4.1.1  Smart Substations

Creating a "strong and smart grid" is the strategic plan for China's smart grids. However, the problems and shortcomings of traditional substations have hampered the development of smart grids. These problems and shortcomings include:

- inconsistency of data due to various of data systems existing in a substation
- repeated data acquisition
- high complexity of design and difficult maintenance due to various devices
- poor interoperability among systems and devices
- complex communication protocol
- lack of conformity test and authorization
- non-standard information, hence difficult to use.

The "strong and smart grid" plan motivates the substation to transform into a strong, reliable, economic, efficient, clean, environmentally friendly, transparent, open, and interactive modern substation. The smart substation has two parts: smart high-voltage equipment and a unified substation information platform.

The smart high-voltage equipment is composed of an organic integration of primary high-voltage equipment and intelligent components, and is characterized by digital measurement, networked control, visualized state, integrated functionalities, and interactive information. Smart high-voltage equipment mainly refers to a smart transformer, smart switching devices, and electronic transformers. The intelligent components are composed of a set of intelligent electronic devices (IEDs), which are state sensing components and intelligent actuators, including all or part of devices for measurement, control, state monitoring, metering, and protection. These components implement the basic functionalities such as measurement, control, and monitoring of the host equipment. In certain cases, the intelligent components are also used for metering and protection.

*Magnetic Field Measurement with Applications to Modern Power Grids*, First Edition.
Qi Huang, Arsalan Habib Khawaja, Yafeng Chen and Jian Li.
© 2020 John Wiley & Sons Ltd. Published 2020 by John Wiley & Sons Ltd.

The requirements and objectives of smart high-voltages equipment are:

- to enhance the reliability of the power grid by making equipment faults predictable
- to reduce the whole lifecycle cost through smart control of the cooling system (energy efficient), integrated design (saving land utilization), and online self-diagnostics (reducing operation and maintenance cost of assets)
- to optimize the utilization of assets.

In the unified substation information platform, the protocol IEC 61850 has been introduced. The IEC 61850, as the future standard for substations, has been chosen to build the communication platform for smart substations. It offers a cost-efficient solution by reducing the need for hardwiring between the switchgear bays. Using Generic Object Oriented Substation Event (GOOSE), communication supervision is a natural and integral part of the communication and is easier to accomplish using GOOSE system enhancements than with hardwired solutions. GOOSE also enables simplified substation wiring. In practice only one ethernet cable is required between the IEDs of a substation and an ethernet switch enables communication between the protection and control IEDs. This can be compared with a hardwired solution where, for each signal, a copper wire is connected from each IED to all the other IEDs in the substation.

Smart substations is a brand new concept and is still under its initial stage. In China, quite a lot of smart substation demonstration projects aimed at deploying and integrating intelligent solutions to enhance the efficiency and reliability of the electricity network are under construction. It is challenging to transition from a legacy substation to a highly efficient, highly reliable, and highly interconnected substation.

### 4.1.2 Gas-insulated Switchgear

High-voltage electrical distribution equipment can receive and distribute electric energy above 1 kV. This equipment is mainly categorized into three types:

- Air-insulated switchgear (AIS) : This is a kind of conventional distribution equipment, the bus bars of which are directly exposed to the air. AIS utilizes air and insulators to realize the insulation between the part of the primary equipment with power and the ground and phase-to-phase insulation.
- Hybrid power distribution system (H-GIS): This is a novel complete equipment system that is based on modularization thinking.
- Gas-insulated switchgear (GIS): This is a shorthand term for gas-insulated metal and enclosed switchgear, which connects the circuit breaker, disconnector, earthing switch, voltage transformer, current transformer, arrester, and bus bar, and encapsulate them in a metal enclosure with cable termination and in-and-out line casing. $SF_6$ is used for the interrupters and insulating medium in the enclosed and sealed enclosure.

Among the three types of high-voltage electrical distribution equipment, GIS-based substations prevail globally [1]. GIS appeared in the late 1950s and was applied in the market in the 1960s. It is a kind of high-voltage electrical distribution equipment. The first 170 kV GIS underground substation was installed in Zurich, Switzerland, by the ABB company [2]. In 1976, the first 500 kV GIS-based substation was installed in

**Figure 4.1** A typical 220 kV GIS-based substation.

Claireville, Canada, the first 800 kV one was installed in South Africa in 1986, while China installed its first 800 kV GIS-based substation in 2005. GIS production of 126 kV and above was realized in China in 2017.

### 4.1.3 GIS-based Substations

With the development of GIS technology, substations have made a great progress. GIS-based substations have been widely constructed. A GIS-based substation is composed of a circuit breaker, disconnector, bus bar, voltage transformer, current transformer, arrester, and cashing. All high-voltage electrical components are sealed in earthed metal cashing (see Figure 4.1). GIS-based substations have some important advantages due to their inherent characteristics and structure:

- Small floor area, small size, light weight. Generally, the floor area of a 220 kV GIS substation accounts for 37% of the area required for conventional equipment. The floor area of a 110 kV GIS substation accounts for 46% of the area of conventional equipment [3].
- Components are sealed and free from environmental disturbance such as pollution, moisture, and salt spray.
- The operating reliability is high becasuse it is oil-free and gas-free.
- Can be installed in urban areas becasue of its good seismic performance.
- High operating safety because SF6 is an inert gas.
- Small energy loss due to the small sensing magnetic field on the GIS surface. Loss caused by eddy effect is small.
- Low noise due to shielding enclosure.
- Short construction period. The components of GIS equipment are versatile. They are assembled in one building block structure and transported to the construction site for local assembly.

Although GIS has high operating reliability, with a fault rate that is 20–40% that of conventional equipment, it still has some shortcomings:

- Flashover failure will occur due to SF6 leakages, external water infiltration, impurity dopant, or aging problems with insulators.
- The enclosed structure of the GIS makes fault location and maintenance difficult.
- The power cut-off range is largely caused by faults in the GIS.

## 4.2 MR-based Electronic Current Transformers

Recently, power networks have undergone a period of rapid change. Power networks used to be systems connecting relatively stable power generation facilities, like large power stations, with a relatively well understood average load, i.e. commercial and domestic consumers. However, sources of renewable energy, such as those relying on wind, solar, tidal or wave power, for example, are dynamic sources by comparison with traditional gas, oil, coal and nuclear power stations. This change in the way some power is generated in the power network acts to destabilize it somewhat. Control of power networks, especially for smart grids, has become an important topic of research as a result [4]. In this context, improvement in non-invasive current measuring devices is desirable as this permits measurement of network control performance. Conventional non-contact current measurement instruments rely on magnetic field readouts where the magnetic flux density generated by the conductor under measurement is interpreted into electric current [5; 6; 7]. Such instruments include current transformer principle based ammeters, fluxgate magnetometer based ammeters, Hall effect based ammeters, and Rogowski coil based ammeters [8]. However, all such instruments are prone to inherited measurement complexities because of the measurement principle of the device. In particularly, Hall effect based ammeters employ a high permeability core to concentrate the maximum flux density into the sensitive axis of the sensor. This restricts operation in instances where the core material saturates in the presence of a large magnetic field. Fluxgate magnetometers present some advantages, such as small size, high sensitivity, and low noise [9]. However, as well as for the high price of fluxgate magnetometers, large drive currents are required to make the core saturated, which increases the power consumption [10; 11]. Similarly, current transformers suffer from the same disadvantage of core saturation in addition to hysteresis losses in the transformer core. Rogowski coil based electronic current transformers require multistage amplification of voltage induced by alternating electric current. These amplifiers require regular inspections and depict a drift with aging [12]. Additionally, Rogowski coils and current transformers can only capture alternating current signals. In the era of smart grids, modern power system applications need broadband low-cost current measurement instruments.

A relatively newer generation of magnetic field sensors has found application in broadband electric current measurement. They are based on the magnetoresistance (MR) effect, remain linear for strong magnetic field stimulus, and can operate at a wide band of frequencies [13]. MR effect based sensors are available in three generations with respect to the advancements in the fabrication process. The first generation employed the anisotropic effect, but these MR sensors require a calibration pulse for aligning the sensor output if exposed to strong magnetic stimulus.

Second-generation magnetic sensors based on the giant magnetoresistance (GMR) effect have been applied for current measurement. For instance, GMR magnetic sensors were used for current measurement of up to 45 A in the work of [14]. However, GMR magnetic sensors suffer from the disadvantage of unipolar output and sensitivity to environment temperature changes [15; 16].

Developed from the tunnel magnetoresistance (TMR) effect, third-generation sensors are more stable than those of the previous two generations. For current measurement, TMR magnetic sensors dominate other variants in terms of miniaturization, low cost, low power consumption, high integration, high response frequency, and high sensitivity characteristics [16; 17]. Meanwhile, the authors' research team have successfully established the appropriateness of TMR magnetic sensors for power system applications in other studies [18; 19; 20; 21].

### 4.2.1 Experimental Research on Hysteresis Effects in MR Sensors

#### 4.2.1.1 Introduction

GMR sensors are generally integrated into a Wheatstone bridge for analog current measurement. In such a configuration, it is found that the hysteresis effects in GMR sensors are the major source of measurement error. Experiments are conducted by measuring hysteresis curves between the applied magnetic field and sensor output voltage with different typical initial magnetization states [22]. GMR sensors have two operation modes: bipolar and unipolar. Bipolar operation means the internal magnetization pattern of the GMR sensor changes direction during the operation range. This mode is inappropriate for direct measurement, as experimental results indicate that the shape of measured curves with the same range as the applied magnetic field may be severely distorted due to different initial magnetization states. Unipolar operation is an intermediate process of bipolar operation with invariant magnetization direction. Its input–output relationship is almost linear so is especially suitable for measurement. The principal measurement error within unipolar operation is DC offset voltage, which is induced by remanence. It can be fixed by saturating the GMR sensor before operation.

From physical point of view, magnetic hysteresis means the internal magnetization pattern of the material does not uniquely depend on an instantaneous applied magnetic field, but is a combined effect with magnetization history [23]. For GMR sensors, this exists in both GMR resistors and flux concentrators. Hysteresis is often desirable in digital system applications because it can provide a lag effect to prevent error switch [24]. However, it is not appreciated in analog measurement applications as it may bring about large measurement error that severely influences the accuracy of the sensor output.

Magnetic hysteresis is characterized by a hysteresis curve, which shows the relationship between magnetization M and applied field H [23]. A pair of closed reversal curves forms a hysteresis loop. For a one-dimensional vector magnetic field sensor, research is mainly focused on hysteresis along the sensitive axis, by discussing the susceptibility, remanence, and coercivity of the major hysteresis loop and minor hysteresis curves for both bipolar operation and unipolar operation. Bipolar operation means the internal magnetization pattern along the sensitive axis changes direction in the range of the external applied magnetic field, and unipolar operation means the direction of the magnetization pattern stays invariant within the operation range. The major hysteresis loop exists only in bipolar operation and is unique. It goes from a negative saturation state to a positive saturation (ascending major curve), and back again (descending major curve). Other hysteresis curves with applied fields varying between two saturation fields are all minor hysteresis curves. Obviously, there is infinite number of minor curves, the shape of which depends on the initial magnetization state of the material, the range of the applied magnetic field, and the direction of variation (ascending or descending). The slope of the hysteresis curve is known as the susceptibility, and it reflects both the sensitivity and the linearity of the GMR sensor. When the external field is removed, the magnetization pattern does not vanish completely. This phenomenon is called remanence and it produces DC offset voltage output to the sensor. The critical applied magnetic field in the reverse direction with respect to the original magnetization pattern that totally removes remanence is called coercivity and it reflects the ability of magnetic material to resist demagnetization [25]. Coercivity exists only in bipolar operation and distinguishes it from unipolar operation.

#### 4.2.1.2 Experimental Setup

To make GMR sensors work properly, an additional conditioning circuit should be added to adjust the output signal to the appropriate level [26]. The Wheatstone bridge (Figure 4.2) can be powered by both current and voltage sources. Here, a 2.048 V ultra-stable voltage reference is selected. The differential output of the Wheatstone bridge is proportional to the voltage supply. This signal is further pre-amplified by an instrumentation amplifier with gain 10 through direct coupling.

Because hysteresis is a rate-independent phenomenon, i.e. the final magnetization pattern is the same no matter how fast te external magnetic field is applied [27], hysteresis curves can be measured in a static way by plotting output voltage V versus applied field H at discrete points. An experimental system for this is setup is shown in Figure 4.3. A solenoid is employed to provide the reference magnetic field. The GMR sensor is placed at the central region of the solenoid where the generated field is uniform. To minimize disturbance induced by the Earth's magnetic field (typically 0.5 Oe), the solenoid is

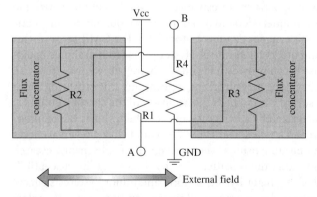

**Figure 4.2** Schematic diagram of a linear output GMR sensor. The two-way arrow denotes the sensitive axis. Terminals A and B are the positive and negative outputs of Wheatstone bridge, respectively. (*Source:* Reprinted with permission from S. Liu et. al., Experimental research on hysteresis effects in GMR sensors for analog measurement applications. *Sensors and Actuators A: Physical*, 182, 72–81, Aug. 2012.)

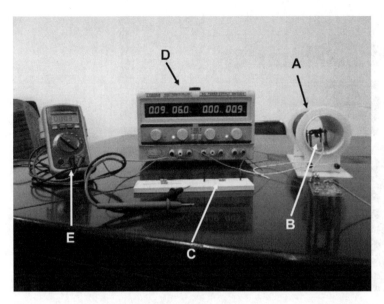

**Figure 4.3** DC magnetic field generation system: A, solenoid; B, GMR sensor with conditioning circuit; C, precision resistor; D, DC power source; E, voltage meter. (*Source:* Reprinted with permission from S. Liu et. al., Experimental research on hysteresis effects in GMR sensors for analog measurement applications. *Sensors and Actuators A: Physical*, 182, 72–81, Aug. 2012.)

carefully oriented to make the sensitive axis of the GMR sensor orthogonal to it. The output signal of the GMR sensor is transmitted through a co-axial cable and then measured by a voltage meter. The solenoid is powered by a DC voltage source that can saturate the GMR sensor in both directions along the sensitive axis. The strength of the generated reference magnetic field is proportional to the current flow within the solenoid, which can be accurately measured. In order to do this, a 1 $\Omega$ precision resistor is connected in series with it. The voltage drop can be captured by a voltage meter. It should be noted that the precision resistor is of relatively large rated power. Consequently, its resistance varies little due to thermal drift during the experiment.

The absolute measurement error of the digital voltage meter is the last digit of its operation range. Hence the relative error is significant when measuring small quantities, including solenoid current flow around remanence, which corresponds to zero applied magnetic field, and sensor output voltage around coercivity, where the measured value approaches zero.

To evaluate the hysteresis influences on dynamic signal measurements, an AC magnetic field generation system is set up as shown in Figure 4.4. The DC power source is replaced by a function generator with programmable voltage output. A two-channel oscilloscope is used to observe the synchronized time-domain waveforms of both the solenoid current flow (applied magnetic field) and the GMR sensor voltage output. Another application of the AC system is to demagnetize the GMR sensor to its equilibrium state, i.e. the natural magnetization state with antiferromagnetic coupling between the ferromagnetic layers (Figure 4.5), by applying a slowly damped AC magnetic field with sufficiently large initial magnitude until zero. This process is called AC demagnetization [27] to distinguish it from its DC counterpart.

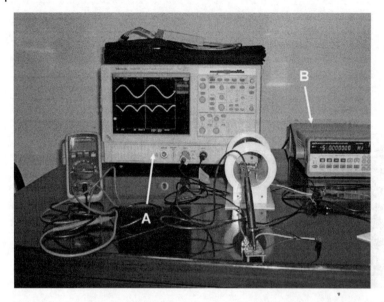

**Figure 4.4** AC magnetic field generation system: A, two-channel oscilloscope; B, function generator. (*Source:* Reprinted with permission from S. Liu et. al., Experimental research on hysteresis effects in GMR sensors for analog measurement applications. *Sensors and Actuators A: Physical*, 182, 72–81, Aug. 2012.)

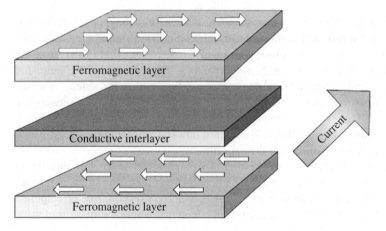

**Figure 4.5** The current flows in the plane of thin film layers. (*Source:* Reprinted with permission from S. Liu et. al., Experimental research on hysteresis effects in GMR sensors for analog measurement applications. *Sensors and Actuators A: Physical*, 182, 72–81, Aug. 2012.)

All ferromagnetic materials are sensitive to thermal influences [28; 29]. The susceptibility has a minus temperature coefficient, i.e. a temperature rise will make the ferromagnetic layers harder to magnetize. Although the Wheatstone bridge configuration can compensate for this, the sensitivity and saturation magnetic field of GMR sensor still suffer from it to some extent and this cannot be neglected in practical applications [30]. This issue is beyond the scope of the following investigations and the ambient temperature is restricted to 27°C during the experimental processes.

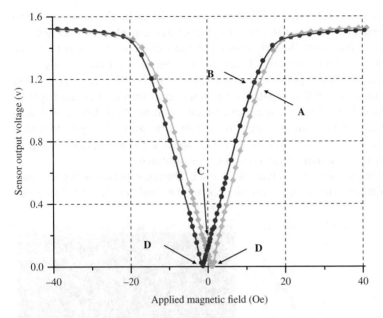

**Figure 4.6** Major hysteresis loop of GMR sensor: A, ascending major curve; B, descending major curve; C, remanence; D, coercivity. (*Source:* Reprinted with permission from S. Liu et. al., Experimental research on hysteresis effects in GMR sensors for analog measurement applications. *Sensors and Actuators A: Physical*, 182, 72–81, Aug. 2012.)

### 4.2.1.3 Hysteresis in Bipolar Operation

The major hysteresis loop provides an overall perspective of the output behaviors of the GMR sensor, which is a good basis for analyzing and evaluating the shapes of minor curves. Figure 4.6 is the measured major hysteresis loop of a typical GMR sensor. The major hysteresis loop consists of two regions: linear and saturation. Only the linear region is suitable for analog measurement applications. The large deviation between the ascending and descending major curves indicates magnetic hysteresis within the major hysteresis loop. For both curves, the upward moving segment has a larger slope (in absolute value) than the downward moving segment, i.e. the susceptibility for the magnetizing process is larger than that for the demagnetizing process. In addition, during both processes the susceptibilities are not absolutely invariant. In other words, there exist nonlinearities within the measurement range of the GMR sensor.

Compared with the range of the linear region, the amount of remanence is large enough that its induced DC offset voltage output cannot be neglected during operation. For a GMR sensor with linear output, coercivity (in absolute value) is proportional to remanence. But it should be noted that despite remanence being minimized in the presence of the coercivity field, it still exists for a small portion after the coercivity field is removed to zero. In other words, the coercivity field cannot demagnetize the GMR sensor thoroughly, but it is smaller than the DC demagnetizing field.

To investigate hysteresis in bipolar operation, experiments primarily focus on steady-state linear response behaviors (after accommodation) of the GMR sensor with respect to applied periodic magnetic fields. The measured hysteresis curves that form the bipolar minor hysteresis loops are closed. For analog measurement applications, the initial magnetization state of the GMR sensor is defined in the absence of the external applied field, i.e. the initial state is uniquely characterized by remanence and can be partially reflected from the DC offset voltage output (different remanence may produce the same offset voltage). Obviously, remanence is minimized after AC demagnetization, and the corresponding magnetization state is called equilibrium. On the other hand, remanence in the negative direction is maximized after the negative saturation magnetic field is removed. These two initial magnetization states are chosen to analyze hysteresis in bipolar minor loops. The range of the applied magnetic field is selected as −3 Oe to 3 Oe. It is relatively small with respect to the whole linear range of the GMR sensor., hence remanence and coercivity are dominant.

The measured loops are shown in Figure 4.7a and c. During experimental processes, accommodation phenomena are indeed observed as the final steady-state hysteresis

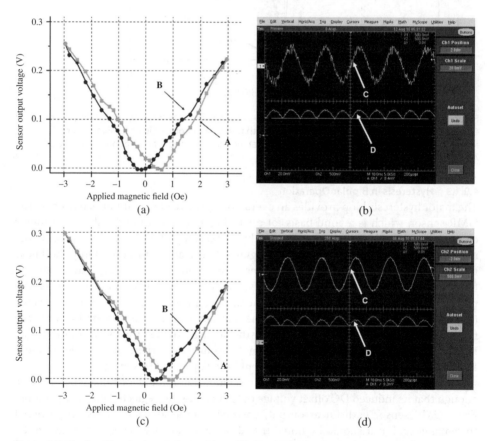

**Figure 4.7** Bipolar minor hysteresis loops: (a) equilibrium initial magnetization state and (c) post-negative saturation initial magnetization state. In both (a) and (c), curves A and B are the ascending minor curve and descending minor curve, respectively. (b) and (d) are the time-domain responses of the GMR sensor when a 50 Hz sine wave magnetic field is applied, which correspond to the hysteresis loops in (a) and (c), respectively. In both (b) and (d) waveform C indicates the applied magnetic field and waveform D indicates sensor output voltage. (*Source:* Reprinted with permission from S. Liu et. al., Experimental research on hysteresis effects in GMR sensors for analog measurement applications. *Sensors and Actuators A: Physical*, 182, 72–81, Aug. 2012.)

loops deviating from the initial magnetizing curves with applied magnetic field ranging from 0 to 3 Oe (not plotted). The bipolar minor hysteresis loop with equilibrium initial state (Figure 4.7a) is not exactly symmetrical because the GMR sensor cannot be demagnetized thoroughly. Here, the initial state has a small portion of remanence in the negative direction due to a larger output voltage at −3 Oe than that at 3 Oe, and a smaller coercivity field (in absolute value) in the negative direction than that in the positive direction. Like the major hysteresis loop, the susceptibility during the magnetizing process (upward moving) in both the ascending and descending minor curves is larger than that during the demagnetizing process (downward moving). This leads to a sensitivity error with the GMR sensor. The joint points between two adjacent half sine waves (waveform D in Figure 4.7b) correspond to the coercivities in Figure 4.7a. It can be seen that the sensor output voltage in the time-domain waveform changes abruptly at coercivity, which indicates a sudden change in susceptibility. Hence, the shape of the minor hysteresis loop around coercivity (Figure 4.7a) is not smooth. This inaccuracy is induced by the experimental system. In Figure 4.7b there exits a lag phase shift between the waveforms of the applied magnetic field and sensor output voltage, the amount of which is determined by coercivity.

The bipolar minor hysteresis loop with a post-negative saturation initial magnetization state (Figure 4.7c) is no longer symmetrical, but is severely distorted. The output voltage at −3 Oe is 50% larger than that at 3 Oe, and the coercivities of both the ascending and descending minor curves lie in the positive region of the applied magnetic field. Consequently, the corresponding time-domain output waveform (Figure 4.7d) for the negative applied field appears to have a higher peak value and a larger time span than that for the positive applied field, from which the original sine wave applied magnetic field can hardly be recognized.

To investigate common properties among these hysteresis loops, the ascending and descending curves are plotted separately (Figure 4.8). It can be seen that the minor hysteresis curves (curves B and C) are parallel with each other, which indicates that the susceptibilities of both the magnetizing and demagnetizing processes are independent of the initial magnetization state and remain almost constant. It is the slight nonlinearities of these curves that close the minor hysteresis loops. The susceptibilities of the major curves (curve A) are slightly larger than those of the minor curves, but these deviations are sufficiently small when compared with the absolute magnitudes of the susceptibilities and the difference between the magnetizing and demagnetizing susceptibilities. Hence, both susceptibilities can be treated as invariant during bipolar operation, regardless of any specific range of the applied magnetic field. This is helpful for developing a mathematical model to estimate the output behaviors of the GMR sensor [31].

The experimental results indicate that the bipolar operation mode of the GMR sensor cannot be used directly for analog magnetic field measurement. First, it is unable to discriminate between the positive and negative applied fields due to the absence of negative voltage output. Second, the output is not unique with respect to the same applied field because the initial magnetization states are infinite. Although some initial states can be known before measurement, a large disturbance magnetic field applied in both directions along the sensitive axis may distort the original shape of the hysteresis loop during operation and induce a change in remanence (DC offset voltage) and coercivity (time delay). Finally, for a particular bipolar minor hysteresis loop, its time delay of coercivity is not fixed, but is magnitude dependent. For example, in Figure 4.7b and d, if

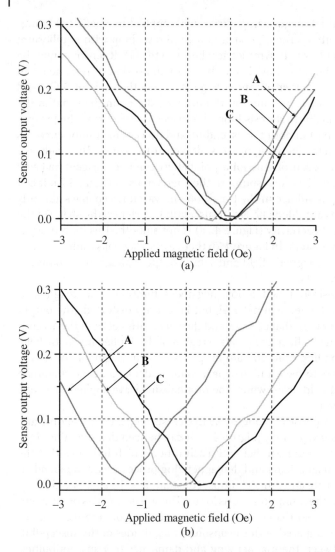

**Figure 4.8** Ascending curves and descending curves of hysteresis loops: (a) ascending curves and (b) descending curves. In each part, curve A is the major hysteresis loop, curve B is the minor hysteresis loop with equilibrium initial magnetization state, and curve C is the minor hysteresis loop with post-negative saturation initial magnetization state. (*Source:* Reprinted with permission from S. Liu et. al., Experimental research on hysteresis effects in GMR sensors for analog measurement applications. *Sensors and Actuators A: Physical*, 182, 72–81, Aug. 2012.)

the frequency of the applied magnetic field is chosen as 5 Hz with amplitude invariant, its time delay is 10 times larger than that for a 50 Hz applied field. To eliminate these drawbacks and make bipolar operation mode possible for analog measurement, a mathematical model for bipolar minor hysteresis loops with different initial magnetization states should be built and the parameters within this model should be identified. This is a rather complex issue and is difficult for hardware realization, which is beyond the scope of this book.

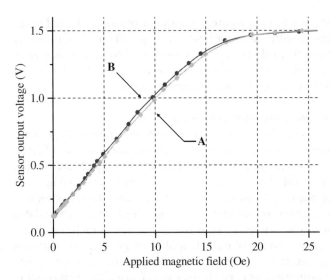

**Figure 4.9** Unipolar minor hysteresis loop: A, unipolar ascending minor curve; B, unipolar descending minor curve. (*Source:* Reprinted with permission from S. Liu et. al., Experimental research on hysteresis effects in GMR sensors for analog measurement applications. *Sensors and Actuators A: Physical*, 182, 72–81, Aug. 2012.)

### 4.2.1.4 Hysteresis in Unipolar Operation

To present the whole output behavior of GMR sensor in positive direction, applied magnetic field ranges from positive saturation to zero and then goes back. The measured curves form unipolar minor hysteresis loop, as shown in Figure 4.9.

In unipolar operation, the susceptibility during the magnetizing process (curve A) is nearly equal to that in the demagnetizing process (curve B). The slight deviation between these two curves indicates magnetic hysteresis, which is rather small compared with that in bipolar operation. According to the manufacturers' report, for each curve, its nonlinearity within the measurement region of the GMR sensor is less than 1.5%, the extent of which depends on the range of the applied magnetic field. Usually, a smaller operation range leads to less deviation. Hence, this error contributes little to the sensors accuracy and can be neglected in most analog measurement applications. In crucial situations, it can be fixed through unipolar hysteresis modeling [32]. For the ascending minor curve in Figure 4.9, if the applied magnetic field decreases at some point in the linear region before reaching positive saturation, the corresponding demagnetizing curve (not plotted) lies within the unipolar minor hysteresis loop and meets the ascending curve at zero magnetic field. The obtained linear unipolar minor hysteresis loop with post-positive-saturation initial magnetization state is especially suitable for analog measurement because its remanence is robust and cannot be changed by any positive applied magnetic field. The induced DC offset voltage output can be easily compensated for by external conditioning circuit. There is an accommodation process in this linear minor loop as the applied field cycles between zero and its maximum. The demagnetizing curve ultimately approaches the descending minor curve in Figure 4.9. This transient process brings about a dynamic nonlinearity error (within 1.5%), which can also be neglected.

In unipolar operation, any negative applied magnetic field is treated as a disturbance, which may change the original operation mode and bring about an additional error. It should be noted that the hysteresis loop in Figure 4.9 does not represent the full range of unipolar operation. Compared with the major hysteresis loop (Figure 4.6), the exact lower limit of the applied magnetic field is s negative coercivity field, rather than zero. The negative applied magnetic field in positive unipolar operation is useless for analog measurement, but can provide a criterion for disturbance detection.

### 4.2.1.5 Unipolar Operation with Negative Initial Remanence

The influence of negative remanence for unipolar operation is maximized when initial magnetization state is chosen as post-negative-saturation. The output behaviors of GMR sensor with applied magnetic fields ranging from 0 to 5 Oe and 0 to 15 Oe are plotted in Figure 4.10 (a) and (b), respectively. For the former case, its initial magnetizing curve follows the path of ascending major hysteresis curve. Then, the subsequent descending and ascending curves form a unipolar minor hysteresis loop. This loop accommodates upwards as applied field cycles, i.e. an increase in positive remanence, which is stronger than that with post-positive-saturation initial state. If applied magnetic field continues to increase, the output voltage exactly goes on to follow the path of ascending major curve until reaching positive saturation, as curve D indicates. During this magnetizing process, the susceptibility changes abruptly at the joint point (5 Oe) of curve C and curve D, from that of unipolar operation to bipolar operation. This phenomenon is independent of accommodation process because it always occurs no matter how many times the applied magnetic field cycles between 0 and 5 Oe. The sensor output performance with 0-15 Oe unipolar operation range (Fig. 10b) is almost the same with 0-5 Oe, except an increase in remanence.

These experimental results indicate that unipolar operation of the GMR sensor is only an intermediate process in bipolar operation whose existence does not influence the original bipolar magnetization pattern. Consequently, the unipolar minor hysteresis loop in Figure 4.9 can be treated as a process that corresponds to the major hysteresis loop. Hence, the complete definition of unipolar operation is given as follows: (i) it always begins with a demagnetizing process, (ii) the total magnetization pattern does not change direction, adn (iii) the susceptibility of the magnetizing hysteresis curve is equal to that of the demagnetizing hysteresis curve.

Since unipolar operation mode is appropriate for analog measurement applications, its measurement error is primarily caused by remanence, which is a combined effect of an initial magnetization state and a maximum applied magnetic field that produces a DC offset voltage output. This offset voltage is not monotonic with respect to the maximum applied field, as shown in Figure 4.11 (with post-negative-saturation initial state). Its descending portion corresponds to a decrease in negative remanence, and the ascending portion corresponds to an increase in positive remanence. The joint point between these two portions relates to the DC demagnetizing field, where the remanence is minimized. At this point, the offset voltage is less than zero, which is caused by a mismatch of GMR resistors in the Wheatstone bridge configuration. This phenomenon is called bridge offset. Its induced error is invariant during operation.

If the lower limit of the range of an applied magnetic field is assumed to be zero, it should be noted that in Figure 4.11 the GMR sensor works in unipolar operation mode in the positive direction only for a maximum applied magnetic field that is larger than the DC demagnetizing field (around 5 Oe), i.e. the remanence is positive. This portion can be divided into two regions. If the maximum applied field lies in region C,

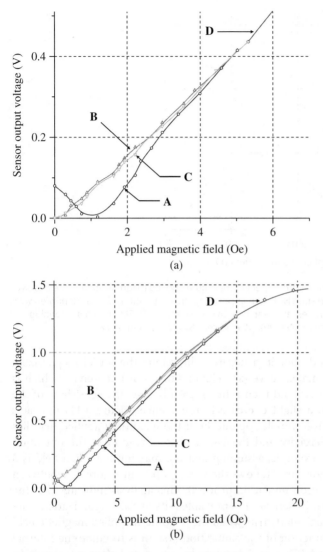

**Figure 4.10** Unipolar operation with post-negative-saturation initial magnetization state. The applied magnetic field ranges from 0 to 5 Oe in (a) and 0 to 15 Oe in (b). In both parts, curve A indicates the initial magnetizing curve, B and C indicate the descending and ascending curves of the unipolar minor hysteresis loop in the linear region, respectively, and D indicates the final magnetizing curve. (*Source:* Reprinted with permission from S. Liu et. al., Experimental research on hysteresis effects in GMR sensors for analog measurement applications. *Sensors and Actuators A: Physical*, 182, 72–81, Aug. 2012.)

the achieved unipolar minor hysteresis loop is linear. Indeed, the output behaviors illustrated in Figure 4.9a and b correspond to the lower and upper limits of region C, respectively. In practical situations, the transient positive disturbance field may lead the maximum field into region D, where the increase in remanence is considerably larger and contributes a lot to measurement error. If the maximum applied field lies in region A, i.e. between zero and the coercivity field (around 1 Oe), the sensor still works

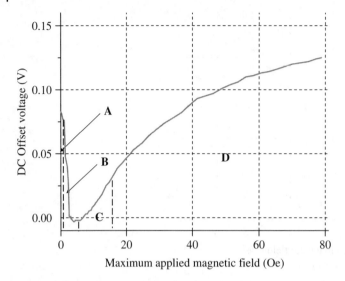

**Figure 4.11** DC offset voltage output with respect to maximum applied magnetic field in the positive direction. The initial magnetization state is chosen as post-negative-saturation. (*Source:* Reprinted with permission from S. Liu et. al., Experimental research on hysteresis effects in GMR sensors for analog measurement applications. *Sensors and Actuators A: Physical*, 182, 72–81, Aug. 2012.)

in its unipolar operation mode but in a negative direction. The initial magnetizing curve follows the ascending major curve (see Figure 4.6) toward coercivity, which is indeed a demagnetizing process, and then returns. In this case, the sensitivity of the GMR sensor is negative and a slight decrease in remanence (Figure 4.11) indicates magnetic hysteresis in negative unipolar operation. If the maximum applied field lies in region B, i.e. between coercivity and the DC demagnetizing field, the operation mode is actually bipolar. For example, assuming that the maximum applied field is 3 Oe, then the descending minor curve follows the same path as in Figure 4.7c between 0 and 3 Oe (neglecting the accommodation process). This analysis indicates that the measurement results of the GMR sensor are unreliable if its initial magnetization state is totally unknown. This is particularly true when the maximum applied magnetic field is small. Hence, one major function of the calibration system is to guide the internal magnetization pattern of the GMR sensor to some known state before measurement applications.

### 4.2.1.6 Calibration Procedure

As mentioned above, a unipolar minor hysteresis loop is most robust when its initial magnetization state is chosen as post-positive-saturation. Then the principal source of disturbance comes from the negative applied magnetic field. In unipolar operation, the input–output relationship is monotonic. Since a small applied negative field does not change operation mode, a threshold voltage between zero and DC offset voltage induced by remanence can be set for its detection. If output voltage is less than threshold voltage, the sensor should start a calibration procedure to re-saturate it in the positive direction before large negative disturbance changes its original remanence. A flowchart for automatic calibration is illustrated in Figure 4.12.

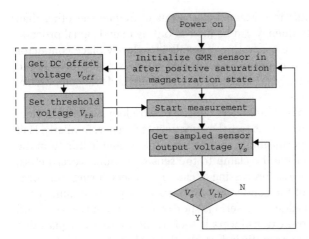

**Figure 4.12** Flowchart of the automatic calibration procedure of a GMR sensor in unipolar operation. (*Source:* Reprinted with permission from S. Liu et. al., Experimental research on hysteresis effects in GMR sensors for analog measurement applications. *Sensors and Actuators A: Physical*, 182, 72–81, Aug. 2012.)

For individual GMR sensor components, the offset voltage after positive saturation may be different for each one. The two steps in the dashed box are designed to get this and are executed only the first time the GMR sensor is powered on. In a complete measurement system, the measured analog quantities are sampled and converted to digital counterparts through an A/D converter for storage and further processing. Typically, a microcontroller is used for controlling the sequence of these commands. The calibration system is integrated in the controller. When the sampled quantity is smaller than threshold voltage, the controller sends a stimulating signal to saturate the GMR sensor through the external coil and then removes it. During this process, the sampled quantities are invalid data. One advantage of this system is that it can provide a calibration service in the presence of a measured magnetic field. After comparing with the threshold voltage, the offset voltage should be subtracted from sampled quantities to obtain the final measurement result.

There exists a post-effect in ferromagnetic materials as the amount of remanence decays with time when the applied magnetic field is small. For GMR sensors, such an effect is not serious. When the applied magnetic field is removed, its internal magnetization state can be held for more than one day. In practical situations, the sensor should be re-saturated regularly to minimize this effect.

### 4.2.2 MR Sensors with Magnetic Shielding

#### 4.2.2.1 Current Transformer with MR Sensor and Shielding Layers

The magnetic field sensing spectrum and increasing sensitivity make this a strong candidate for wide range electric current measurements from a few to hundreds of amperes of transient and steady-state signal frequencies [33]. However, when employed in exposed environments, particularly for large electric current measurement, there is a constant risk of exposure to magnetic interference and superposition (crosstalk) from nearby equipment. Consequently, these disturbances are superimposed on a magnetic field radiating from the conductor under test. Magnetic crosstalk of frequencies other than the one at the measured signal frequency can be filtered out by digital signal processing techniques, but when the source generating crosstalk is a closely place conductor operating at the same frequency it requires sophisticated offline signal processing to curb such disturbances. Consequently, existing measurement methods rely on physical structures for noise mitigation.

As an example, Rogowski coil based ammeters consist of wire loops wound in electrically opposite directions to cancel out external magnetic interference. Similarly, Hall sensor based ammeters use high permeability coils to focus the magnetic flux from the conductor under measurement inside the clamp to the sensing region. Nevertheless, all such devices require auxiliary signal processing to cater for offsets during amplification, and for the shape and size of the conductor with respect to position, number of turns, and material of the coil. In addition, a serious shortcoming of Rogoswki coil and transformer based electronic current transducers is their response to AC signals due to the operating principle of electromagnetic induction. This makes such arrangements saturated with large DC bias when they are presented in an AC current source [34; 35].

Stimulated by the shortcomings of existing devices and method, a novel method and design has been proposed to provide the basis for simple yet efficient wide-range current measurement. It requires measurement of the magnetic field generated from a current-carrying conductor at some distance. Magnetic sensors based on the MR effect are sensitive and prone to significant measurement errors in operational conditions where strong magnetic noise can be predicted, therefore a solution to this can be to shield the sensing unit from external noise where such noise is anticipated. However, an efficient shielding structure requires to be ascertained. Despite the fact that the working principle remains same, the shape, size, and material of the shielding layers need to be adjusted for use in different applications.

Electric current measurement by interpretation of magnetic fields finds its application in various fixed conductor arrangements. A typical gas-insulated switchgear-based substation equipment commonly consists of an embedded Rogowski coil based module for electric current measurement (reconstruction) using the voltage induced in the coil. Such a system is commonly implemented in a gas-insulated substation of a modern substation switching yard. To ensure noise immunity for magnetic measurements, a multi-layered shield using mu-metal for a gas-insulated substation has been conceptualized in steps and evaluated by finite element analysis (FEA) using a three-dimensional model of the conceived prototype shown in Figure 4.13. Keeping the circular tube-like shape of the gas-insulated substation, where a conductor through the center filled with SF6 gas, a number of geometries were evaluated in the software to determine a feasible and efficient substitute for Rogowski coil based electronic current transformers in a gas-insulated substation. The mu-metal based shield requires any disturbance to be attenuated in the external region close to the sensor. Since the sensor employed is sensitive along one axis, it requires shielding along the sensitive direction to curb the impact of noise from these dimensions. To get the same effect, Mu-metal is arranged and tested in single to multiple layers to determine the optimal design. Finally, an arrangement with three sheets of Mu-metal arranged in an arc manifested adequate results in presence of strong magnetic noise. The semicircular arc-like layers provide attenuation to the magnetic flux density approaching the sensor from the external space close to the sensor, thus protecting the sensitive axis of the sensor. The larger outer sheet attenuates most

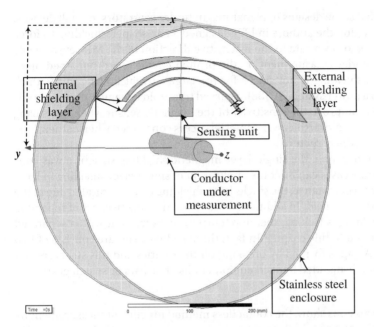

**Figure 4.13** Conceptual design of an electronic current transformer with a MR sensor and shielding layers.

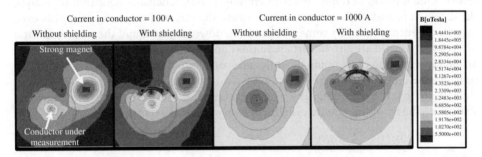

**Figure 4.14** Illustration of shielding efficacy towards the penetration of a strong magnetic stimulus.

of the external flux, whereas the inner shielding layers provide added protection for the flux density penetrating the outer layer. The FEA of the design reveals external magnetic flux attenuation from the shielding structure in Figure 4.14.

### 4.2.2.2 Separated Double-layer Magnetic Shielding Based Current Transformer

Algorithms based on an array structure of magnetic sensors have been proposed to distinguish the targeted magnetic field from the crosstalk. Most of the research is carried out on condition that the interference source is a filamentary current parallel to the targeted current. Nevertheless, the magnetic environment is much more complex in a power system [19]. The position, the modality, the number of the interfering magnetic flux densities, and the direction of its magnetic strength are all unknown, suggesting that the dilemma cannot be solved from the algorithm level. Magnetic shielding to decrease the external interference therefore appears to be an effective method.

Work in [36] studied some designs of planar magnetic concentrators for Hall devices. Based on the work in [36], the authors in [37] designed a bar-shaped shielding to force the magnetic noise vector to rotate to the insensitive direction of the MR sensor while the electric current under measurement on the printed circuit board remained unaffected. However, the shielding is not suited for the large current flowing in the thick copper conductor in a power system. In [38], a curved trapezoidal magnetic flux concentrator was designed to improve the sensitivity of the magnetic sensor when measuring multi-conductor current. Nevertheless, no interference is considered when authenticating the effectiveness of this structure.

In all the studies above, only the single-layer magnetic shielding structure was considered. However, some cases can limit the application of single-layer shielding in large current measurement. For instance, the single-layer shielding can be saturated due to the strong interfering magnetic field. In addition, research on the effectiveness of separated the double-layer shielding is lacking, therefore current measurement with a separated double-layer is proposed in this chapter. In fact, the shielding structure consists of two cylindrical cavities. A gap is formed by breaking off the cavities and this structure can be easily installed and dismantled in fixed equipment like bus bars or switch gears.

### A. Design Overview

As the foregoing discussions show, the contactless method for current measurement by a magnetic sensor faces the drawback that it cannot separate the targeted magnetic field from the superposition of the interfering magnetic field on the detected results of the sensor. One method of doing this is to employ a comprehensive algorithm to analyze the detected results of the sensor then estimate the targeted magnetic field. However, the interference commonly comes from an unknown position, and the magnitude and direction of its field are also uncertain. In this context, magnetic shielding appears to be an effective method of damping the influence contributed by the external noise. Traditionally, shielding efficiency (SE) is defined as a ratio of the magnetic flux density without shielding to that with shielding. The lager the value of SE, the more effective the shielding structure. The SE of the infinite cylindrical cavity is calculated by (4.1):

$$SE = 20lg(1 + \frac{\mu_r t}{2R})(dB) \tag{4.1}$$

where $\mu_r$ is the permeability of the shielding material, $t$ is the thickness of the material, and $R$ is the average of the radius of the inner wall and the outside wall of the shielding. However, the linearity range of current commercial sensor output cannot satisfy these requirements when exposed to a strong magnetic interfering environment. One single shielding sometimes shows this shortcoming due to the saturation of the magnetic material. For the same reason, double shielding layers have been proposed to satisfy the requirements for effective shielding. Supposing that the two shielding materials are utilized with space between them, the thickness of which is $l$. The length of the space between the two shielding layers is $l_2$. The SE for double shielding is defined as [39]

$$SE = \sum_{n=1}^{2} A_n + B_n + R_n \tag{4.2}$$

where $A_n$, $B_n$ and $R_n$ are absorption loss, reflection loss, and multiple reflection loss. This requires that the shielding layers must be consistent without gaps, while, in industrial applications, a separated shielding structure is more convenient for installation.

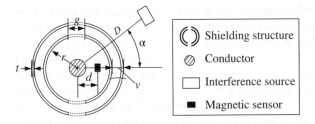

**Figure 4.15** The separated double-layer shielding structure with a strong interference source. (*Source: Reprinted with permission from Y. Chen et. al. Separated double-layer magnetic shielding with magnetic sensor for large current measurement. IEEE ISGT Conference (Asia), May 2018.*)

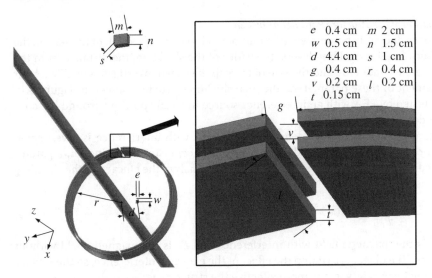

**Figure 4.16** Emulations system and its dimensions by FEA. (*Source: Reprinted with permission from Y. Chen et. al., Separated double-layer magnetic shielding with magnetic sensor for large current measurement, IEEE ISGT Conference (Asia), May 2018.*)

Research on separated double-layer magnetic shielding to dissipate the strong external magnetic noise should therefore be studied [40].

The structure designed in this chapter is shown in Figure 4.15. The separated double-layer shielding consists of four curved pieces of high permeability metal with the conductor under measurement at the center. After some simulation, the final design of the structure is as shown in Figure 4.16. To reach the requirement of miniaturization demanded by the development of the smart grid, the shielding width of $l$ is 2 cm. The two inner pieces of metal are on the same circle with a radius of 7.6 cm. The outside pieces are on another circle with a radius of 7.95 cm. A gap of 4 mm between the two parts of the shielding is incorporated. According to the work presented in [41], a complete cylindrical shielding around the conductor under measurement fails to support the shielding as desired. On the contrary, the effect of shielding would take place inside the shielding because the magnetic field would be tangential and continuous at the surface of the shielding. Discontinuous shielding due to a gap is therefore proposed to concentrate the magnetic field around the region of the gap. The results in [42] illustrate that the shielding ratio of the infinite cylinder is dependent on the thickness of the shielding material and the permeability: the higher the permeability and the thicker the shielding, the higher the shielding ratio. For a magnetic field below 100 kHz, high permeability materials are utilized to separate the route of the interfering magnetic field, concentrating the magnetic lines of flux to pass through the shielding. Having performed as an effective shielding material, mu-metal is employed to attenuate the interference when electric current ranging in power distribution frequency from 50 to 60 Hz is measured. For the same effect, mu-metal is adopted in this chapter. Meanwhile, considering the commercially available mu-metal, the thickness $t$ is set to 1.5 mm. Therefore the interval $v$ between the two layers is 2 mm. The distance $d$ from the sensor to the conductor under measurement is set to 4.4 cm.

### B. Investigation of the Structure Parameters by FEA

The influence due to six parameters (the number of layers, the position of the sensor, the interval between the shielding layers, the width of the shielding, the distance from the shielding to the conductor, and the size of the gap) has been investigated by FEA. After the assessment of these parameters, the final shielding structure shown in Figure 4.16 has been tested by FEA with an interference source from any position around the conductor under measurement.

Here, targeted magnetic flux density without and with interference is investigated on condition that there is a shielding structure. Hence, instead of using SE, a parameter named shielding uncertainty (SU) is defined to analyze the efficacy of the shielding structure as follows:

$$SU = |\frac{B_r - B_0}{B_0}| \times 100\% \tag{4.3}$$

where $B_r$ is the magnetic field with interference and $B_0$ is the magnetic field without interference. The object is to damp the effect of the external interference, so the smaller the shielding uncertainty is, the more effective the structure is.

To demonstrate the advantage of separated double-layer shielding over single-layer, an emulation system and its dimensions have been designed in accordance with the structure shown in Figure 4.16. The material of the shielding is mu-metal. The current under measurement of 100 A flows in a copper conductor with a finite length of 0.2 m. A rectangle region with dimensions of $4 \times 5$ mm is designed to represent the magnetic sensor.

To study the effect due to the number of shielding layers when there is a strong interference and there is no strong interference separately, seven situations have been conducted by FEA, as shown in Figure 4.17. A magnetic material NdFe35 with dimensions of $2 \times 1.5 \times 1$ cm is adopted to intimate the interference perpendicularly above the gap of the shielding layer. The distance $D$ from the conductor is 24 cm. The strength of NdFe35 is 0.28 Tesla, with its magnetic field radiating in three axis directions. The magnetic flux density generated by a filament current of 1000 A near the shielding is approximately $12 \times 10^{-4}$ Tesla, much smaller than the magnetic strength of NdFe35. In addition, the magnetic strength of NdFe35 radiates in all three axis directions, thus employing NdFe35 to represent the strong magnetic noise whose modality is unknown is proper.

The simulation results are depicted in Figure 4.17, from which it can be seen that the magnetic field around the conductor under measurement is affected by magnetic

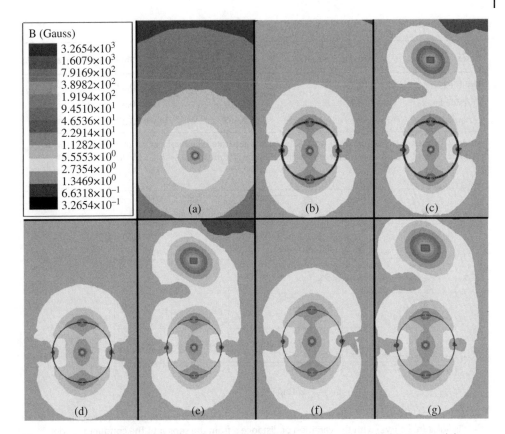

**Figure 4.17** A color map of the magnetic distribution for seven cases: (a) no shielding around the targeted current, (b) separated double-layer shielding without an interference source, (c) separated double-layer shielding with an interference source, (d) single-layer shielding of the inner layer without an interference source, (e) single-layer shielding of the inner layer with an interference source, (f) single-layer shielding of the outside layer without an interference source, and (g) single-layer shielding of the outside layer with an interference source. (*Source:* Reprinted with permission from Y. Chen et. al., Separated double-layer magnetic shielding with magnetic sensor for large current measurement, *IEEE ISGT Conference (Asia)*, May 2018.)

shielding and the external interference. A more precise conclusion can be extracted from Table 4.1: the shielding uncertainty reaches 0.33% with separated double layers compared with only one single layer, which demonstrates the efficacy of separated double-layer shielding. In addition, by comparing the results of cases when there is an outside layer and or just an inner layer, it can be concluded that a closer shielding layer does not guarantee smaller shielding uncertainty.

As suggested by the color map of the magnetic distribution depicted in Figure 4.17, the magnetic field at an adjacent position to the conductor under measurement remains relatively stable even with the effect of a strong interference source. Seven cases are tested when the distance $d$ from the sensor at the test region to the conductor changes from 1.9 to 6.9 cm in steps of 0.5 cm by FEA simulations to find the optimal position of the sensor. The results of the magnetic flux density with and without noise are depicted in Figure 4.18. The magnitude of the magnetic flux density declines as the position of

**Table 4.1** Shielding uncertainty versus layer mumbers.

| Cases | Without noise (Gs) | With noise (Gs) | Shielding uncertainty (%) |
| --- | --- | --- | --- |
| Double layer | 3.065 | 3.055 | 0.33 |
| Outside layer | 2.619 | 2.604 | 0.57 |
| Inner layer | 2.491 | 2.407 | 3.37 |

*Source:* Reprinted with permission from Y. Chen et. al., Separated double-layer magnetic shielding with magnetic sensor for large current measurement, *IEEE ISGT Conference (Asia)*, May 2018.

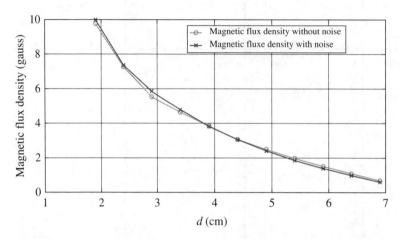

**Figure 4.18** Simulation results of the magnetic flux density when there is no noise (blue curve) and there is noise (red curve) with the variation of distance *d* from the sensor to the conductor under measurement. (*Source:* Reprinted with permission from Y. Chen et. al., Separated double-layer magnetic shielding with magnetic sensor for large current measurement, *IEEE ISGT Conference (Asia)*, May 2018.)

the sensor becomes farther away from the conductor. Furthermore, it can be seen that the curves of the magnetic flux density in two cases are close when *d* is around 4 cm, which indicates that the shielding uncertainty is not linearly related to the distance *d*. When *d* is less than 4.4 cm, the difference between the magnetic field with and without interference decreases with increasing *d*. Contrary to this, when *d* is larger than 4.4 cm, the difference between the magnetic field with and without interference increases with increasing *d*. When *d* is 3.9 cm and 4.4 cm, the shielding uncertainty is less than 0.4%. The minimum shielding uncertainty reaches 0.33% when the distance from the sensor to the conductor is 4.4 cm.

The interval of *v* between the two shielding layers can influence the shielding uncertainty [40]. Emulations are thus performed to investigate the effect due to the value of *v*. The inner layer remains unchanged while the distance from the outside shielding to the conductor increases so that the interval increases. It can be seen from Figure 4.19 that when the interval is 2, 5, 8, and 10 mm, the shielding uncertainty is less than 1%. The minimum shielding uncertainty of 0.33% is calculated when the interval is 2 mm.

To design a device with small dimensions, the optimal width *l* of the shielding has been analyzed. The simulated magnetic flux density with and without noise is depicted

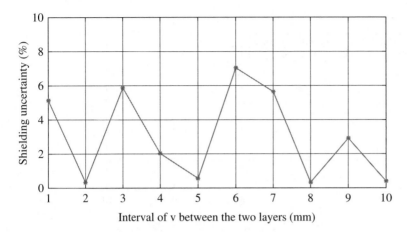

**Figure 4.19** Shielding uncertainty versus the interval *v* between the inner and outer layers. (*Source:* Reprinted with permission from Y. Chen et. al., Separated double-layer magnetic shielding with magnetic sensor for large current measurement, *IEEE ISGT Conference (Asia)*, May 2018.)

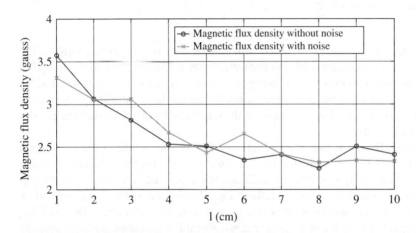

**Figure 4.20** Simulation results of the magnetic flux density when there is no noise and when there is noise with variation of the width *l* of the shielding layer. (*Source:* Reprinted with permission from Yafeng Chen et. al., Separated double-layer magnetic shielding with magnetic sensor for large current measurement, *IEEE ISGT Conference (Asia)*, May 2018.)

in Figure 4.20. It can be seen that a larger area of the shielding does not confirm smaller shielding uncertainty for this structure. As a whole, the magnetic flux density damps in accordance with the increase in *l*. When *l* is 2 cm and 7 cm, separately, the curve representing the magnetic flux density with noise and the curve representing the magnetic flux density without noise almost coincide with one point. When the width of the shielding is 2 cm, the shielding uncertainty is 0.33%, and when *l* is 7 cm, the shielding uncertainty is 0.21%. To make the measurement device smaller, the width is set to 2 cm.

In the work reported in [40], the distance from the shielding to the conductor plays a significant role in the shielding effectiveness of the separated double-layer magnetic shielding. To investigate the contribution due to the distance *r* from the inner shielding to the conductor in the structure proposed in this chapter, simulations have been tested

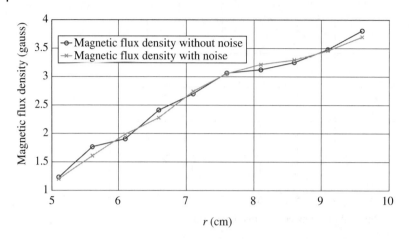

**Figure 4.21** Simulation results of the magnetic flux density when there is no noise and when there is noise with variation of the shielding position of *r*. (*Source:* Reprinted with permission from Y. Chen et. al., Separated double-layer magnetic shielding with magnetic sensor for large current measurement, *IEEE ISGT Conference (Asia)*, May 2018.)

for *r* increasing from 5.1 cm up to 9.6 cm while other parameters are unchanged. The results are presented in Figure 4.21. For the overall trend, the variation in the direction of the magnetic field magnitude is contrary to that of *r*. When *r* is 7.1, 7.6, 8.6, and 9.1 cm, the difference between the two curves is small. The minimum shielding uncertainty of 0.33% is calculated when the distance from the inner shielding to the conductor under measurement is 7.6 cm.

The results by FEA show that the strengths of the magnetic field at the test region are 4.499 and 4.521 Gs, individually, with and without strong noise NeFe35 around. The shielding uncertainty is 0.49%. In addition, the separated structure with the gap is designed for the convenience of installation in places such as bus bars and switch gears. Figure 4.22 is a color map of the magnetic field distribution at one of the shielding gap when the gap is 4 mm. It can be seen that mu-metal concentrates the high-strength magnetic field around the gap region, which is consistent with the conclusion in [41].

Simulations are conducted by changing the size of the gap. From the results presented in Figure 4.23, the relationship between the gap size and the shielding uncertainty is disproportionate. The maximum shielding uncertainty reaches 11.14%. When the gap is 2 and 4 mm, the uncertainty is less than 1%. The minimum shielding uncertainty of 0.33% is calculated when the size of the gap is 4 mm.

According to the results above, the parameters of the shielding structure are set as in Figure 4.16. In the real industrial environment, the interference source can come from random directions. Because the designed shielding structure is symmetrical, the effect due to the position of the noise has been investigated by changing the angle of $\alpha$ shown in Figure 4.15 from 0° to 180° in steps of 10°. The magnetic field with interference in comparison with the magnetic field without interference is shown in Figure 4.24. Supposing that the magnetic flux density is the mathematical exception value when there is no noise, the standard deviation is 0.1259. Two other situations when the magnitude of the targeted current is 500 A and 1000 A, and the standard deviations are 0.0661 and

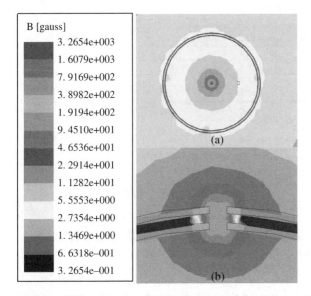

Figure 4.22 Color map of magnetic field distribution: (a) a complete cylindrical shielding and (b) shielding structure with the gap. (*Source:* Reprinted with permission from Y. Chen et. al., Separated double-layer magnetic shielding with magnetic sensor for large current measurement, *IEEE ISGT Conference (Asia)*, May 2018.)

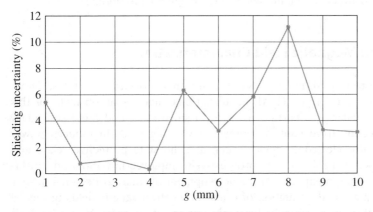

Figure 4.23 Shielding uncertainty in relation to the gap size of *g*. (*Source:* Reprinted with permission from Y. Chen et al., Separated double-layer magnetic shielding with magnetic sensor for large current measurement, *IEEE ISGT Conference (Asia)*, May 2018.)

0.1023, respectively, authenticate the effectiveness of this shielding structure for current measurement up to 1000 A.

Accelerated by the demands of smart grids, providing accurate information about electric current by devices with small size, low cost, and simple installation has emerged as an active research field. However, knowledge about interference is generally uncertain. Magnetic shielding therefore appears to be an effective method to damp the influence of the interference. Here, a separated double-layer magnetic shielding structure is designed with the gap. The parameters of this structure are optimally confirmed by FEA.

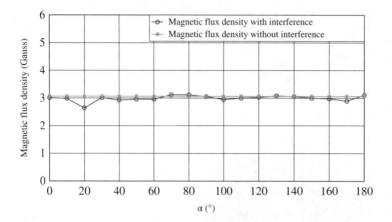

**Figure 4.24** Shielding uncertainty in relation to the position of the interference. (*Source:* Reprinted with permission from Y. Chen et al., Separated double-layer magnetic shielding with magnetic sensor for large current measurement, *IEEE ISGT Conference (Asia)*, May 2018.)

The minimum shielding uncertainty between the cases without and with interference is 0.33%. In addition, the standard deviations between the cases without and with interference from random places are 0.1259, 0.0661, and 0.1023, respectively, when the current under measurement is 100, 500, and 1000 A. These results prove the feasibility of the shielding structure to attenuate the influence of the magnetic interference.

## 4.3 Broadband Magnetic Field Characterization

The transformation from traditional power system into smart gird is very fast. However, electromagnetic interference (EMI) in substations might be an obstacle for this development. This challenge is particularly critical since more and more secondary systems are transferring from control rooms to switching yards in modern power systems [43]. Instantly, electronic devices such as solid-state protective relays and microprocessor-based control units are progressively employed in GIS-based substations to satisfy the requirements due to the development of smart grids [44; 45]. As a result, the measurement and evaluation of transient electromagnetic fields produced by switching or lightning become increasingly important in order to ensure EMI compatibility. Against this background, studies on advanced tools that can characterize and evaluate the EMI level have become a hot research field.

### 4.3.1 Transient Magnetic Field Events

Under transient conditions, the secondary systems installed in switching yards endure a poor EMI environment composed of a transient magnetic field (TMF) and a transient electric field (TEF). The measurement and evaluation of the TMF are more important than for the TEF. First, it is more difficult to shield a magnetic field compared with an electric field. Since the standards for the smart grid are still far from completion [46], many onsite cabinets in the substations, traditionally located in control room but now in the switching yard, are not properly shielded against TMF. Second, overcurrent is

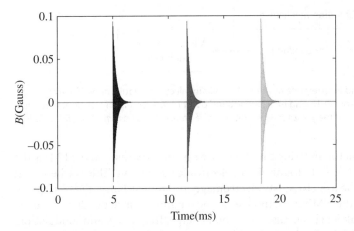

**Figure 4.25** The measured TMF waveform when there is a switching operation of an isolating switch on a bus bar in a 220 kV substation.

more of a menace than overvoltage [47] because overvoltage is generally well protected in substations. Hence, the disturbances caused by TMF can be much more disastrous than those caused by TEF.

TMF events usually include situations when the electrical equipment in the power system is struck by lightning, disconnecting switches, resulting in power network faults or failure of switching operations in substations. Figure 4.25 shows the measured transient magnetic field waveform recorded during a switching operation of an isolating switch on a bus bar in a 220 kV substation (3 m below the bushing of a transmission line connected to the bus bar).

The measurement of EMI is not easy since EMI generally appears in a very wide frequency band [43; 48]. Commercially available EMI meters generally can only measure the interference level without the functionality of high-frequency waveform analysis. In order to identify the actual power system faults, it is necessary to obtain the waveform of the transient interference for analysis. A search coil can be used to measure high-frequency EMI [49], but this solution does not work well for low-frequency EMI and is not easy for miniaturization to perform point measurement. It is imperative to find a solution which can realize broadband point measurement of EMI [50].

### 4.3.2  Evaluation of TMF Event Impact on Electronic Equipment

In addition to the interference level of a TMF, it is necessary to evaluate the effect of the TMF on the victim electronic equipment and its circuits, which are typically secondary system circuits [51] such as the communication or control devices used in substations. An ideal instrumentation device for measuring TMF in substations should possess the following capabilities: (1) measurement of TMF with high spatial resolution, (2) sufficient bandwidth and dynamic range, and (3) correlation of the EMI to the power system events for long-term quantitative evaluation of EMI.

Figure 4.26 depicts how the EMI affects the victim electronic equipment. Generally, the EMI problem involves the source, the propagation route, and the victim electronic equipment. All three factors and their associated parameters affect the EMI of the victim electronic equipment.

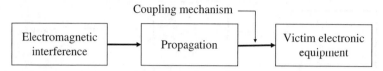

**Figure 4.26** Propagation and interference of the TMF on victim electronic equipment. (*Source:* Reprinted with permission from Q. Huang et al., Broadband point measurement of transient magnetic interference in substations with magnetoresistive sensors, *IEEE Transactions on Magnetics*, 50, 1–5, July 2014.)

The most important factor affecting the ability of a system to withstand the EMI is the coupling mechanism of the TMF to the victim electronic equipment. This is where EMI shielding strategies can be used to satisfy electromagnetic compatibility (EMC) requirements. The coupling of the TMF to victim electronic equipment is dependent on many parameters, such as peak level, average level, total energy, frequency, and modulation, etc. Most of the parameters are determined by the source. However, the propagation path (media) may greatly change the amplitude, phase, and direction of the TMF [52], depending on the physical shape and material properties of the media.

As described above, the secondary devices are immersed in the radiating electromagnetic fields from high-voltage equipment. Some of the parts in the secondary devices serve as an antenna or a loop to induce a voltage that superpositions on the normal working circuit. Since the circuits in the secondary device do not generally serve as an amplifying circuit, the antenna effect can be neglected. The induction severely affects the normal operation of the secondary electronic system. The induced voltage in the victim electronic equipment and its circuits is proportional to the derivative of the flux, and it is defined as the *received disturbance level* (RDL) at the victim electronic equipment:

$$\varepsilon(t) = \frac{d\phi(t)}{dt} = \vec{A} \bullet \frac{d\vec{B}(t)}{dt} \tag{4.4}$$

where $\vec{A}$ is the area of the victim electronic equipment projected onto the direction of the local incident magnetic field, $\phi(t)$ is the flux, and $\vec{B}(t)$ is the local transient magnetic field vector. The induced voltage is the most important parameter in evaluating the effect of the TMF on the secondary system in the substation. The response of the victim circuit $v(t)$ to the disturbance can be evaluated with the following equation:

$$v(t) = L^{-1}(s\hat{g}(s)\vec{A} \bullet \vec{B}(s)) \tag{4.5}$$

where $L^{-1}$ is the inverse Laplace transform operator, $\vec{B}(s)$ is the Laplace transform of the measured magnetic field vector $\vec{B}(t)$, and $\hat{g}(s)$ is the transfer function between induced voltage $\hat{g}(s)$ and output $v(t)$.

From (4.4), the effect of the TMF on the victim electronic equipment can be directly evaluated by calculating the derivative of $\vec{B}(t)$. Nevertheless, in practice, it is not so convenient to obtain the derivative of the TMF due to noise and the analog-digital quantizing error. For engineering applications, one can first perform the Fourier transform of the measured magnetic field waveform and then find the frequency components and their effects by computing $\omega\vec{B}(\omega)$, i.e.

$$\varepsilon = \vec{A} \bullet \sum_{\omega} |\omega\vec{B}(\omega)| \tag{4.6}$$

where $\omega$ is the angular frequency and $\vec{B}(\omega)$ is the magnetic flux density at frequency $\omega$.

## 4.4 Broadband Point Measurement of the TMF in Substations with MR Sensors

In this section, the effect of sensor size will be first studied, and then a measurement system with small MR sensor head will be designed. The designed system will be tested and applied in the detection of the local spatial magnetic field of a substation.

### 4.4.1 Effect of sensor size

The traditional solution to most power system TMF evaluation is the search coil. This solution may have very broad band of frequency response, with especially good performance in the high frequency band. However, it does not have a similar performance when the TMF is at low frequency. The disadvantage of this solution is the size: it is almost impossible to realize point measurement.

To demonstrate the importance of point measurement, numerical simulations have been performed to illustrate the effect of the area of the sensor head on the measurement accuracy. Suppose a circular sensor with a diameter $D$ is placed in the magnetic field produced by a current carrying conductor, as shown in Figure 4.27. In Figure 4.27a, the sensor is placed vertically below the conductor; in Figure 4.27b, the sensor is moved to 45° while the distance between the center of the sensor head and the current carrying conductor, $r$, remains unchanged. The current flows into the paper in Figure 4.27b. According to the Biot–Savart law, the magnetic field generated at the point at distance $r$ from a conductor carrying current $i$ should be:

$$B = \frac{\mu_0 i}{2\pi r} \tag{4.7}$$

where $\mu_0$ is the permeability constant ($4\pi \times 10^{-7}$ H.m$^{-1}$).

Numerical simulation was carried out based on a magnetic search-coil sensor with $D = 1$ cm. The changes in the magnetic field vectors $\vec{B}_{ctr}$ (the magnetic field at the center point of the sensor head), $\vec{B}_{meas}$ (the average magnetic field over the sensor circular area, which is the signal measured by the sensor in practice), $\vec{B}_{min}$ (the magnetic field

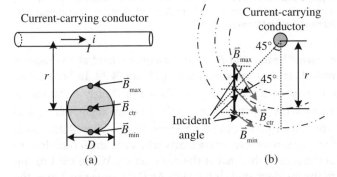

(a)          (b)

**Figure 4.27** Effect of sensor size in the measurement of magnetic field: (a) sensor vertically below the conductor and (b) sensor below the conductor at 45°.(*Source:* Reprinted with permission from Q. Huang et al., Broadband point measurement of transient magnetic interference in substations with magnetoresistive sensors, *IEEE Transactions on Magnetics*, 50, 1–5, July 2014.)

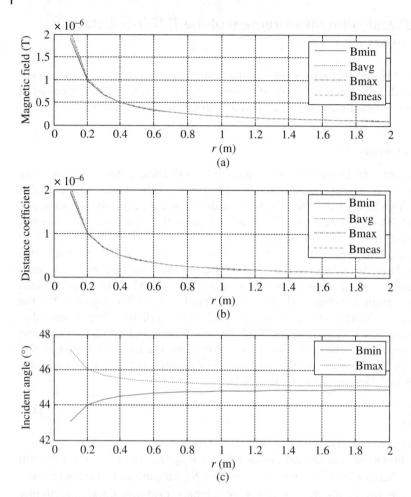

**Figure 4.28** Analytic results of the effect of sensor size in the measurement of the magnetic field: (a) the magnetic field at different points when sensor is vertically below the conductor, (b) the magnetic field at different points when the sensor is below the conductor at 45°, and (c) the incident angle of the magnetic field at the farthest point and the nearest point. (*Source:* Reprinted with permission from Q. Huang et al., Broadband point measurement of transient magnetic interference in substations with magnetoresistive sensors, *IEEE Transactions on Magnetics*, July 2014.)

at the farthest point of the sensor head), and $\vec{B}_{max}$ (the magnetic field at the nearest point of the sensor head) with distance $r$ are simulated in Figure 4.28. In both cases, the magnitude of $\vec{B}_{meas}$ agrees well with that of $\vec{B}_{ctr}$, as shown in Figure 4.28a and b, respectively. However, a noticeable discrepancy between the incident angles of $\vec{B}_{min}$ and $\vec{B}_{max}$ at different points on the sensor plane can be observed from Figure 4.28c. Such a discrepancy is dependent on the sensor size, particularly when $r$ is small (i.e. close to the field source). The smaller the sensor, the smaller the discrepancy. When $r = 1$ m and $D = 1$ cm, the discrepancy of the incident angle is 0.4052°. As $D$ decreases to 3 mm, the discrepancy diminishes to 0.1216°. Thus, $\vec{B}_{meas}$ agrees with $\vec{B}_{ctr}$ more as $D$ decreases. In order for the sensor measurement to truly reflect the actual magnetic field at the point

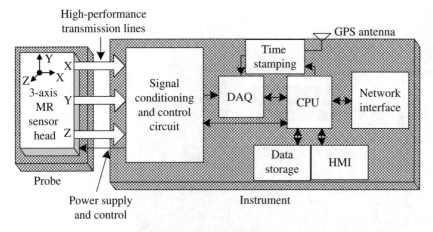

**Figure 4.29** System architecture of the TMF point measurement system. (*Source:* Reprinted with permission from Q. Huang et al., Broadband point measurement of transient magnetic interference in substations with magnetoresistive sensors, *IEEE Transactions on Magnetics*, 50, 1–5, July 2014.)

(i.e. $\vec{B}_{ctr}$), the sensor size should be minimized so that $\vec{B}_{min}$ and $\vec{B}_{max}$ are as close to each other as possible. Therefore, the point measurement (i.e. small sensor size) is critical for obtaining accurate measurement of the TMF with high spatial resolution. Three-axis MR sensors are a promising candidate for achieving this goal because of their compact size. In addition, a commercially available three-axis MR sensor can be fabricated into a microelectromechanical system (MEMS) fluxgate with a very small size ($\approx 3$ mm or even less) [53] and have a dynamic range from 120 $\mu$Gs to 6 Gs. All these characteristics are favorable for designing a point measurement system for a spatial TMF in substations, which requires a broad frequency band (e.g. EMI caused by DC, power-frequency or high-frequency transient) and large dynamic range (e.g. the current might be tens times of that under normal conditions).

## 4.4.2 Design of a Point Measurement System

Figure 4.29 shows the system architecture of a measurement system. A three-axis MR sensor (HMC1043 with a packaged size of $3 \times 3 \times 1.5$ mm) installed on a small printed circuit board (PCB) serves as the probe for measuring the TMF. In order not to affect the spatial distribution of the magnetic field to be measured, the ancillary circuits are assembled as another module, leaving only the MR sensor head on the probe. The measured weak signals are transmitted to the instrument by differential transmission in a high-performance, well-shielded and high-frequency transmission line. The signal conditioning circuits filter and amplify the received signals before data acquisition (DAQ). The data storage, network interface, and human-machine interface (HMI) are properly designed for convenient use in the field. The signal processing and analysis algorithms are implemented on the central processing unit (CPU). Currently, a global positioning system (GPS) is widely used in power grids to perform synchronous phasor measurements [54]. A GPS antenna is installed with this system to provide time-stamped data in order to facilitate correlating the EMI to the timing of the power system outages and finding the sources of interference.

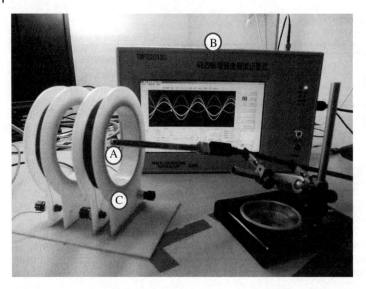

**Figure 4.30** Prototype of a TMF point measurement system: A, sensor head; B, associated instrumentation; C, solenoid. (*Source:* Reprinted with permission from Q. Huang et al., Broadband point measurement of transient magnetic interference in substations with magnetoresistive sensors, *IEEE Transactions on Magnetics*, 50, 1–5, July 2014.)

Shown in Figure 4.30 is a prototype for a TMF point measurement system. The whole system has a bandwidth from DC to 5 MHz, covering most transient EMI phenomena in substations, and a measurement range of 6 Gs (with a resolution 0.02 Gs). The main specifications are:

- measurement range: −6 Gs ∼ 6 Gs
- solution: 0.02 Gs
- sampling frequency: 40 MHz
- bandwidth: DC, 5 MHz
- storage depth: 32 MB/channel
- recording times: 2000 (sampling at 40 MHz for 100 ms for every recording)
- GPS timing accuracy: 1 $\mu$s

### 4.4.3 Laboratory Testing of the Measurement System

A series of tests has been designed to test the performance of the designed system. The test is generally conducted in the setup shown in Figure 4.31. A magnetic generator (actually a current generator in sinusoidal, pulse or damped oscillated waveform) is connected to a coil (to amplify the magnitude a solenoid may be used). The current and magnetic waveshapes are generally recorded simultaneously in the test.

Experimental characterization shows that the developed system exhibits good linearity (>99.7%) and relatively small error (<3%), as shown in Figure 4.32. In the characterization experiment, the magnetic field reference is generated by a Helmholtz coil using a current generator. The results here are measured at 50 Hz.

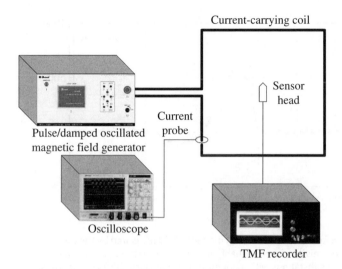

**Figure 4.31** Experiment setup of a laboratory test. (*Source:* Reprinted with permission from Q. Huang et al., Broadband point measurement of transient magnetic interference in substations with magnetoresistive sensors, *IEEE Transactions on Magnetics*, 50, 1–5, July 2014.)

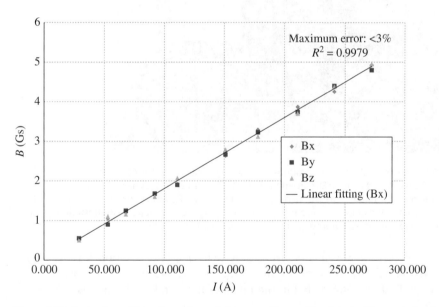

**Figure 4.32** Evaluation of linearity and measuring error. (*Source:* Reprinted with permission from Q. Huang et al., Broadband point measurement of transient magnetic interference in substations with magnetoresistive sensors, *IEEE Transactions on Magnetics*, 50, 1–5, July 2014.)

A test was carried out to investigate the system performance in measuring the TMF. A standard lightning surge generator (SG-5009G) was connected to standard magnetic field test equipment compliant with IEC 61000-4-9 [55]. The current waveform was observed by an oscilloscope while the magnetic field waveform was measured by the system described in this chapter. The results are shown in Figure 4.33. It can be

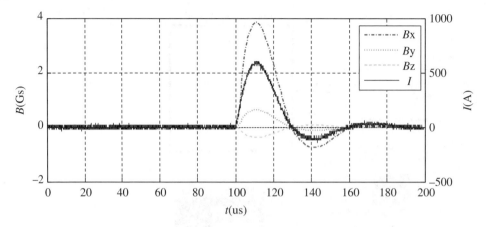

**Figure 4.33** A current pulse waveform and the resulting magnetic field measured by the developed system. (*Source:* Reprinted with permission from Q. Huang et al., Broadband point measurement of transient magnetic interference in substations with magnetoresistive sensors, *IEEE Transactions on Magnetics*, 50, 1–5, July 2014.)

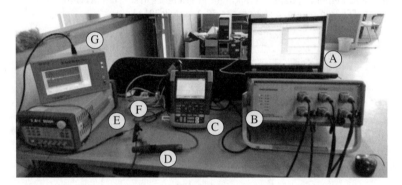

**Figure 4.34** Experimental setup: A, computer; B, traveling wave generator; C, oscilloscope for displaying the measured current; D, current probe; E, TMF measurement system; F, solenoid; G, oscilloscope for displaying the measured magnetic field. (*Source:* Reprinted with permission from Q. Huang et al., Broadband point measurement of transient magnetic interference in substations with magnetoresistive sensors, *IEEE Transactions on Magnetics*, 50, 1–5, July 2014.)

seen that the current waveform's rising time (6.65 $\mu$s) and duration time (17.4 $\mu$s) are followed closely by those of the magnetic field waveform (6.76 $\mu$s and 18.1 $\mu$s, respectively). Both waveforms fall in the range of the standard, i.e. 6.4 $\mu$s $\pm$ 30% and 16 $\mu$s $\pm$ 30%.

In order to test the capability of the developed measurement system to capture the TMF produced by a traveling wave when a switching operation or lightning occurs, a traveling wave was simulated and amplified to produce a transient current and transient magnetic field. The developed system was then used to measure the generated TMF. The experimental setup is shown in Figure 4.34. A charging operation of a transmission line is simulated with power system computer-aided design (PACAD) software [56]. The simulated output is connected to a power amplifier (denoted as a traveling wave generator in the figure) to generate a transient current of up to 10 A. The amplified current is connected to a small solenoid to generate TMF. The sensor head is placed close to (<1

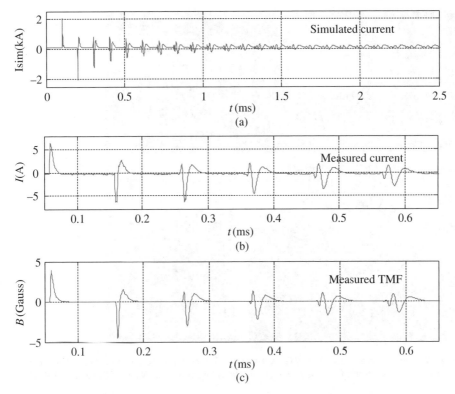

**Figure 4.35** Measurement results for a magnetic field generated by a simulated traveling wave current passing a power amplifier: (a) the simulated traveling current, (b) the output current measured by a current probe, and (c) the magnetic field measured by the developed system. (*Source:* Reprinted with permission from Q. Huang et al., Broadband point measurement of transient magnetic interference in substations with magnetoresistive sensors, *IEEE Transactions on Magnetics*, 50, 1–5, July 2014.)

cm) the solenoid to measure the TMF. The simulated traveling current (Figure 4.35a), output current measured by a current probe (Figure 4.35b), and emanated magnetic field waveform measured by the sensor (Figure 4.35c) are plotted in Figure 4.35. It can be seen that the proposed TMF measurement system can measure the TMF caused by traveling current wave fairly well.

### 4.4.4 Onsite Testing

In order to study the performance of the developed system in onsite applications, it was used to measure magnetic fields in various circumstances, including a power frequency steady state, a distorted waveform in a reactor, inside a high-voltage direct current valve house, and switching operations. Figure 4.36 demonstrates the application of the developed system in a typical GIS-based substation.

Figure 4.37 shows a typical waveform of the TMF for a switching operation measured by the developed system. When a switching operation was applied to charge a bus bar in the 110 kV substation, the TMF waveform was measured at 3 m directly below the bushing of the transmission line connected to the bus bar. The measured traveling waveform was verified by a similar PSCAD setup to that described above.

**Figure 4.36** Field test in a typical 220 kV GIS substation. (*Source:* Reprinted with permission from Q. Huang et al., *Innovative testing and measurement solutions for smart grid*, Wiley-IEEE Press, April 2015.)

Figure 4.38 presents the numerical differentiation and fast Fourier transform (FFT) analysis results of the zoomed-in part of the measured waveform (Figure 4.37d). By using the FFT method, it can be estimated that the induced voltage is 1.43 V cm$^{-2}$ with the dominant frequency at 3.66 MHz (with peak magnetic field 6.2183 Gs), as shown in Figure 4.38b. From the numerical differentiation result (Figure 4.38a), the evaluated maximum induced voltage is 1.71 V cm$^{-2}$. Taking into account the measurement noise in the signal, these two values can be regarded as almost on the same level. This can be further verified by time frequency analysis. Figure 4.39 shows the short-time Fourier transform (STFT) analysis of the same window of signal (with Kaiser Window, $N = 7$, $\beta = 0.8$). It is indicated that at 48.213 ms ($\frac{dB}{dt}$ reaches s maximum at 48.210 ms, as shown in Figure 4.38a), there is a maximum component of frequency 3.75 MHz and magnitude 6.5674 Gs. The corresponding induced voltage is 1.55 V cm$^{-2}$. This level of induced voltage may cause significant interference in analog and digital circuits. it is therefore of great application value to have a measurement system that can evaluate the effect of the TMF on secondary devices in substations.

## 4.5 Noise and External Field Protection

A single magnetic sensor can be employed for large current measurement when placed in the close vicinity of a conductor. However, the magnetic field is complex in modern substations. Higher sensitivity limits the application of MR sensors in environments

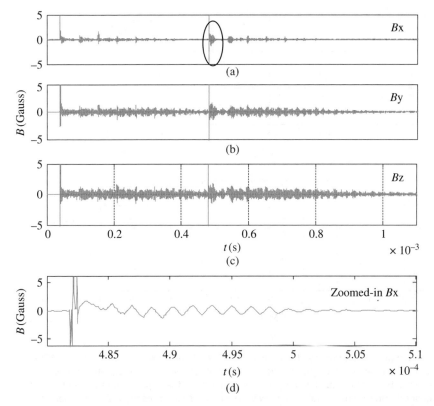

**Figure 4.37** The measured TMF components in (a) the *x* axis, (b) the *y* axis, and (c) the *z* axis during a charging bus bar in a 110 kV substation. The highlighted waveform in (a) is magnified in in (d). (*Source:* Reprinted with permission from Q. Huang et al., Broadband point measurement of transient magnetic interference in substations with magnetoresistive sensors, *IEEE Transactions on Magnetics*, 50, 1–5, July 2014.)

with nearby magnetic noise sources. For instance, the magnetic field from a nearby conductor is superimposed and the sensor fails to distinguish the magnetic field generated by the current under measurement from the magnetic interference [57], the information for which is generally unknown. Therefore, various methods have been designed to eliminate the external magnetic interference and noise from different sources.

Methods of magnetic sensor array have been proposed. In [58], the authors designed a sensor unit composed of three Hall effect sensors to measure AC and DC current in a circular conductor and this design can calibrate the position of the conductor. Nevertheless, the work is based on the assumption that there is no other interfering magnetic field around the magnetic sensors. However, due to the superposition of magnetic field from nearby noise sources, the measurement uncertainty rises to significant levels. In [59], the authors presented a new algorithm based on spatial discrete Fourier transform (DFT), which calculates the current to be measured by placing solid state magnetic sensors in a circular array with a current flowing conductor under the measurement at the center. In [60], the authors further refined the method presented in [59] to study the effect of the position of the interference current on the error at different angles. In

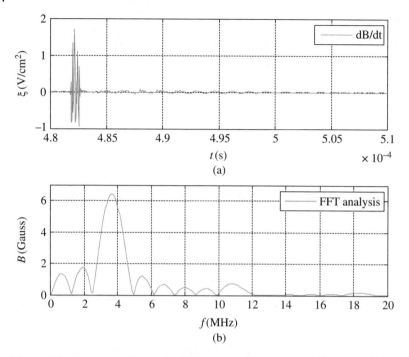

**Figure 4.38** Comparison of numerical differential and FFT analyses for evaluating the effect of TMF. (a) Numerical differentiation of the magnetic field, and (b) spectrum analysis of magnetic field by FFT.(*Source:* Reprinted with permission from Qi Huang, et al, Broadband point measurement of transient magnetic interference in substations with magnetoresistive sensors, *IEEE Transactions on Magnetics*, 50, 1–5, July 2014.)

[61] and [62] the authors presented algorithms that were able to calculate the intensity of the DC flowing in a rectangular bus bar with a circular sensor array. In [63], a new principle for simultaneous measurement of ACs flowing in bus bars was proposed for multi-conductor systems. The principle proposed was tested in laboratory experiments with Hall sensors that showed shortcomings in high consumption of power and poor linearity.

On the other hand, high permeability materials can be used as concentrators or to shield the external interference. Work in [36] studied some designs of planar magnetic concentrators for Hall devices. However, magnetic concentrators may suffer from the problem of saturation. Based on the work in [36], the authors in [37] designed a bar-shaped shielding to force the magnetic noise vector to rotate to the insensitive direction of the MR sensor while the electric current under measurement on the PCB remains unaffected. However, the shielding is not suitable for the large current flowing in a thick copper conductor in a power system. In [38], a curved trapezoidal magnetic flux concentrator was designed to improve the sensitivity of the magnetic sensor when measuring multi-conductor current. Nevertheless, no interference is considered when authenticating the effectiveness of this structure. For all these systems only single-layer magnetic shielding was considered. However, some cases, for example when the shielding is saturated, can limit the application of single-layer shielding in large current measurement.

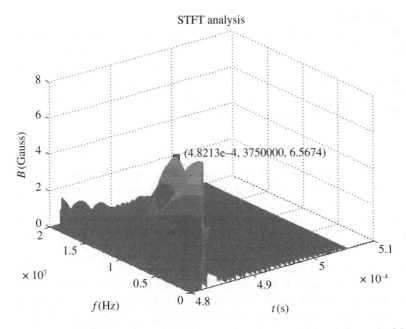

**Figure 4.39** The STFT analysis result of the windowed measured signal: the peak of the signal is denoted as (time, frequency, magnetic field strength). (*Source:* Reprinted with permission from Q. Huang et al., Broadband point measurement of transient magnetic interference in substations with magnetoresistive sensors, *IEEE Transactions on Magnetics*, 50, 1–5, July 2014.)

### 4.5.1 MR Sensor Array Based Interference-rejecting Current Measurement Method

The most frequent source of an interfering field is a nearby conductor [60]. A novel method for a wide range of electric current measurement with TMR sensors has been proposed to reduce the effect from an adjacent current-carrying conductor, which is verified by numerical simulations, FEA of the field, and laboratory experiments in the presence of strong magnetic disturbance. The tested current under measurement ranges from 100 to 1000 A. The method is demonstrated with a circular disc consisting of four TMR magnetic sensors encircling the conductor under measurement. A mathematical model consisting of transcendental equations is then developed. The mathematical model is inspired by the study presented in [60]. However, the algorithm is improved to minimize the number of magnetic sensors to four and this method can estimate the electric current in the presence of an interference conductor placed at unknown locations. The validity of this method is first tested by numerical analysis. From the FEA, it shows that the accuracy of the proposed method proves the potential of this method for practical measurement of current varying from 100 A from 1000 A with relative error less than 3% at 50 Hz from 0.12 m distance in the presence of an interfering flux density up to 16.67 Gs also at a frequency of 50 Hz. A simple experiment, with some unavoidable limitations, shows that the worst case error is less than 5%.

### 4.5.1.1 Mathematical Model

As a first step, it is vital to establish a mathematical framework that lays out a relation between electric current, external magnetic noise, number, and position of magnetic sensors. According to Ampere's law, the magnetic flux density generated by an electric current flowing in a conductor is directly proportional to the electric current magnitude [37]. When the length $L$ of a conductor is infinite, the magnetic flux density generated by the filamentary electric current can be expressed as [64; 65]

$$B = \frac{\mu I}{2\pi r} \tag{4.8}$$

where $\mu$ is the magnetic permeability of free space, $I$ is the electric current magnitude, and $r$ is the distance between the test point and the electric current source. Equation (4.8) is used to approximate the magnetic flux density generated by the electric current flowing in a conductor when the length $L$ of the conductor is much longer than the distance $D$ from the sensor to the conductor. In this case, $L$ can be considered infinite, therefore when the magnetic flux density is sampled, the electric current of $I$ can be calculated by equation (4.8).

In practical applications, magnetic flux density from current flowing in the nearby equipment superimposes on flux from the conductor under measurement. For instance, let us consider a case when the interference current is flowing in parallel to the filamentary current under measurement. Meanwhile, the sensitive direction of a TMR magnetic sensor, which is parallel to the tangent line of the conductor under measurement, needs to be considered because any magnetic flux density along the sensitive axis of the sensor in a certain range can be effectively detected. As shown in Figure 4.40, $B_\alpha$ and $B_r$ are two components of the magnetic flux density of $B_{noise}$, which is the superimposed disturbance. $B_\alpha$ is parallel to the sensitive direction of the TMR magnetic sensor. $B_r$ is parallel to the radial direction of the current under measurement. $B_0$ is the magnetic flux density generated by the current under measurement. Assuming that the directions of the two currents flowing in the conductors $I_1$ and $I_2$ are opposite, then

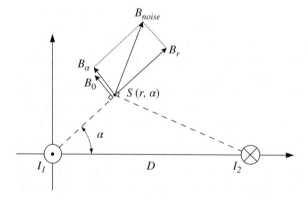

**Figure 4.40** Magnetic flux density generated by current under measurement ($I_1$) and interference current ($I_2$) at the test point. (*Source:* Reprinted with permission from Y. Chen et al., A novel interference-rejecting current measurement method with TMR magnetic sensor array, *IET Science Measurement Technology*, 50, 1–5, Sep. 2018.)

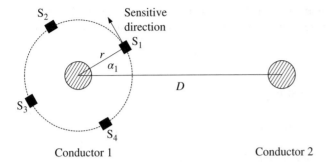

**Figure 4.41** Four sensors are used for magnetic flux density measurement. (*Source:* Reprinted with permission from Y. Chen et al., A novel interference-rejecting current measurement method with TMR magnetic sensor array, *IET Science Measurement Technology*, 50, 1–5, Sep. 2018.)

magnetic flux density B at the test point can be calculated based on Ampere's law as following

$$B = \frac{\mu I_1}{2\pi r} + \frac{\mu I_2 (D \cos \alpha - r)}{2\pi (D^2 + r^2 - 2Dr \cos \alpha)} \tag{4.9}$$

Utilizing the magnetic flux density detected by the TMR magnetic sensor, electric current from conductor under measurement can be determined by (4.9). However, in a real industrial environment, the measurement results of the sensor will be affected by the nearby interference current whose current magnitude and position are unknown and therefore unresolvable by (4.9). For this effect, an effective measurement method needs to be adopted to estimate electric current in the presence of unknown magnetic disturbance.

This work minimizes the number of magnetic sensors to four and estimates the electric current in the presence of an interference conductor placed at unknown locations. As shown in Figure 4.41, four TMR magnetic sensors, $S_1, S_2, S_3$, and $S_4$, are installed in a circular pattern around the conductor. A TMR sensor measures the magnetic field only in one direction, which is shown in Figure 4.41 as the sensitive direction. Because of this, the detected results of the magnetic sensor are not affected by the magnetic field perpendicular to the conductor cross-section or the magnetic field along the radial direction of the conductor cross-section. Figure 4.41 shows the case when a current-carrying conductor acting as the interference source is parallel to the current-carrying conductor of interest.

Using the equation reported in [60], we can denote the distance from the conductor under measurement to the sensor as $r$ and the distance between conductor under measurement and interference conductor as $D$. The magnetic field for $r < D$ can be calculated from the magnetic scalar potential and is expressed as

$$H_\alpha = \frac{I_1}{2\pi r} - \frac{I_2}{2\pi r} \sum_{m=1}^{+\infty} \left(\frac{r}{D}\right)^m \cos(m\alpha) \tag{4.10}$$

$$H_r = -\frac{I_2}{2\pi D} \sum_{m=1}^{+\infty} \left(\frac{r}{D}\right)^{m-1} \sin(m\alpha) \tag{4.11}$$

Keeping in view the sensitive direction of the TMR magnetic sensors, the impact due to the angle between the sensitive direction and the magnetic vector is considered.

Therefore, a system of transcendental equations to calculate the magnetic flux density at four test points, which represent the four sensors, can be expressed as

$$
\begin{cases}
B_1 = \dfrac{\mu I_1}{2\pi r} - \dfrac{\mu I_2}{2\pi r} \displaystyle\sum_{m=1}^{+\infty} \left(\dfrac{r}{D}\right)^m \cos(m\alpha_1) \\[2ex]
B_2 = \dfrac{\mu I_1}{2\pi r} - \dfrac{\mu I_2}{2\pi r} \displaystyle\sum_{m=1}^{+\infty} \left(\dfrac{r}{D}\right)^m \cos(m\alpha_2) \\[2ex]
B_3 = \dfrac{\mu I_1}{2\pi r} - \dfrac{\mu I_2}{2\pi r} \displaystyle\sum_{m=1}^{+\infty} \left(\dfrac{r}{D}\right)^m \cos(m\alpha_3) \\[2ex]
B_4 = \dfrac{\mu I_1}{2\pi r} - \dfrac{\mu I_2}{2\pi r} \displaystyle\sum_{m=1}^{+\infty} \left(\dfrac{r}{D}\right)^m \cos(m\alpha_4)
\end{cases}
\tag{4.12}
$$

where

$$
\alpha_i = \alpha_1 + \frac{\pi}{2}(i-1),\ i = 2, 3, 4
\tag{4.13}
$$

In equation (4.12), there are four unknown variables, i.e. $I_1$, $I_2$, $D$, and $\alpha_1$ that require to be processed iteratively using the magnetic flux density ($B_1$ to $B_4$) sampled from each of the TMR magnetic sensors. It is important to note that with the four equations, the estimation process using (4.12) not only returns the electric current from the conductor under measurement but also estimates other important parameters such as the external interference current of $I_2$.

#### 4.5.1.2 Simulations

Before performing simulations by FEA and laboratory experiments, numerical simulations were conducted to demonstrate the validity of equation (4.12) for calculating the targeted current. Supposing that the targeted current and the interference current are filamentary, the method is first evaluated to estimate the source current based on Ampere's law. $B_i(i = 1, 2, 3, 4)$ representing the magnetic flux density detected by each TMR magnetic sensor is calculated based on the Ampere circuit law as

$$
B_i = \frac{\mu I_1}{2\pi r} + \frac{\mu I_2(D \cos \alpha_i - r)}{2\pi (D^2 + r^2 - 2Dr \cos \alpha_i)}
\tag{4.14}
$$

where

$$
\alpha_i = \alpha_1 + \frac{\pi}{2}(i-1),\ i = 2, 3, 4
\tag{4.15}
$$

Then $B_i$ is substituted in equation (4.12) to calculate the targeted current $I_1$. The calculated results for $I_1$ are represented as $\tilde{I}_1$. The four unknown variables ($I_1$, $I_2$, $\alpha_1$, and $D$) are determined by employing a built-in algorithm to solve the system of nonlinear equations. To solve the transcendental equations, the initial values of the four unknown variants are set as the testing parameters to calculate the theoretical magnetic field that will be detected by the magnetic sensor. Considering the range of tested current from 100 to 1000 A, the initial values of the targeted current $I_1$ and interference current $I_2$ are set to 500 A, which approximates to the average of the current range. Since the circle is symmetric and the four sensors are placed uniformly, the range of $\alpha_1$ is set from $0°$ to $90°$ in order to simplify the analysis process. Thus, the initial value of $\alpha_1$ is set to $90°$, which is the maximum of the angle. From equation (4.9), the distance $D$ between the interference current and the current under measurement is inversely proportional

**Table 4.2** Relative error (%) of $I_1$ by numerical simulation for $D = 0.5$ m.

| $I_1(A)$ | | 100 | 500 | 1000 |
|---|---|---|---|---|
| $I_2(A)$ | 100 | $2.84 \times 10^{-14}$ | $1.59 \times 10^{-13}$ | $2.27 \times 10^{-14}$ |
| | 280 | $1.42 \times 10^{-14}$ | $2.27 \times 10^{-14}$ | $4.09 \times 10^{-12}$ |
| | 460 | $3.84 \times 10^{-13}$ | $3.59 \times 10^{-11}$ | $1.14 \times 10^{-14}$ |
| | 640 | $5.68 \times 10^{-14}$ | $1.51 \times 10^{-12}$ | $1.48 \times 10^{-13}$ |
| | 820 | $3.92 \times 10^{-10}$ | $8.25 \times 10^{-12}$ | $6.82 \times 10^{-14}$ |
| | 1000 | $4.62 \times 10^{-11}$ | $1.25 \times 10^{-11}$ | 0 |

*Source:* Reprinted with permission from Y. Chen et al., A novel interference-rejecting current measurement method with TMR magnetic sensor array, *IET Science Measurement Technology*, Sep. 2018.)

to the magnetic field. When the distance is long enough, the effect of the interference can be neglected. Therefore, the initial value of $D$ is set at 0.5 m, which is a relatively closer distance, whereas $r$ is set to 0.06 m taking into consideration the practicalities of the circuit design in the laboratory.

The algorithm is tested for three cases, i.e. 100 A, 500 A, and 1000 A for the current under measurement, with the interference current changing from 100 to 1000 A with a step of 180 A.

The calculated results of the targeted current $I_1$ are shown in Table 4.2. The calculated error analyzes the accuracy of this method. $\tilde{I}_1$ is the estimated result of the current under measurement $I_1$. As defined in equation (4.16), relative error between the estimated current result and the reference current value is used to analyze the accuracy of this method. $\tilde{I}_1$ is the estimated result of the targeted current under measurement while $I_1$ is the real magnitude of the current under measurement.

$$\text{relative error} = |\frac{\tilde{I}_1 - I_1}{I_1}| \times 100\% \tag{4.16}$$

From the results shown in Table 4.2, when the targeted current is 100 A and the interference current is 820 A, the relative error of $3.92 \times 10^{-10}\%$ is highest, but still less than $4 \times 10^{-10}\%$, which demonstrates the validity of equation (4.12) to calculate the targeted current with high precision.

The proposed method can calculate the electric current under measurement. In addition, the method is capable of calculating the interference current generated at locations near the conductor under measurement. To demonstrate this, we tested cases when $I_2$ was 100, 500, and 1000 A while $I_1$ changed from 100 to 1000 A in steps of 180 A. As shown in Figure 4.42, the results show that the relative error by equation (4.16) is around 0.4%. This error is due to the equation, therefore the results verify the feasibility of equation (4.12) to estimate the interference current.

In practical scenarios, the positions of the conductor under measurement as well as the conductor generating the interfering magnetic field are uncertain. Thanks to the symmetry of the circle, we tested this method when the angle position $\alpha_1$ of the interference current changed from 0° to 90° in steps of 30° to simplify the analysis when $r = 0.06$ m, as shown in Figure 4.41 for $I_1 = 500$ A and $I_2 = 500$ A. According to equation (4.8), the magnetic flux density is inversely proportional to the distance $r$ between the test point

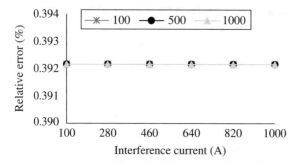

**Figure 4.42** Relative error of $I_2$ based on the Ampere circuit law. (*Source:* Reprinted with permission from Y. Chen et al., A novel interference-rejecting current measurement method with TMR magnetic sensor array, *IET Science Measurement Technology*, 50, 1–5, Sep. 2018.)

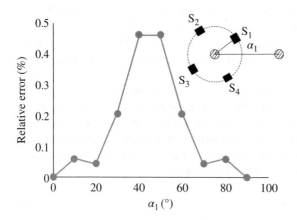

**Figure 4.43** Relative error for $I_1$ when $\alpha_1$ changes. (*Source:* Reprinted with permission from Y. Chen et al., A novel interference-rejecting current measurement method with TMR magnetic sensor array, *IET Science Measurement Technology*, 50, 1–5, Sep. 2018.)

and the current. For conductor 2, the distance from the test point to the current flowing in conductor 2 is $D - r$, therefore a relatively shorter distance, $D = 0.12$ m, is considered in this case. From the results shown in Figure 4.43, when the angle of the interference current changes, the maximum error for $I_1$ is less than 0.5%. When $\alpha_1$ is 0° and 90°, the relative errors are $4.55 \times 10^{-14}$ and $1.25 \times 10^{-13}$, respectively.

In addition, we tested the worst case when the current under measurement was influenced by a strong interference current, i.e. $I_1 = 100$ A and $I_2 = 1000$ A, with the interference placed at five random places. The positions $(\alpha, D)$ in the polar coordinate system of interference current as shown in Figure 4.41 are given in Table 4.3 along with the results of the relative error. The maximum error for $I_1$ is 0.23%, which further demonstrates that this method is effective to figure out the current under measurement even when the interference current is randomly placed.

**Table 4.3** Relative error (%) for $I_1$ for interference generated at five random places.

| $\alpha_1$ | 30 | 45 | 75 | 67.5 | 70 |
|---|---|---|---|---|---|
| $D$ | 0.18 | 1 | 0.6 | 0.7 | 0.4 |
| Relative error | 0.01 | 0.23 | $6.75 \times 10^{-07}$ | $2.84 \times 10^{-13}$ | $1.19 \times 10^{-7}$ |

*Source:* Reprinted with permission from Y. Chen et al., A novel interference-rejecting current measurement method with TMR magnetic sensor array, *IET Science Measurement Technology*, Sep. 2018.

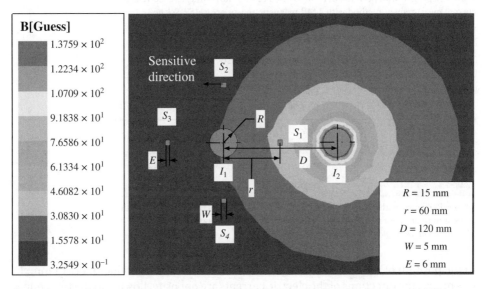

**Figure 4.44** Simulation model and magnetic field distribution when $I_1 = 100$ A and $I_2 = 1000$ A in ANSYS Maxwell. (*Source:* Reprinted with permission from Y. Chen et al., A novel interference-rejecting current measurement method with TMR magnetic sensor array, *IET Science Measurement Technology*, 50, 1–5, Sep. 2018.)

Numerical simulations provide a mathematical analysis based on Ampere's law. However, it does not consider factors like the permeability of free air with the Earth's magnetic field interference, or the magnetic field distribution pattern from each of the conductors. In practical measurement, the detected magnetic field by the sensor will be a little different due to factors such as the precision of the sensor, so simulations need to be conducted to test the validity of the method before applying it to practical current measurement. Because of these considerations, the method was validated by ANSYS Maxwell 16.0, which provides a graphic user interface to analyze electromagnetic field at low frequencies using FEA. In Maxwell, we constructed two cylindrical copper conductors set to a finite length of 3000 mm, with four rectangle areas representing the sensing region of the TMR magnetic sensors (S1–S4), for which the dimensions of 5 × 6 mm are shown in Figure 4.44. For the four sensors, the sensitive direction is along the counterclockwise direction as presented in Figure 4.44. The distance $D$ from the interference to the conductor under measurement is 120 mm. The angle $\alpha_1$, which is defined as shown in Figure 4.41, is set to 0. The distance $r$ from the conductor under measurement

**Table 4.4** Relative error (%) of $I_1$ for FEA simulations in Maxwell for $D = 0.12$ m.

| | $I_2$ (A) | | |
|---|---|---|---|
| $I_1$ (A) | 100 | 500 | 1000 |
| 100 | 0.3164 | 1.2334 | 2.3796 |
| 500 | 0.134 | 0.3176 | 0.5468 |
| 1000 | 0.0786 | 0.1703 | 0.2849 |

*Source:* Reprinted with permission from Y. Chen et al., A novel interference-rejecting current measurement method with TMR magnetic sensor array, *IET Science Measurement Technology*, Sep. 2018.

**Table 4.5** Magnitude of the magnetic field at four sensor positions by FEA.

| Magnitude of $S_1$–$S_4$ by FEA (Gs) | Theoretical magnitude (Gs) | Difference (Gs) |
|---|---|---|
| 35.696 | 36.67 | 0.974 |
| 2.944 | 3.33 | 0.386 |
| 6.845 | 7.78 | 0.935 |
| 2.944 | 3.33 | 0.386 |

*Source:* Reprinted with permission from Y. Chen et al., A novel interference-rejecting current measurement method with TMR magnetic sensor array, *IET Science Measurement Technology*, Sep. 2018.

to the position of the sensor is 60 mm. The radius of both the conductors is 15 mm. In this model, the directions of the electric currents are opposite, flowing in the two copper conductors separately. The four sensing regions are placed on the same circumference uniformly.

Three cases were tested: $I_1 = 100$, 500, and 1000 A. For each case $I_2$ changes from 100 to 500 A and then to 1000 A. By means of FEA in ANSYS Maxwell, the magnetic field distribution in the simulation region is as shown in Figure 4.44 when $I_1$ is 100 A and $I_2$ is 1000 A. The magnetic flux densities detected at each of the sensing regions with different current magnitudes were put into equation (4.12). The targeted current $I_1$ and relative error for $I_1$ were then processed in MATLAB. From the results in Table 4.4, for the worst case when the magnitude of $I_1$ is 100 A and the magnitude of $I_2$ is 1000 A, the error of 2.3796% is highest, but still remains less than 2.5%. Compared with other errors, this larger error is caused by the strong interference current nearby, which can affect the distribution of current in the conductor under measurement.

Comparing with the numerical simulations, the relative error between the targeted current and the calculated error from equation (4.12) becomes larger. Due to the mesh setting in Ansys Maxwell and the finite length of the conductor, a discrepancy exists between the simulated magnetic field at the rectangular areas and the theoretical values by Ampere's law. For instance, when $I_1$ is 100 A and $I_2$ is 1000 A, the magnitudes of the magnetic fields at the four rectangular areas and the difference between the simulation magnitude and the theoretical magnitude are presented in Table 4.5. The relative

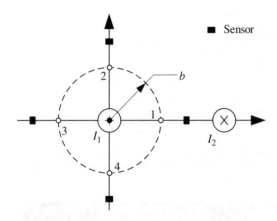

**Figure 4.45** Displacement due to the targeted conductor at four places (1, 2, 3, 4) when $I_1 = 100$ A and $I_2 = 100$ A in ANSYS Maxwell; the four places are all 1 mm away from the center. (*Source*: Reprinted with permission from Y. Chen et al., A novel interference-rejecting current measurement method with TMR magnetic sensor array, *IET Science Measurement Technology*, 50, 1–5, Sep. 2018.)

**Table 4.6** Relative error due to conductor displacement.

| Places | 1 | 2 | 3 | 4 |
|--------|--------|--------|--------|--------|
| RE (%) | 3.3154 | 3.4897 | 3.4144 | 3.4066 |

*Source*: Reprinted with permission from Y. Chen et al., A novel interference-rejecting current measurement method with TMR magnetic sensor array, *IET Science Measurement Technology*, Sep. 2018.

error comes from the difference between the simulated magnetic field and the theoretical magnitude.

In addition, the center of the targeted conductor cross-section should be at the center of the sensor array. However, displacement of the conductor may still exist, therefore four cases are tested when the conductor is placed at the four positions (1, 2, 3, and 4) shown in Figure 4.45. The four places are the intersections of the two axes and a circle with a radius of 1 mm. Other dimensions remain the same while the magnitudes of $I_1$ and $I_2$ are both set to 100 A. As the results presented in Table 4.6 show, the relative error due to the displacement of the conductor reaches around 3%, indicating that the influence contributed by the uncertainty placement of the conductor should be properly handled.

### 4.5.1.3 Experimental Validation

Comprehensive experiments were performed in laboratory to authenticate the effectiveness of the proposed method. As shown in Figure 4.46, the electric current source of $I_1$ and interference current of $I_2$ are generated by a many-to-two-turn transformer especially constructed for our application. In particular, the turn ratio of the primary and secondary copper coil is 138/2. The electric current is flowing in a thick copper conductor, which is connected to the primary and secondary sides of the transformer to form a short circuit. The radius of the conductor is 14 mm. In this experiment, one side

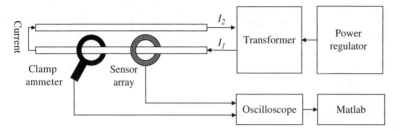

**Figure 4.46** Block diagram for laboratory experiment. (*Source:* Reprinted with permission from Y. Chen et al., A novel interference-rejecting current measurement method with TMR magnetic sensor array, *IET Science Measurement Technology*, 50, 1–5, Sep. 2018.)

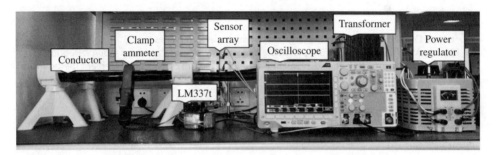

**Figure 4.47** Experimental equipment. (*Source:* Reprinted with permission from Y. Chen et al., A novel interference-rejecting current measurement method with TMR magnetic sensor array, *IET Science Measurement Technology*, 50, 1–5, Sep. 2018.)

of the conductor is used as the conductor under measurement while the other side is used as the interference conductor. Therefore, the electric current source and interference current equal in current magnitude but opposite in direction. In the experiment, the distance of the two conductors is 15 cm. The output current at the secondary side of the transformer is controlled by a power regulator, with which the current can reach an amplitude of 1000 A. All the outputs sampled in the oscilloscope, MDO3012 from Tektronix, are calculated in MATLAB to determine the source current. The amplitude precision of MDO3012 is less than 1.5%. A clamp ammeter TK-65 from Tektronix with an output of 1 mV/A is applied as a reference of the source current. For alternating current varying from 100 and 1000 A, the measurement error of TK-65 is less than 0.3%. To eliminate the effect due to the vibration of the laboratory environment to a negligible level, the equipment has been fixed. The error of the displacement of cables is within +/− 1 mm. For sensors, the placement error is within +/− 0.01 mm. The installation of the experimental equipment is shown in Figure 4.47.

As shown in Figure 4.48, four TMR magnetic sensors, model TMR2104, are installed at a distance of 6 cm from the conductor under measurement on a circular PCB uniformly with a rated sensitivity of 3.1 mV/V/Gs when the detected magnetic flux density is in a range of 80 Gs. The sensor can work properly from −40 to 125° C. In this range, from the datasheet supported by the manufacturer, the offset varies from −1.88 mV to −1.51 mV. In this context, we consider that the temperature in the laboratory experiments has a negligible influence on the robustness of the sensor. Instrumental amplifiers INA333 with a gain factor of 2 applied on the same PCB amplify the output of the sensors to

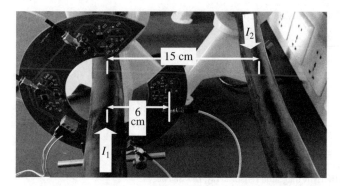

**Figure 4.48** Four TMR sensors on a PCB with a conductor under measurement in the center. (*Source:* Reprinted with permission from Y. Chen et al., A novel interference-rejecting current measurement method with TMR magnetic sensor array, *IET Science Measurement Technology*, 50, 1–5, Sep. 2018.)

**Table 4.7** Results for laboratory experiments.

| Reference current (A) | Estimated current (A) | Error (%) |
| --- | --- | --- |
| 100 | 103.83 | 3.83 |
| 500 | 520.72 | 4.14 |
| 1000 | 1034.67 | 3.47 |

*Source:* Reprinted with permission from Y. Chen et al., A novel interference-rejecting current measurement method with TMR magnetic sensor array, *IET Science Measurement Technology*, Sep. 2018.

put the output in a satisfactory detection range. A linear tuneable filter-regulated DC power supply board, LM337t, is applied to supply the voltage of 5 V to the TMR magnetic sensors and amplifiers. The TMR magnetic sensors provide an output of 31 mV/Gs.

By adjusting the power regulator, three cases were tested with reference currents of 100, 500, and 1000 A. When the data of the magnetic flux densities ($B_1$ to $B_4$) sampled in the oscilloscope were put into equation (4.12), the estimated current and error for $I_1$ were processed by the designed algorithm. From the experimental results shown in Table 4.7, when the magnitudes of $I_1$ and $I_2$ are both 500 A, the error of 4.14% is the maximum, which validates that this method has potential for usage in practical scenarios.

Due to the limitation of the environmental conditions, the measured magnetic flux density deviates from the theoretical magnitude. The length of the conductor is 104 cm, which is not long enough compared with the distance from the conductor to the sensor of 6 cm. The conductor is not exactly straight so it is not strictly in the center of the sensor array. In addition, the magnetic flux density generated from other equipment in teh laboratory also affects the measurement of the TMR magnetic sensors. As shown in Table 4.8, when the reference current is 500 A, the maximum error of measured magnetic flux density and magnetic flux density based on Ampere's law reaches 4.95%. This relative interference error is close to that found in [60]. In their experiment, the magnitudes of the targeted current and interference current are the same, so their relative

**Table 4.8** Measured magnetic flux density and theoretical magnetic flux density for each sensor.

| Measured B (Gs) | Theoretical B (Gs) | Error (%) |
|---|---|---|
| 28.65 | 27.54 | 4.03 |
| 15.13 | 14.42 | 4.95 |
| 11.53 | 11.95 | 3.54 |
| 15.13 | 14.42 | 4.91 |

*Source:* Reprinted with permission from Y. Chen et al., A novel interference-rejecting current measurement method with TMR magnetic sensor array, *IET Science Measurement Technology*, Sep. 2018.

interference error is same as the relative error defined by equation (4.16), demonstrating the validity of the method. For the effect of the measurement deviation, the relative error in the laboratory environment is larger than the error in simulations by means of FEA of Maxwell and numerical simulations based on the Ampere circuit law.

In conclusion, a novel interference-rejecting method to measure large current flowing in the long conductor at a fixed place with a circular TMR magnetic sensor array has been verified. Measurement for a large electric current is a significant research field in power systems. For this reason, the proposed method here is validated to be effective for the peak magnitude of the source current changing from 100 to 1000 A, with interference current changing from 100 to 1000 A. In order to eliminate the unwanted magnetic field generated by nearby current noise, a system of four transcendental equations is constructed to calculate the current under measurement. The magnetic flux densities at test points are detected by a four-sensor array with a conductor under measurement in the centre. The relative error between the calculated results and the targeted magnitude by equation (4.12) can be as small as $4 \times 10^{-10}\%$, proving the validity of equation (4.12) to calculate the source current. When tested by means of FEA in ANSYS Maxwell, the error is less than 2.5% for the worst case scenario when the source current is 100 A and the interference current is 1000 A. The experimental data validate the efficacy of our method for electric currents of up to 1000 A. In addition, the interference current generated at random locations was calculated with error less than 0.4% in numerical simulations.

The results also show that other practical factors can affect the estimated current. The uncertainty contributions due to the placement of +/− 1 mm of the conductor could be considered in the laboratory experiments. The length of the measured conductor, which is 104 cm, can affect the detected results of the TMR magnetic sensor for the equations employed on condition that the length of the conductor is infinite compared with the distance from the sensor to the conductor under measurement. In this context, methods to reduce the uncertainty contributions need to be studied in future research.

### 4.5.2 Adaptive Filter Algorithm Based Current Measurement

With progress in digital signal processors, adaptive filters have been applied in various fields to remove noise in signal processing [66; 67]. Adaptive filters can be used in applications where some parameters of a system are not known, like the unknown targeted current and the uncertainty of the interference current mentioned above. In [68], a method aimed at rejecting magnetic interference was employed based on the Kalman filter. The authors designed an optimal steady-state filtering and a sub-optimal steady-state filtering based on Kalman filtering for non-contact current measurement. The application of optimal filtering was limited because it relied on information about the characteristics of the system status, which were unknown in real scenarios. Sub-optimal filtering was proposed to compensate this shortcoming. However, only currents up to a few amperes were tested and the error reached up to 5%.

Here, an adaptive filter is used with a circular magnetic sensor array constituting of three nodes to estimate the magnetic field under measurement accurately. This current measurement scheme is small and easy to install, benefiting from the miniaturization of the magnetic sensor and non-use of magnetic shielding. In addition, the adaptive-filtering algorithm is free from the complexity of spatial DFT analysis. Analyzing the results from simulations when the measurement range of the peak current, $I$, is from 500 to 1500 A and the interference current ranges from $I = -250$ A to $I = +50$ A, proves that this method is effective to dampen the interference to a negligible level.

### 4.5.2.1 Mathematical Principle

According to the Biot–Savart law, the magnetic flux density generated by the filamentary current can be calculated by

$$B(t) = \frac{\mu_0 I(t)}{4\pi R}(\sin\ \theta_1 - \sin\ \theta_2) \tag{4.17}$$

where $\mu_0$ is the permeability of free space, $I(t)$ is the filamentary current magnitude, $R$ is the distance from the current under measurement to the test point, and $\theta_1$ and $\theta_2$ are the angles of the line connected by the current terminal and the test point and the line passing through the test point and perpendicular to the current.

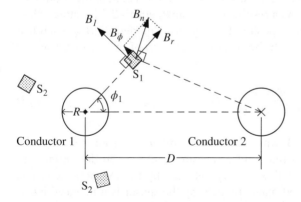

**Figure 4.49** Magnetic flux density generated by the current under measurement flowing in conductor 1 and the interference current flowing in conductor 2. (*Source:* Reprinted with permission from Y. Chen et al., A novel adaptive filter for accurate measurement of current with magnetic sensor array, *IEEE PES General Meeting*, Aug. 2018.)

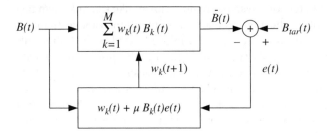

**Figure 4.50** The process to estimate the magnetic field under measurement. (*Source:* Reprinted with permission from Y. Chen et al., A novel adaptive filter for accurate measurement of current with magnetic sensor array, *IEEE PES General Meeting*, Aug. 2018.)

An MR effect based sensor can detect the magnetic flux density along its sensitive direction. For this effect, the measured result will be affected because the magnetic flux density generated by nearby interference current can superpose on the magnetic field under measurement. Usually, this situation happens when interference magnetic flux density is generated by a current-carrying conductor parallel to the conductor under measurement at a fixed distance, like bus bars. Therefore, a situation is considered where the nearby interference current is parallel to the filamentary current under measurement, as shown in Figure 4.49. The electric current $I_1(t)$ flowing in conductor 1 is under measurement while conductor 2 is the interference conductor with electric current $I_2(t)$. $D$ is the distance from conductor 1 to conductor 2. $\phi$ is the angle between the line connecting the two conductors and the line connecting the sensor and the targeted conductor. $B_\phi$ and $B_r$ are two components of the interference magnetic flux density of $B_n$. $B_\phi$ is parallel to the sensitive direction of the magnetic sensor, while $B_r$ is along the radial direction of the current under measurement. $B_1$ is the magnetic flux density under measurement at $S_1$ that represents a magnetic effect based sensor. For equation (4.17), the length of the conductor is often considered as infinite when the length of the conductor is much longer than $R$ in practice. As a result, $\sin\theta_1 - \sin\theta_2$ equals 2. Therefore, when the directions of $I_1(t)$ and $I_2(t)$ are opposite, the magnetic flux density $B_i$ at each test point can be expressed as follows with the sensitive direction of the magnetic sensor considered:

$$B_i(t) = \frac{\mu_0}{2\pi R}\left(I_1(t) + \frac{(\alpha\cos\phi_i - 1)}{\alpha^2 + 1 - 2\alpha\cos\phi_i}I_2(t)\right) \tag{4.18}$$

where $\phi_i = \phi_1 + \frac{2\pi}{N}(i-1), i = 1, 2, 3$, where $N$ is the number of sensors and $\alpha$ is the ratio of $D$ and $R$. The electric current under measurement can be determined by equation (4.18) when the magnetic flux density detected by the magnetic sensor is obtained. Nevertheless, the detected magnetic field by the sensor is influenced by an unknown interference magnetic field in practical industrial engineering.

To damp the effect of the external interference magnetic field, the authors proposed to use a classical adaptive-filtering algorithm, the least mean square (LMS) algorithm, as shown in Figure 4.50. The mathematical model is constructed based on the theory of the adaptive-filtering algorithm [69]. The calculation expression of the LMS algorithm can be expressed as

$$
\begin{cases}
\overline{B}(t) = \sum_{k=1}^{M} w_k(t)B_k(t) \\
w_k(t+1) = w_k + \mu B_k(t)e(t) \\
e(t) = B_{tar}(t) - \overline{B}(t)
\end{cases}
\tag{4.19}
$$

where $\overline{B}(t)$ is the processed output of the magnetic field with filtered weights, $w_1(t)$, $w_2(t),\ldots, w_M(t)$, at time $t$. $M$ is the number of filter taps, $\mu$ is a small constant called the learning rate, and $B_{tar}(t)$ is the actual signal to be measured, which is usually unknown in real problems. Let $B(t) = [B_1(t),B_2(t),\ldots, B_M(t)]$, $W(t) = [w_1(t),w_2(t), w_M(t)]$. The stimulus vector $B(t)$ can arise in one of two fundamentally different ways, one spatial and the other temporal: (1) the $M$ elements of $B(t)$ are detected by different magnetic sensors in space and (2) the $M$ elements of $B(t)$ represent the set of present and $(M-1)$ past values generated by some excitations that are uniformly spaced in time.

For the spatial way, more power consumption is concomitant as more magnetic sensors are applied. Meanwhile, hundreds of sensors are needed to confirm the searching steps of this algorithm, which is impractical for the installation of a sensor array. Therefore, the temporal way is used to solve the problem here.

In the adaptive-filtering algorithm, to refine the filtering procession an actual targeted signal $B_{tar}(t)$ is employed to form an error signal due to the feedback of the adaptive filter. However, for current measurement, the actual current signal under measurement is unknown. The work in [67] has verified that it is possible to use the delayed observed signal instead. For this effect, the authors proposed using an average of all sensor outputs to replace $B_{tar}(t)$, which can be explained by the LMS.

Keeping in mind the power consumption, three sensors are adopted. Based on (4.18) and (4.19), equations can be obtained to estimate the magnetic field signal under measurement as follows:

$$
\begin{cases}
\overline{B_1}(t) = \sum_{k=1}^{M} w_{1k}(t)B_1(t) \\
\overline{B_2}(t) = \sum_{k=1}^{M} w_{2k}(t)B_2(t) \\
\overline{B_3}(t) = \sum_{k=1}^{M} w_{3k}(t)B_3(t) \\
e(t) = \frac{B_1(t)+B_2(t)+B_3(t)}{3} - \overline{B_i}(t) \\
w_{ik}(t+1) = w_{ik}(t) + \mu B_i(t)e(t), i = 1, 2, 3
\end{cases}
\tag{4.20}
$$

where $B_1(t)$, $B_2(t)$, and $B_3(t)$ are the magnetic fields detected by each sensor at time $t$, and $\overline{B_1}(t)$, $\overline{B_2}(t)$, and $\overline{B_3}(t)$ are the corresponding filtered magnetic fields. Using the LMS, the estimated value of the magnetic flux density under measurement at time $t$ can be determined by

$$
B_c(t) = \frac{\overline{B_1}(t) + \overline{B_2}(t) + \overline{B_3}(t)}{3}
\tag{4.21}
$$

According to equation (4.17), the estimated current is in proportion to the magnetic flux density. Therefore, the relative error (RE) of the estimated current and the actual current under measurement is equivalent to the RE of the estimated magnetic flux density and the actual magnetic flux density under measurement. Therefore, for rest of this section, we will focus on the RE of the estimated magnetic field and the actual magnetic field under measurement, as follows:

$$RE = |\frac{\overline{B_c} - \overline{B_0}}{\overline{B_0}}| \times 100\% \tag{4.22}$$

where $\overline{B_c}$ is the estimated magnetic flux density and $\overline{B_0}$ is the peak value of the magnetic flux density under measurement.

### 4.5.2.2 Numerical Simulations

The adaptive filter for current measurement is first validated with numerical simulations when alternating current is considered. From equation (4.18), the ratio of $\alpha$ contributes to the result of the magnetic field. Therefore, simulations are performed to analyze the effect of $\alpha$ on the estimated results of magnetic field under measurement in three cases when the peak value, $I_1$, of the current under measurement of $I_1$ and the interference current, $I_2$, of $I_2(t)$ is 500, 1000 and 1500 A. The ratio changes with $R$ while $D$ stays at 0.12 m. In all the simulations, $\phi_1$ is 0. The magnetic flux density of $B_i$ calculated based on equation (4.18) is adopted as the detected magnitude at each test point. Utilizing the calculated magnetic flux density of $B_i$, the estimated magnetic flux density of $B_c$ can be calculated by equations (4.20) and (4.21). The root mean square errors (RMSEs) of the estimated magnetic flux density and the actual magnetic flux density under measurement are shown in Figure 4.51, from which the minimum RMSE for all three cases is calculated when the ratio is 2. This proves that a ratio of 2 is appropriate to minimize the RMSE.

To determine a valid value of $R$, numerical simulations are performed in the next step. In these simulations, the peak value, $I_1(t)$, of the current under measurement of $I_1(t)$ and the peak value, $I_2(t)$, of the interference current of $I_2(t)$ are 500 A. The ratio of $D$ and $R$ is set as 2. The behavior of RMSE is as shown in Figure 4.52 when $R$ changes. It can be seen that when $R$ is 0.06, the RMSE is least.

To authenticate the efficacy of employing the adaptive-filtering algorithm, three cases are tested when the peak value $I_1$ of $I_1(t)$ is 500, 1000 and 1500 A with interference current $I_2$ of $I_2(t)$ changing from $I_1 = -250$ A to $I_1 = +50$ A. Meanwhile, $R$ is 0.06 m, $D$ is 0.12 m, and $\phi_1$ is 0. The processed results are shown in Table 4.9, where $B_c$ is the estimated magnetic flux density while $B_0$ is the magnetic flux density under measurement. From Table 4.9, the maximum RMSE of 0.8944 is calculated, which infers that the processed magnetic flux density waveform is highly consistent with the magnetic flux density waveform under measurement. For the maximum RMSE case when $I_1(t) = 1500\sin(100\pi t)$ and $I_2(t) = 1550\sin(100\pi t)$ A, the waveforms are depicted in Figure 4.53, showing that the two curves almost coincide.

Further simulations are performed to test the effectiveness of the adaptive filter for current measurement. When the peak value $I_1$ of $I_1(t)$ is 500, 1000 and 1500 A with interference current $I_2$ of $I_2(t)$) changing from $I_1 = -250$ A to $I_1 = +50$ A in steps of 50 A, the RE is shown in Table 4.10, from which it can be seen that the maximum RE is 2.685%.

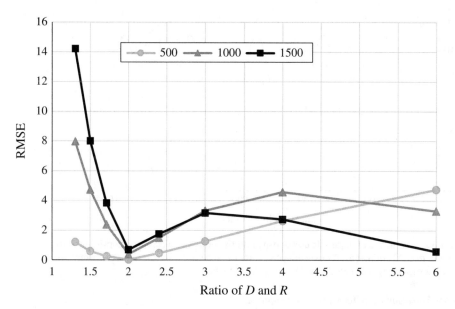

**Figure 4.51** RMSE changes with the ratio of *D* and *R* when $I_1 = I_2$ = 500, 1000, and 1500 A. (*Source:* Reprinted with permission from Y. Chen et al., A novel adaptive filter for accurate measurement of current with magnetic sensor array, *IEEE PES General Meeting*, Aug. 2018.)

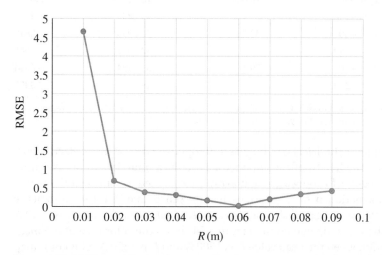

**Figure 4.52** RMSE versus *R* when $I_1 = I_2$ = 500 A. (*Source:* Reprinted with permission from Y. Chen et al., A novel adaptive filter for accurate measurement of current with magnetic sensor array, *IEEE PES General Meeting*, Aug. 2018.)

### 4.5.2.3 Finite Element Analysis

Conditions for numerical simulations are deficient of practical factors like the pragmatic length of the conductor. Thus, we have performed simulations by means of FEA. The designed model is shown in Figure 4.54, in which *R* is the distance from the conductor to the sensor, *D* is the distance between the two conductors, and *r* is the radius of the

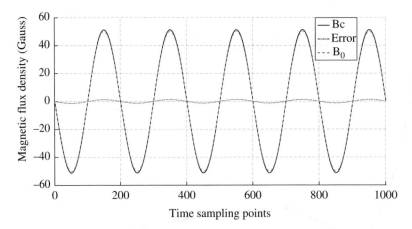

**Figure 4.53** Waveforms of magnetic flux density and the error signal. (*Source:* Reprinted with permission from Y. Chen et al., A novel adaptive filter for accurate measurement of current with magnetic sensor array, *IEEE PES General Meeting*, Aug. 2018.)

**Table 4.9** The results at different current magnitudes.

| $I_1$ (A) | $I_2$ (A) | $B_0$ (Gs) | $B_c$ (Gs) | RMSE |
|---|---|---|---|---|
| 500 | $I_1 = +50$ | 16.6667 | 16.6307 | 0.0364 |
| | $I_1 = -250$ | | 16.2835 | 0.2808 |
| 1000 | $I_1 = +50$ | 33.3333 | 34.1935 | 0.5812 |
| | $I_1 = -250$ | | 32.5371 | 0.5888 |
| 1500 | $I_1 = +50$ | 50 | 51.3424 | 0.8944 |
| | $I_1 = -250$ | | 49.5974 | 0.3372 |

*Source:* Reprinted with permission from Y. Chen et al., A novel adaptive filter for accurate measurement of current with magnetic sensor array, *IEEE PES General Meeting*, Aug. 2018.

conductor. Two copper cylindrical conductors with opposite directions of electric current were designed. The length of the conductor is set to a finite length of 3000 mm. Three rectangular regions forming a circular array are designed to represent the detective region of the magnetic sensor. $E$ is the length of the sensor and $W$ is the width of the sensor. According to the numerical simulations above, when the peak value $I_1$ is 1500 A and $I_2$ is 1400 A, the RE is the maximum. Figure 4.54 shows the color map of the magnetic field distribution when the peak value $I_1$ is 1500 A and $I_2$ is 1550 A. This color map shows the superposition of the interference magnetic field on the magnetic field under measurement in a more intuitive way.

To demonstrate the effectiveness of the adaptive-filtering algorithm applied to reduce the magnetic interference, cases are tested when the peak value of the current under measurement is 500 A while the peak value of the external interference current changes from 250 to 550 A in steps of 150 A. The results are shown in Table 4.11. The maximum RE for magnetic flux density at peak value is 3.55% while the RMSE is 0.81 for this situation, which proves the capability of this method for practical electric current measurement applications.

**Table 4.10** RE between the practical and estimated values.

| $I_1$ (A) | $I_1 = +50$ | $I_1$ | $I_1 = -50$ | $I_1 = -100$ | $I_1 = -150$ | $I_1 = -200$ | $I_1 = -250$ |
|---|---|---|---|---|---|---|---|
| | | | | $I_2$ (A) | | | |
| 500 | 0.216 | 0.134 | 0.249 | 0.542 | 0.994 | 1.586 | 2.299 |
| 600 | 1.694 | 1.278 | 0.761 | 0.157 | 0.521 | 1.261 | 2.053 |
| 700 | 2.502 | 1.797 | 1.052 | 0.275 | 0.524 | 1.34 | 2.166 |
| 800 | 2.691 | 1.86 | 1.018 | 0.171 | 0.675 | 1.517 | 2.352 |
| 900 | 2.661 | 1.797 | 0.935 | 0.078 | 0.773 | 1.613 | 2.442 |
| 1000 | 2.581 | 1.736 | 0.896 | 0.062 | 0.765 | 1.582 | 2.389 |
| 1100 | 2.53 | 1.728 | 0.93 | 0.137 | 0.65 | 1.431 | 2.204 |
| 1200 | 2.523 | 1.772 | 1.024 | 0.279 | 0.462 | 1.197 | 1.928 |
| 1300 | 2.538 | 1.844 | 1.151 | 0.461 | 0.226 | 0.91 | 1.591 |
| 1400 | 2.604 | 1.965 | 1.328 | 0.692 | 0.057 | 0.575 | 1.205 |
| 1500 | 2.685 | 2.101 | 1.517 | 0.935 | 0.353 | 0.227 | 0.805 |

*Source:* Reprinted with permission from Y. Chen et al., A novel adaptive filter for accurate measurement of current with magnetic sensor array, *IEEE PES General Meeting*, Aug. 2018.

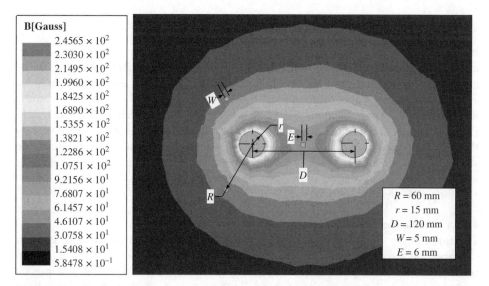

**Figure 4.54** Color map of the magnetic distribution. (*Source:* Reprinted with permission from Y. Chen et al., A novel adaptive filter for accurate measurement of current with magnetic sensor array, *IEEE PES General Meeting*, Aug. 2018.)

### 4.5.2.4 Comments

MR effect based sensors realize non-contact measurement for electric current. However, the interference magnetic flux density generated by nearby current is an important factor in the results measured by this method, therefore we have tried to damp the effect of the interference based on the adaptive-filtering algorithm. Here, a novel adaptive filter is proposed to reject the magnetic interference with a sensor array. The current under

**Table 4.11** RE and RMSE in FEA simulations.

| $I_2$ (A) | $I_1 = -250$ | | $I_1 = -100$ | | $I_1 = +50$ | |
|---|---|---|---|---|---|---|
| | RE | RMSE | RE | RMSE | RE | RMSE |
| 500 | 1.5 | 0.19 | 1.61 | 0.41 | 0.26 | 0.14 |
| 1000 | 0.71 | 0.07 | 0.93 | 0.19 | 1.54 | 0.49 |
| 1500 | 1.71 | 0.19 | 3.55 | 0.81 | 3.34 | 1.13 |

*Source:* Reprinted with permission from Y. Chen et al., A novel adaptive filter for accurate measurement of current with magnetic sensor array, *IEEE PES General Meeting*, Aug. 2018.

measurement can then be determined by reconstruction when the magnetic flux density at the test point is confirmed. Without the magnetic shielding of the metal, the measurement device is small and can be installed conveniently. Furthermore, the equations of the adaptive-filtering algorithm are easy to construct. With numerical simulations and the FEA method, we have demonstrated that the adaptive filter is feasible to reduce the interference effect when the peak value of the current under measurement $I$ ranges from 500 to 1500 A with the parallel interference current ranging from $I = -250$ A to $I = +50$ A, providing a novel idea for current measurement for practical applications.

### 4.5.3 Current Measurement Under Strong Interference

In [70], the authors proposed a three-phase current sensor based on the anisotropic magnetoresistance (AMR) effect and utilized a sensor placement strategy to reduce the effect of the magnetic interference. The same method has been extended to propose an improved strategy applicable to real-world scenarios where the targeted current is much smaller in comparison with the two other interfering currents. When the algorithm in [70] is applied in situations when the interfering currents are much larger than the targeted current, a lager error can be caused by summing up the magnetic strengths detected by two sensors with opposite sensitive direction. Therefore, a strategy is proposed to damp the effect of the strong magnetic interference, combining magnetic shielding and sensor placement.

#### 4.5.3.1 Mathematical Background
The three-phase current system is usually applied in power systems due to its high operating performance and high efficiency. The mathematical model of the algorithm to calculate the three-phase current is shown in Figure 4.55. $I_1$, $I_2$ and $I_3$ comprise a typical geometrical arrangement of three-phase wires in medium voltage/low voltage substations, $r_i$ ($i = 1, 2$) is the distance from $I_i$ to sensor $S_{ij}$ ($i = 1, 2; j = 1, 2$), and $d$ is the distance between the two currents. The main idea is based on the assumption that the contribution to the magnetic field from magnetic interference can be reduced to a negligible level by summing up the detected magnetic field of two adjacent magnetic sensors ($S_{i1}$ and $S_{i2}$) with opposite sensitive directions. In this scheme, uniaxial sensors, which can only detect the magnetic flux density along one sensitive direction (shown as the arrows in Figure 4.55), are used.

The mathematical model from prior work is briefly discussed so that a firm grounding can be established for the contribution of this chapter. From [71], calculation of

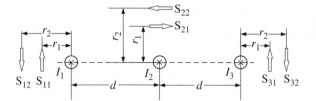

**Figure 4.55** Magnetic sensor array configuration for a three-phrase current system. (*Source:* Reprinted with permission from Y. Chen et al., An interference-rejection strategy for measurement of small current under strong interference with magnetic sensor array, *IEEE Sensors Journal*, 19, 692–700, 2018.)

the phase current is given based on the Biot–Savart law. Due to the symmetry of the three-phase structure and the sensors, the calculation processes for $I_1$ and $I_3$ are similar so the calculation for $I_1$ is discussed. In theory, the magnetic fields detected by the sensors should be equal to the values based on the Biot–Savart law as follows:

$$\begin{cases} B_{11} = \frac{\mu I_1}{2\pi r_1} + \frac{\mu I_2}{2\pi(r_1+d)} + \frac{\mu I_3}{2\pi(r_1+2d)} \\ B_{12} = -\frac{\mu I_1}{2\pi r_2} - \frac{\mu I_2}{2\pi(r_2+d)} - \frac{\mu I_3}{2\pi(r_2+2d)} \\ B_{21} = \frac{\mu I_2}{2\pi r_1} + \frac{\mu r_1 I_1}{2\pi(r_1^2+d^2)} + \frac{\mu r_1 I_3}{2\pi(r_1^2+d^2)} \\ B_{22} = -\frac{\mu I_2}{2\pi r_2} - \frac{\mu r_2 I_1}{2\pi(r_2^2+d^2)} - \frac{\mu r_2 I_3}{2\pi(r_2^2+d^2)} \end{cases} \tag{4.23}$$

where $B_{ij}$ is the magnetic field along the sensitive direction of $S_{ij}$.

For instance, when $I_2$, $I_3$, $B_{11}$, and the dimensions are known, $I_1$ can be confirmed based on the equation related to $B_{11}$. However, since the information on the currents is unknown, we propose to estimate the currents by

$$\begin{cases} \tilde{I}_1 = \frac{2\pi r_1 r_2 (B_{11}+B_{12})}{\mu(r_2-r_1)} \\ \tilde{I}_2 = \frac{2\pi r_1 r_2 (B_{21}+B_{22})}{\mu(r_2-r_1)} \end{cases} \tag{4.24}$$

where $\tilde{I}_1$ is the estimated value for $I_1$ and $\tilde{I}_2$ is the estimated value for $I_2$.

From equations (4.23) and (4.24), for $\tilde{I}_1$, the error between the estimated current and the true value of the current under measurement is:

$$error = \left| \frac{\tilde{I}_1 - I_1}{I_1} \right| \times 100\%$$

$$= \left| \frac{r_1 r_2 I_2}{(r_1 + d)(r_2 + d)I_1} + \frac{r_1 r_2 I_3}{(r_1 + 2d)(r_2 + 2d)I_1} \right| \times 100\% \tag{4.25}$$

For $\tilde{I}_2$, the error between the estimated current and the true value of the targeted current is

$$error = \left| \frac{\tilde{I}_2 - I_2}{I_2} \right| \times 100\%$$

$$= \left| \frac{I_1 + I_3}{I_2} \left[ \frac{r_1^2 r_2}{(r_1^2 + d)(r_2 - r_1)} + \frac{r_1 r_2^2}{(r_2^2 + d)(r_2 - r_1)} \right] \right| \times 100\% \tag{4.26}$$

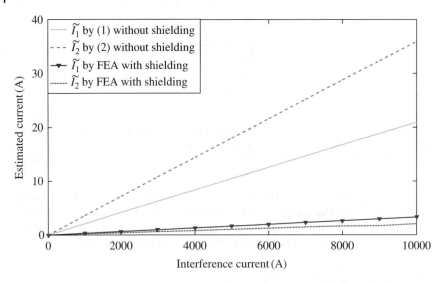

**Figure 4.56** Estimated currents with and without shielding. (*Source:* Reprinted with permission from Y. Chen et al., An interference-rejection strategy for measurement of small current under strong interference with magnetic sensor array, *IEEE Sensors Journal*, 19, 692–700, 2018.)

From equations (4.25) and (4.26), it is clear that the error will be affected by the magnitudes of the currents. If the parameters of the measurement system are confirmed, when the targeted current is much smaller than the other two currents, the error becomes large. When the targeted current is zero at some moment while the other two interfering currents are not, the error cannot be calculated by equations (4.25) and (4.26). Under this situation, the estimated error equals the estimated currents, i.e. equation (4.24). To clarify this point, a graph (see Figure 4.56) is used to depict the estimated currents based on (4.23) and (4.24). It can be seen that when $I_1$ is zero while $I_2$ and $I_3$ change from 0 to 10,000 A simultaneously, the theoretically estimated magnitude $\tilde{I}_1$ by equation (4.23) changes from 0 to 20.9447 A. When $I_2$ is zero while $I_1$ and $I_3$ change from 0 to 10,000 A simultaneously, the theoretically estimated magnitude $\tilde{I}_2$ by (4.24) changes from 0 to 35.9611 A.

### 4.5.3.2 Design Overview

To compensate for the estimated error for the targeted current being large when the interfering current is much stronger, an improved strategy has been proposed, which employs a curved piece of magnetic shielding as shown in Figure 4.57. In this strategy, the magnetic shielding should have two functions. One is to damp the effects contributed by the interference [72], which can also be damped by summing up the magnetic fields detected by two sensors with opposite directions. The other is to avoid the effect on the magnetic flux lines generated by the targeted current inside the shielding [40]. The primary purpose of the proposed structure is to reduce the effect on the magnetic field generated by $I_2$, therefore the magnetic shielding is designed as a circular arc. The center of the arc is at the center of the conductor under measurement, trying to prevent the magnetic flux lines from being influenced by the strong influence due to the shielding. This is demonstrated by simulations of FEA as shown in Figure 4.58. In the simulations,

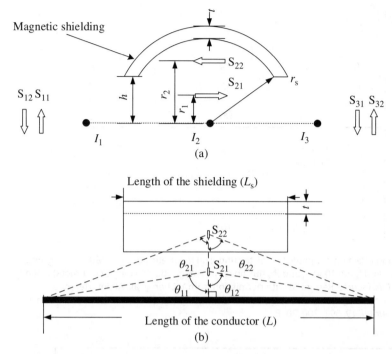

**Figure 4.57** Proposed current measurement system with magnetic shielding: (a) view along the conductor section and (b) side view perpendicular to the circular conductor section. (*Source:* Reprinted with permission from Y. Chen et al., An interference-rejection strategy for measurement of small current under strong interference with magnetic sensor array, *IEEE Sensors Journal*, 19, 692–700, 2018.)

$R$ is 0.005 m, $h$ is 0.02 m, $d$ is 0.3 m, $t$ is 0.0002 m, and $r_s$ is 0.036 m. The material of the conductor is copper. Mu-metal is employed as the magnetic shielding layer. The distance from position A to $I_2$ is 10.5 mm and the distance from position B to $I_1$ is 10.5 mm.

First, the effect on the magnetic field generated by $I_2$ is investigated. $I_1$ and $I_3$ are 0 A. $I_2$ changes from 1000 to 10000 A. The magnetic fields at position A are shown in Figure 4.59. It should be noted that the magnetic field without magnetic shielding is close to the magnetic field with shielding and they are linear. The error between the magnetic field without shielding and the magnetic field with shielding is 3.35% when $I_2$ is 10,000 A, proving that the effect of the shielding on the magnetic field inside the shielding is small. When the position of A changes from 6 to 35 mm, the magnetic field at position A with and without magnetic shielding is investigated. $I_2$ is 1000 A. From Figure 4.60, the minimum error between the magnetic field without and with shielding is only 2.80%, therefore the curved arc forming the shielding has a small effect on the magnetic field inside it.

Simulations are then performed to investigate the effect on the magnetic field near $I_1$ due to the shielding. $I_1$, $I_2$ and $I_3$ change from 1000 to 10000 A. The magnetic fields at position B are depicted in Figure 4.59. It can be seen that when exposed to the interference from $I_2$ and $I_3$, the magnetic fields at position B without and with shielding are close. The error between the magnetic field with shielding and the magnetic field without

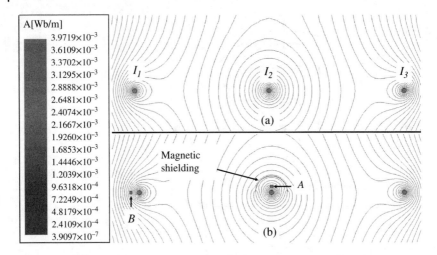

**Figure 4.58** Shielding effect on magnetic flux distribution: (a) magnetic flux lines without magnetic shielding when $I_1$, $I_2$, and $I_3$ are 1000 A, and (b) flux lines with magnetic shielding of mu-metal when $I_1$, $I_2$, and $I_3$ are all 1000 A. (*Source:* Reprinted with permission from Y. Chen et al., An interference-rejection strategy for measurement of small current under strong interference with magnetic sensor array, *IEEE Sensors Journal*, 19, 692–700, 2018.)

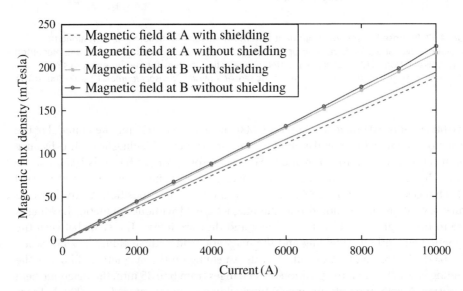

**Figure 4.59** Magnetic field in relation to current under measurement ($I_2$). (*Source:* Reprinted with permission from Y. Chen et al., An interference-rejection strategy for measurement of small current under strong interference with magnetic sensor array, *IEEE Sensors Journal*, 19, 692–700, 2018.)

shielding is 3.67% when $I_2$ is 10,000 A. In addition, the flux lines are shown in Figure 4.58 when $I_1$, $I_2$ and $I_3$ are 1000 A.

From the results, the error between the magnetic field without and with shielding is less than 3.7% and can be further eliminated by summing up the magnetic flux density detected by two sensors with opposite directions. Equations based on the Bior–Savart

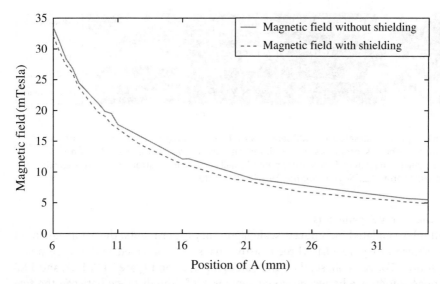

**Figure 4.60** Magnetic field in relation to position A. (*Source:* Reprinted with permission from Y. Chen et al., An interference-rejection strategy for measurement of small current under strong interference with magnetic sensor array, *IEEE Sensors Journal*, 19, 692–700, 2018.)

law can also be utilized to estimate the targeted current. With the length of the conductor taken into consideration, the estimated currents $\tilde{I}_1$ and $\tilde{I}_2$ are

$$\tilde{I}_1 = \frac{B_{11} + B_{12}}{A_1} = \frac{A_1 I_1 + B_1 I_2 - C_1 I_3}{A_1} \tag{4.27}$$

$$\tilde{I}_2 = \frac{B_{21} + B_{22}}{A_2} = \frac{A_2 I_2 + C_2(I_1 - I_3)}{A_2} \tag{4.28}$$

$$\begin{cases} A_1 = \frac{\mu}{4\pi}\left(\frac{\sin\theta_{11} - \sin\theta_{12}}{r_1} - \frac{\sin\theta_{21} - \sin\theta_{22}}{r_2}\right) \\ B_1 = \frac{\mu}{4\pi}\left(\frac{\sin\theta_{11} - \sin\theta_{12}}{r_1+d} - \frac{\sin\theta_{21} - \sin\theta_{22}}{r_2+d}\right) \\ C_1 = \frac{\mu}{4\pi}\left(\frac{\sin\theta_{11} - \sin\theta_{12}}{r_1+2d} - \frac{\sin\theta_{21} - \sin\theta_{22}}{r_2+2d}\right) \end{cases} \tag{4.29}$$

$$\begin{cases} A_2 = \frac{\mu}{4\pi}\left(\frac{\sin\theta_{11} - \sin\theta_{12}}{r_1} - \frac{\sin\theta_{21} - \sin\theta_{22}}{r_2}\right) \\ C_2 = \frac{\mu}{4\pi}\left(r_1\left(\frac{\sin\theta_{11} - \sin\theta_{12}}{r_1^2+d^2}\right) - r_2\left(\frac{\sin\theta_{21} - \sin\theta_{22}}{r_2^2+d^2}\right)\right) \end{cases} \tag{4.30}$$

where $\theta_{ij}(i=1,2;j=1,2)$ is the angle shown in Figure 4.57b.

Therefore, for $\tilde{I}_1$, the error is

$$error = \left|\frac{\tilde{I}_1 - I_1}{I_1}\right| \times 100\% = \left|\frac{B_1 I_2 - C_1 I_3}{A_1 I_1}\right| \times 100\% \tag{4.31}$$

For $\tilde{I}_2$, the error is

$$error = \left|\frac{\tilde{I}_2 - I_2}{I_2}\right| \times 100\% = \left|\frac{C_2 I_1 - C_2 I_3}{A_2 I_2}\right| \times 100\% \tag{4.32}$$

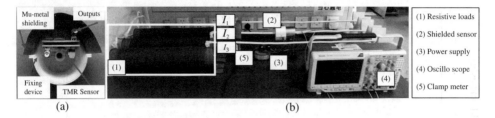

**Figure 4.61** Experimental setup: (a) TMR sensor with shielding and (b) three-phase current system with shielded magnetic sensors. (*Source:* Reprinted with permission from Y. Chen et al., An interference-rejection strategy for measurement of small current under strong interference with magnetic sensor array, *IEEE Sensors Journal*, 19, 692–700, 2018.)

### 4.5.3.3 Laboratory Experiments

The method is validated by reduced-scale laboratory experiments, the arrangement of which is shown in Figure 4.61. Three resistive loads are connected to the three-phase current source. For each phase, the RMS values for $I_1$, $I_2$, and $I_3$ are 14.7, 14.7, and 14.3 A. The phase difference between two currents is 120°. The distance between the two conductors is 15 cm and the radius of each conductor is 5 mm. A clamp meter is utilized to measure the current as the reference with an output of 1 mV/A. Two TMR sensors are employed. In [70], an AMR sensor is utilized. Compared with the AMR sensor, the TMR sensor does not need the polarization to be restored to maintain the sensor properties, does not need to compensate the offset due to resistor mismatch, and is free from the problem of recalibration contributed by the temperature variation [70; 71; 73]. At a laboratorial temperature of 25°, the employed TMR has a linear output when the range of the applied magnetic field is within ±80 Gs, therefore the maximum magnitude of the current should be smaller than 420 A. Otherwise, the output of the sensor is not linear and the relationship between the sensor output and the detected magnetic field is hard to find. In the circuit design, differential amplifiers are used. Due to the effectiveness of the amplifiers, the common mode signal is regarded as suppressed. Mu-metal is employed with an initial permeability of 60 mH/m and maximum permeability of 280 mH/m. The thickness of the shielding is 0.2 mm and the length is 10 cm. In this experimental setup, $r_s$ is 36 mm and $h$ (see Figure 4.57) is 2 mm. The whole structure of the shielded sensors is as shown in Figure 4.61a. The distances from the conductor to each sensor are 10.5 and 17.5 mm, respectively. The power supply for the sensor is 3.00 V. Placement uncertainty for the shielded sensor is within 0.1 mm. The outputs of the sensor and the clamp meter were recorded on an oscilloscope.

From the results in Figures 4.59 and 4.60, it can be concluded that the shielding has little effect on the magnetic field inside it. This conclusion is also demonstrated by laboratorial experiments. In the experiments, only $I_2$ is employed, the RMS value of which is 14.7 A. Due to the limitations of the device, position A (see Figure 4.58) at 6.5, 10.5, and 13.5 mm with and without shielding is tested. The results of the magnetic field are compared with the simulated results by FEA methods. All the results are presented in Table 4.12. From the results, the minimum error between the magnetic field without and with shielding is 2.52% by FEA simulations. For the experiments, the error is larger, but the minimum error is 3.25%. From the datasheet of the TMR, when the temperature is 25° C and the supply voltage is 3 V, the sensitivity of the sensor is 3.227 mV/V/Gs.

**Table 4.12** Comparison of estimated magnetic field.

| Methods | | Position (mm) | | |
|---|---|---|---|---|
| | | 5.5 | 10.5 | 15.5 |
| FEA simulations | Without shielding | 0.4437 mT | 0.2813 mT | 0.1882 mT |
| | With shielding | 0.4315 mT | 0.2742 mT | 0.1829 mT |
| | Error | 2.75% | 2.52% | 2.82% |
| Experiment | Without shielding | 0.4623 mT | 0.2856 mT | 0.1908 mT |
| | With shielding | 0.4436 mT | 0.2761 mT | 0.1846 mT |
| | Error | 4.04% | 3.33% | 3.25% |

*Source:* Reprinted with permission from Y. Chen et al., An interference-rejection strategy for measurement of small current under strong interference with magnetic sensor array, *IEEE Sensors Journal*, 182, 72–81, 2018.

Since the sensitivity of the sensor will be influenced by the temperature and the supply voltage, calibration for the sensitivity is done to improve the accuracy of the sensor.

The output of the sensor is voltage, which represents the magnetic field, and calibration for the sensor measurement is performed to determine the relationship between the current and the output voltage of the TMR sensor. In addition, a geomagnetic field of 0.03–0.05 mT will affect the sensor. Nevertheless, since the magnetic lines of the geomagnetic field are uniformly distributed, the geomagnetic field can be eliminated by summing up the sensed results of the two sensors with opposite sensitive directions. The effect of the geomagnetic field can be further dampened by measurement calibration. Experiments are tested when only one conductor carries current while other two phase currents are zero. The relationship can be described by

$$\tilde{I}_i = \frac{S_{ei1} S_{ei2} (V_{i1} - V_{i2})}{S_{ei2} - S_{ei1}} \tag{4.33}$$

where $S_{ei1}$ and $S_{ei1}$ ($i = 1, 2, 3$) are coefficients between the currents and the output voltages of the sensors, respectively.

Then two cases are tested to prove the feasibility of the method for the three-phase current system, and to demonstrate that the method is able to solve the problem that the estimated error for the targeted current is large when the interfering current is much stronger.

Case A: Experiments are carried out to demonstrate the improved strategy applied in the three-phase alternating current system. The estimated results from equation (4.33) for the three-phase currents are presented in Table 4.13. It can be seen that in the worst case scenario the error is 1.61%.

Case B: Situations are tested when one phase current is zero while other two are energized. For instance, when $I_1$ is zero, $I_2$ and $I_3$ are 14.7 and 14.3 A, respectively. For each phase, when they are zero, the estimated results are 0.11, 0.19, and 0.10 A, respectively.

From the investigation of the two cases, the feasibility of this method is demonstrated by the laboratorial experiments.

**Table 4.13** Estimated RMS currents.

| | | Case A | | Case B |
|---|---|---|---|---|
| Phase | Reference value (A) | Estimated value (A) | Error (%) | Estimated value (A) |
| $I_1$ | 14.7 | 14.84 | 0.95 | 0.11 |
| $I_2$ | 14.7 | 14.9 | 1.36 | 0.19 |
| $I_3$ | 14.3 | 14.53 | 1.61 | 0.1 |

*Source:* Reprinted with permission from Y. Chen et al., An interference-rejection strategy for measurement of small current under strong interference with magnetic sensor array, *IEEE Sensors Journal*, 182, 72–81, 2018.

**Table 4.14** Comparison of estimated results by different methods.

| | | Phase | | |
|---|---|---|---|---|
| Methods | | $I_1$ | $I_2$ | $I_3$ |
| FEA simulations | Without shielding | 0.146 A | 0.228 A | 0.146 A |
| | With shielding | 0.038 A | 0.049 A | 0.038 A |
| Experiment | Without shielding | 0.28 A | 0.49 A | 0.29 A |
| | With shielding | 0.11 A | 0.19 A | 0.10 A |

*Source:* Reprinted with permission from Y. Chen et al., An interference-rejection strategy for measurement of small current under strong interference with magnetic sensor array, *IEEE Sensors Journal*, 182, 72–81, 2018.

#### 4.5.3.4 Validation by FEA Simulations

Due to the limited conditions of the laboratorial experiments, the tests of the method are supplemented by FEA simulations. Using FEA simulations, larger interfering currents can be set to demonstrate the feasibility of the method. Mu-metal is employed as the shielding material. For the magnetic shielding, $r_s$ is 36 mm and $h$ is 2 mm. $r_s$ is the distance from the shielding layer to the targeted conductor and $h$ is the distance from the end of the shielding piece to the line connected by the three centers of the conductors. The length of the copper conductor is 1 m with a radius of 5 mm. The length of the shielding is set to 10 cm.

First, a comparison of the estimated results is made using the design outlined in this chapter in FEA simulations and experiments. The parameters are set the same as in the laboratory experiments. In each method, the true value of the targeted current is zero at some moment while the other two phases are not zero. For instance, since $I_1(t)$ is $14.7\sqrt{2}\,\sin(100\pi t - 2\pi/3)$, $I_2(t)$ is $14.7\sqrt{2}\,\sin(100\pi t)$, and $I_3(t)$ is $14.3\sqrt{2}\,\sin(100\pi t + 2\pi/3)$, when $t$ is 0, $I_2$ is zero but the estimated result for $I_2$ is $5.2\pm10^{-2}$ A by the method of [70]. Additionally, in the tests of FEA simulations and experiments, the sensors with and without magnetic shielding are investigated. When the sensors are not shielded, the method is the same as in [71]. The results are shown in Table 4.14. It can be seen that the estimated value with shielding becomes smaller compared with that without shielding, demonstrating the effectiveness of the shielding.

First, simulations for direct current are tested when $I_1$ is zero while $I_2$ and $I_3$ flow in the same direction and have the same magnitude, varying from 0 to 10,000 A.

The simulated magnitude for $I_1$ is calculated by equation (4.34). Then similar simulations for $I_2$ are tested. In equation (4.34), $B_{i1}$ and $B_{i2}$ are simulated by FEA at sensor positions $S_{i1}$ and $S_{i2}$, representing the magnetic fields detected by the magnetic sensors. The results are shown in Figure 4.56. It can be seen that the estimated magnitude of the targeted current with magnetic shielding becomes much smaller compared with the theoretically current estimated by equations (4.23) and (4.24) without magnetic shielding. The maximum drift from the estimated current to the targeted current of $I_1$, which is zero, is 3.29 A. The maximum drift from the estimated current to the targeted current of $I_2$, which is zero, is 2.07 A:

$$\tilde{I}_i = \frac{2\mu r_1 r_2 (B_{i1} - B_{i2})}{\mu(r_2 - r_1)} \tag{4.34}$$

where $i = 1, 2, 3$.

This strategy is not only suitable for the direct current measurement system, but can also be applied in the alternating current measurement system. Tests for alternating current are then performed. Simulations are tested when $I_1(t)$ is $I_m \sin(100\pi t)$ A and $I_2(t)$ and $I_3(t)$ are $I_m \sin(100\pi t + \pi/2)$ A. $I_m$ changes from 1000 to 10,000. $I_1(t)$ is the current to be measured. At time 0 s when $I_1$ is 0 zero, $I_2$ and $I_3$ are at peak values, and the results of the estimated value $\tilde{I}_1$ with the magnetic shielding around $I_2$ and without shielding are depicted in Figure 4.62. It can be seen that the estimated value $\tilde{I}_1$ with magnetic shielding around $I_2$ is closer to its real magnitude than that without shielding. The maximum drift is 7.22 A from the measured current to the targeted current of $I_1$, which is zero. Simulations are then performed when $I_2(t)$ is $I_m \sin(100\pi t)$ A and $I_1(t)$ and $I_3(t)$ are $I_m \sin(100\pi t + \pi/2)$ A. $I_m$ changes from 1000 to 10,000. $I_2(t)$ is the current to be measured. At time 0 s when $I_2$ is zero, $I_1$ and $I_3$ are at peak values, and the results of the estimated value $\tilde{I}_2$ with magnetic shielding around $I_2$ and without shielding are depicted in Figure 4.62. The estimated value $\tilde{I}_2$ with magnetic shielding around $I_2$ is closer to its real magnitude than that without shielding. Compared with the direct current simulations, the drift becomes larger due to the eddy effect.

#### 4.5.3.5 Discussion on the Displacement of Sensor Installation

In this method, the two adjacent sensors should be parallel. However, displacement of the sensor may happen when installing the sensors. According to the manufacturer's technique, the angular error is considered within $\pm2°$. FEA simulations are therefore tested to investigate the effect of the angular displacement, which is shown in Figure 4.63. The parameters are set in accordance with the previous tests in Section 4.5.3.4. Results for the currents, the peak value of which is 10000 A, are presented in Table 4.15, where 1, −1 means that the angular displacement for $S_{i1}$ is 1° and −1° for $S_{i2}$. 0, 0 represents the situation where there is no angular displacement of the sensors. From the results, the estimated results are doubled. Especially for alternating currents, when the angular displacement is 2°, the estimated results are about double those without angular displacement. Therefore, the maximum angular displacement should be limited within 1°.

The simulations above demonstrate the effectiveness of this design to improve the measurement accuracy when the targeted current is much smaller than the interfering current. In addition, this design is has been tested for application in a three-phrase alternating current system where $\tilde{I}_1$ is $10000 \sin(100\pi t - 2\pi/3)$, $\tilde{I}_2$ is $10000 \setminus \sin(100\pi t)$,

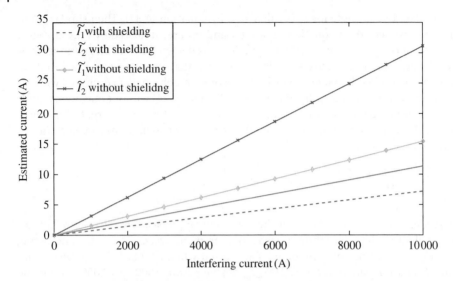

**Figure 4.62** FEA simulations for alternating current with and without magnetic shielding. (*Source:* Reprinted with permission from Y. Chen et al., An interference-rejection strategy for measurement of small current under strong interference with magnetic sensor array, *IEEE Sensors Journal*, 19, 692–700, 2018.)

**Table 4.15** Estimated currents with sensor displacement.

| Angular displacement (°) | DC | | AC | |
|---|---|---|---|---|
| | $\tilde{I}_1$ | $\tilde{I}_2$ | $\tilde{I}_1$ | $\tilde{I}_2$ |
| 0, 0 | 3.29 A | 2.07 A | 7.22 A | 11.36 A |
| 1, 1 | 3.99 A | 2.39 A | 8.35 A | 14.32 A |
| 1, −1 | 3.99 A | 2.36 A | 7.49 A | 14.31 A |
| −1, 1 | 4.03 A | 2.38 A | 7.98 A | 14.31 A |
| −1, −1 | 4.03 A | 2.42 A | 8.16 A | 14.32 A |
| 2, 2 | 4.41 A | 3.41 A | 14.32 A | 20.27 A |
| 2, −2 | 4.99 A | 3.36 A | 14.04 A | 20.32 A |
| −2, 2 | 4.98 A | 3.39 A | 13.98 A | 18.47 A |
| −2, −2 | 5.01 A | 3.34 A | 14.31 A | 20.27 A |

*Source:* Reprinted with permission from Y. Chen et al., An interference-rejection strategy for measurement of small current under strong interference with magnetic sensor array, *IEEE Sensors Journal*, 182, 72–81, 2018.

and $\tilde{I}_3$ is $10000 \sin(100t + 2/3)$. The measured results from equation (4.34) are shown in Figure 4.64. It can be seen that the estimated currents from equation (4.34) deviate from the true currents. Accuracy can be improved by correlating the current and the magnetic field when there is a piece of curved magnetic shielding layer. Simulations are first tested to find the relationship. From the Biot–Savart law, the magnetic field is proportional to

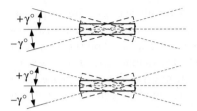

**Figure 4.63** FEA simulations for alternating current with and without magnetic shielding. (*Source:* Reprinted with permission from Y. Chen et al., An interference-rejection strategy for measurement of small current under strong interference with magnetic sensor array, *IEEE Sensors Journal*, 19, 692–700, 2018.)

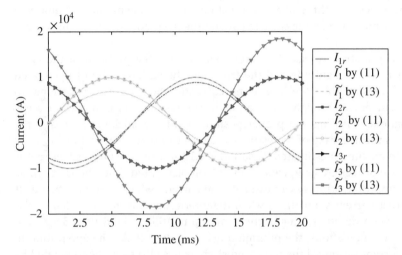

**Figure 4.64** Comparison of the estimated currents by equations (4.34) and (4.36) in FEA simulations. (*Source:* Reprinted with permission from Y. Chen et al., An interference-rejection strategy for measurement of small current under strong interference with magnetic sensor array, *IEEE Sensors Journal*, 19, 692–700, 2018.)

the current. By data fitting, the coefficient between the current $I$ and magnetic field $B$ can be confirmed as

$$B = mI \tag{4.35}$$

where $m$ is a scale coefficient. The current flowing in each conductor can therefore be confirmed as

$$I_i = \frac{m_{i1}m_{i2}(B_{i1} - B_{i2})}{m_{i2} - m_{i1}} \tag{4.36}$$

where $m_{i1}$ and $m_{i2}$ ($i = 1, 2$) are the coefficients between $I_i$ and the detected magnetic fields of sensors $S_{i1}$ and $S_{i2}$, respectively.

The results are shown in Figure 4.64, where $I_{1r}$, $I_{2r}$, and $I_{3r}$ are the reference currents of $I_1$, $I_2$, and $I_3$, respectively. At the peak value of $I_1$, the error between the estimated current from equation (4.36) and the reference current is 0.795%. At the peak value of $I_2$, the error between the estimated current from equation (4.36) and the reference current is 0.206%. At the peak value of $I_3$ the error between the estimated current from equation (4.36) and the reference current is 0.15%. Those results demonstrate the potential of the design for current measurement in cases when the current under measurement is much smaller than the interfering current. In addition, based on the main idea that the error can be reduced by summing up the detected magnetic fields by a sensor with opposite direction, the accuracy of the shielded sensor array in a three-phase alternating current system can be improved when the relationship between the current and the magnetic field with shielding is confirmed.

Those results demonstrate the potential of the design for current measurement in cases when the current under measurement is much smaller than the interfering current. In addition, based on the idea that the error can be reduced by summing up the detected magnetic fields by sensors with opposite directions, the accuracy of the shielded sensor array in a three-phase alternating current system can be improved when the relationship between the the current and the magnetic field with shielding is confirmed.

Inspired by [70], a strategy has been proposed to solve the problem of the error between the estimated magnitude and the true value being large when the targeted current is much smaller than the interfering currents. The error is proportional to the interfering currents. Through the analysis of the results, the improved strategy with some magnetic shielding and proper positioning of the shielding layer and the sensors increases the accuracy of the estimated current in that case. From laboratory experiments, the maximum estimated deviation is 0.19 A when the targeted current is zero but other two phases are energized. From the supplemented simulations, it can be noted that this improved strategy can estimate the current, which is zero, with a drift of 0.21 A in direct current simulations when the interfering currents are 1000 A. For the worst case, when the interfering currents are up to 10 kA, the drift is 3.3 A. For alternating current simulations, the minimum drift is only 0.28 A. The application in a three-phase current system of the improved strategy has also been demonstrated by FEA and laboratory experiments. In FEA simulations, the maximum error between the peak value of the estimated current and the true value is 0.795% for current up to 10 kA. From laboratory experiments to validate application in a three-phase current system, the error between the targeted current and the estimated magnitude is less than 1.61%. This easily installed structure with high accuracy proves it is possible to solve the addressed problem and demonstrates its potential application for current measurement in smart grids.

# Bibliography

1 Q. Huang, S. Jing, J. Li, D. Cai, J. Wu, and W. Zhen, "Smart substation: State of the art and future development," *IEEE Transactions on Power Delivery*, vol. 32, no. 2, pp. 1098–1105, 2017.

2 U. Riechert and W. Holaus, "Ultra high-voltage gas-insulated switchgear – a technology milestone," *International Transactions on Electrical Energy Systems*, vol. 22, no. 1, pp. 60–82, 2012.

3 J. Zhang, "Application of the gas-insulated switchgear in the substation," *Guangdong Electric Power*, no. 6, pp. 41–42, 45, 2001.

4 S. Amin and B. Wollenberg, "Toward a smart grid: power delivery for the 21st century," *IEEE Power and Energy Magazine*, vol. 3, no. 5, pp. 34–41, 2005.

5 R. Bhadra and A. Acharyya, "A proposed DC line current measurement technique based on current induced magnetic field sensing using n-channel enhancement-type MOSFET," in *Michael Faraday IET International Summit*, 2015, pp. 1–4.

6 S. Williamson and J. W. Ralph, "Finite-element analysis for nonlinear magnetic field problems with complex current sources," *IEE Proceedings A – Physical Science, Measurement and Instrumentation, Management and Education – Reviews*, vol. 129, no. 6, pp. 391–395, 2008.

7 T. W. Wieckowski, "Electric-field and magnetic-field sensors," *Electronics Letters*, vol. 29, no. 11, pp. 968–970, 1993.

8 A. Cataliotti, D. D. Cara, A. E. Emanuel, and S. Nuccio, "Current transformers effects on the measurement of harmonic active power in LV and MV networks," *IEEE Transactions on Power Delivery*, vol. 26, no. 1, pp. 360–368, 2011.

9 P. Ripka, P. Kejik, P. Kaspar, and K. Draxler, "Precise DC current sensors," in *Quality Measurements: The Indispensable Bridge between Theory and Reality*. IEEE Instrumentation and Measurement Technology Conference, 1996, pp. 1479–1483 vol.2.

10 F. Primdahl, "The fluxgate magnetometer," *Journal of Physics E: Scientific Instruments*, vol. 12, no. 4, p. 241, 2001.

11 M. F. Snoeij, V. Schaffer, S. Udayashankar, and M. V. Ivanov, "Integrated fluxgate magnetometer for use in isolated current sensing," *IEEE Journal of Solid-State Circuits*, vol. 51, no. 7, pp. 1684–1694, 2016.

12 J. G. Webster, *The Measurements*, Instrumentation and Sensors Handbook. CRC Press, 1999.

13 H. Reeg, M. Schwickert, and Darmstadt , "Sensor studies for DC current transformer application," *Proceedings of IBIC, Monterey, CA, USA*, 2014.

14 P. T. Yan, T. N. C. Fai, and L. R. W. Hong, "Extending the GMR current measurement range with a counteracting magnetic field," *Sensors*, vol. 13, no. 6, p. 8042, 2013.

15 R. S. Popovic, P. M. Drljaca, and C. Schott, "Bridging the gap between AMR, GMR, and Hall magnetic sensors," in *International Conference on Microelectronics*, 2002, pp. 55–58 vol.1.

16 E. Quandt, M. Lohndorf, A. Ludwig, M. Ruhrig, and J. Wecker, "TMR sensor," Patent, 2007, US10405934.

17 W. Y. Lee, C. M. Park, B. R. York, and A. M. Zeltser, "Magnetoresistive sensor with high TMR and low RA," Patent, 2009, US20090257152 A1.

18 A. H. Khawaja and Q. Huang, "Characteristic estimation of high voltage transmission line conductors with simultaneous magnetic field and current measurements," in *Instrumentation and Measurement Technology Conference Proceedings*, 2016, pp. 1–6.

19 A. H. Khawaja, Q. Huang, and L. Lian, "Experimental study of tunnel and anisotropic magnetoresistive sensor for power system magnetic field measurement applications," in *IEEE International Conference on Smart Instrumentation, Measurement and Applications*, 2016.

20 A. H. Khawaja and Q. Huang, "Estimating sag and wind-induced motion of overhead power lines with current and magnetic-flux density measurements," *IEEE Transactions on Instrumentation & Measurement*, vol. 66, no. 5, pp. 897–909, 2017.

21 A. H. Khawaja, Q. Huang, J. Li, and Z. Zhang, "Estimation of current and sag in overhead power transmission lines with optimized magnetic field sensor array placement," *IEEE Transactions on Magnetics*, vol. 53, no. 5, pp. 1–10, 2017.

22 S. Liu, Q. Huang, Y. Li, and W. Zhen, "Experimental research on hysteresis effects in GMR sensors for analog measurement applications," *Sensors and Actuators A: Physical*, vol. 182, pp. 72–81, 2012.

23 E. D. Torre, *Magnetic Hysteresis*. Hoboken, NJ: Wiley-IEEE Press, 2000.

24 J. M. Anderson and A. Pohm, "Ultra-low hysteresis and self-biasing in GMR sandwich sensor elements," *IEEE Transactions on Magnetics*, vol. 37, no. 4, pp. 1989– 1991, 2001.

25 S. Kasap, *Principles of Electronic Materials and Devices*. Beijing: Tsinghua University Press, 2007.

26 J. P. Sebasti, J. A. Lluch, and J. R. L. Vizcaino, "Signal conditioning for GMR magnetic sensors: Applied to traffic speed monitoring GMR sensors," *Sensors & Actuators A: Physical*, vol. 137, no. 2, pp. 230–235, 2007.

27 E. D. Torre, *Magnetic Hysteresis*. Hoboken, NJ: Wiley-IEEE Press, 1999.

28 D. Wang, J. Anderson, and J. M. Daughton, "Thermally stable, low saturation field, low hysteresis, high GMR CoFe/Cu multilayers," *IEEE Transactions on Magnetics*, vol. 33, no. 5, pp. 3520–3522, 1997.

29 P. Ripka, *Improving the Accuracy of Magnetic Sensors*. Springer Link, 2008.

30 C. Reig, M.-D. Cubells-Beltran, and D. R. Muoz, "Magnetic field sensors based on giant magnetoresistance (GMR) technology," *Applications in Electrical Current Sensing*, vol. 9, no. 10, pp. 7919–7942, 2009.

31 M. Sablik, "A phenomenological model for giant magnetoresistance hysteresis in magnetic discontinuous multilayer films," *IEEE Transactions on Magnetics*, vol. 33, no. 3, pp. 2375–2385, 1997.

32 I. Jedlicska, R. Weiss, and R. Weigel, "Linearizing the output characteristic of GMR current sensors through hysteresis modeling," *IEEE Transactions on Industrial Electronics*, vol. 57, no. 5, pp. 1728–1734, 2010.

33 J. Sánchez, D. Ramrez, J. Amaral, S. Cardoso, and P. P. Freitas, "Electrical ammeter based on spin-valve sensor," *Review of Scientific Instruments*, vol. 83, no. 10, p. 105113, 2012. [Online]. Available: http://aip.scitation.org/doi/abs/10._1063/1.4759020

34 V. Skendzic and B. Hughes, "Using Rogowski coils inside protective relays," in *66th Annual Conference for Protective Relay Engineers*, 2013, Conference Proceedings, pp. 1–10.

35 L. A. Kojovic, "Practical aspects of Rogowski coil applications to relaying," Power System Relaying Committee, Report, 2010.

36 P. M. Drljača, F. Vincent, P.-A. Besse, and R. S. Popović, "Design of planar magnetic concentrators for high sensitivity Hall devices," *Sensors & Actuators A: Physical*, vol. 97-98, no. 01, pp. 10–14, 2002.

37 X. Sun, P. T. Lai, and P. W. T. Pong, "A novel bar-shaped magnetic shielding for magnetoresistive sensors in current measurement on printed circuit boards," *IEEE Transactions on Magnetics*, vol. 50, no. 1, pp. 1–4, 2014.

38 K. Zhu and P. W. T. Pong, "Curved trapezoidal magnetic flux concentrator design for improving sensitivity of magnetic sensor in multi-conductor current measurement," in *International Symposium on Next-Generation Electronics*, 2016, pp. 1–2.

39 K. Chang, *Electromagnetic Shielding*. New York: Springer, 1990.

40  S. Ren, S. Guo, X. Liu, and Q. Liu, "Shielding effectiveness of double-layer magnetic shield of current comparator under radial disturbing magnetic field," *IEEE Transactions on Magnetics*, vol. 52, no. 10, pp. 1–7, 2016.

41  K. Wassef, V. V. Varadan, and V. K. Varadan, "Magnetic field shielding concepts for power transmission lines," *IEEE Transactions on Magnetics*, vol. 34, no. 3, pp. 649–654, 1998.

42  T. Rikitake, *Magnetic and Electromagnetic Shielding*. Tokyo: Springer, 1987.

43  C. Wiggins, D. E. Thomas, F. Nickel, and T. S. S. Wright, "Transient electromagnetic interference in substations," *IEEE Transactions on Power Delivery*, vol. 9, no. 4, pp. 1869–1884, 1994.

44  F. Li, W. Qiao, H. Sun, H. Wan, J. Wang, Y. Xia, Z. Xu, and P. Zhang, "Smart transmission grid: Vision and framework," *IEEE Transactions on Smart Grid*, vol. 1, no. 2, pp. 168–177, 2010.

45  M. Camp, H. Garbe, and F. Sabath, "Coupling of transient ultra wide band electromagnetic fields to complex electronic systems," in *International Symposium on Electromagnetic Compatibility*, vol. 2, 2005, pp. 483–488.

46  Q. Yu and R. J. Johnson, "Smart grid communications equipment: EMI, safety, and environmental compliance testing considerations," *Bell Labs Technical Journal*, vol. 16, no. 3, pp. 109–131, 2011.

47  L. Grcev and F. Menter, "Transient electromagnetic fields near large earthing systems," *IEEE Transactions on Magnetics*, vol. 32, no. 3, pp. 1525–1528, 1996.

48  D. E. Thomas, C. Wiggins, F. Nickel, C. Ko, and S. Wright, "Prediction of electromagnetic field and current transients in power transmission and distribution systems," *IEEE Transactions on Power Delivery*, vol. 4, no. 1, pp. 744–755, 1989.

49  J. L. Guttman, J. Niple, R. Kavet, and G. Johnson, "Measurement instrumentation for transient magnetic fields and currents," in *IEEE International Symposium on Electromagnetic Compatibility*, vol. 1, 2001, pp. 419–424.

50  Q. Huang, X. Wang, W. Zhen, and P. W. T. Pong, "Point measurement of transient magnetic interference in substations with broadband magneto-resistive sensors," *IEEE Transactions on Magnetics*, vol. 50, no. 7, pp. 1–5, 2014.

51  D. Nitsch, M. Camp, F. Sabath, J. L. ter Haseborg, and H. Garbe, "Susceptibility of some electronic equipment to HPEM threats," *IEEE Transactions on Electromagnetic Compatibility*, vol. 46, no. 3, pp. 380–389, 2004.

52  J. Holtz, "On the spatial propagation of transient magnetic fields in AC machines," in *Conference Record of the 1995 IEEE Industry Applications Conference*, vol. 1, pp. 90–97.

53  A. B. E. Dallago, P. Malcovati, M. Marchesi, and G. Venchi, "A fluxgate magnetic sensor: From PCB to micro-integrated technology," *IEEE Transactions on Instrumentation and Measurement*, vol. 56, no. 1, pp. 25–31, 2007.

54  A. Phadke and J. Thorp, *Synchronized Phasor Measurements and their Applications*. New York: Spinger, 2008.

55  IEC, *IEC 61000-4-9 Electromagnetic Compatibility (EMC) – Part 4–9: Testing and Measurement Techniques – Pulse Magnetic Field Immunity Test*. Geneva: International Electrotechnical Commission.

56  *PSCAD: User guide*. Manitoba-HVDC Research Center, 2010.

57  J. T. Scoville and P. I. Petersen, "A low-cost multiple Hall probe current transducer," *Review of Scientific Instruments*, vol. 62, no. 3, pp. 755–760, 1991.

**58** J. Y. C. Chan, N. C. F. Tse, and L. L. Lai, "A coreless electric current sensor with circular conductor positioning calibration," *IEEE Transactions on Instrumentation & Measurement*, vol. 62, no. 11, pp. 2922–2928, 2013.

**59** R. Bazzocchi and L. D. Rienzo, "Interference rejection algorithm for current measurement using magnetic sensor arrays," *Sensors & Actuators A: Physical*, vol. 85, no. 1, pp. 38–41, 2000.

**60** R. Weiss, R. Makuch, A. Itzke, and R. Weigel, "Crosstalk in circular arrays of magnetic sensors for current measurement," *IEEE Transactions on Industrial Electronics*, vol. 64, no. 6, pp. 4903–4909, 2017.

**61** L. D. Rienzo and R. B. A. Manara, "Circular arrays of magnetic sensors for current measurement," *IEEE Transactions on Instrumentation and Measurement*, vol. 50, no. 5, pp. 1093–1096, 2001.

**62** R. Bazzocchi, A. Manara, and L. D. Rienzo, "Spatial DFT analysis from magnetic sensor circular array data for measuring a DC current flowing in a rectangular busbar," vol. 3, no. 1, pp. 1194–1197, 2000.

**63** G. Antona, L. Rienzo, and R. Toboni, "Processing magnetic sensor array data for AC current measurement in multiconductor systems," *IEEE Transactions on Instrumentation and Measurement*, vol. 50, no. 5, pp. 1289–1295, 2001.

**64** Y. P. Tsai, K. L. Chen, and N. Chen, "Design of a Hall effect current microsensor for power networks," *IEEE Transactions on Smart Grid*, vol. 2, no. 3, pp. 421–427, 2011.

**65** D. K. Cheng, *Field and Wave Electromagnetics*. Beijing: Tsinghua University Press, 2007.

**66** M. Andrews, "Adaptive filtering prediction and control [book reviews]," *IEEE Transactions on Acoustics Speech & Signal Processing*, vol. 33, no. 1, pp. 337–338, 2003.

**67** M. Kolinova, A. Prochazka, and M. Mudrova, "Adaptive FIR filter use for signal noise cancelling," *Neural Networks for Signal Processing VIII Proceedings of the IEEE Signal Processing Society Workshop*, pp. 496–505, 1998.

**68** J. Wang, Y. S. Geng, J. H. Wang, and Z. X. Song, "*Electric current measurement using magnetic sensor array based on Kalman filtering*," Automation of Electric Power Systems, 2005.

**69** S. S. Haykin, *Neural Networks and Learning Machines*. Beijing: China Machine Press, 2009.

**70** A. Bernieri, L. Ferrigno, M. Laracca, and A. Rasile, "An AMR-based three-phase current sensor for smart grid applications," *IEEE Sensors Journal*, vol. 17, no. 23, pp. 7704–7712, Dec. 2017.

**71** R. S. Popovic, P. M. Drljaca, and C. Schott, "Bridging the gap between AMR, GMR, and Hall magnetic sensors," in *23rd International Conference on Microelectronics. Proceedings (Cat. No.02TH8595)*, vol. 1, May 2002, pp. 55–58 vol.1.

**72** S. Celozzi and R. Araneo, *Electromagnetic Shielding*. New York: Springer, 1990.

**73** X. Wang, Q. Huang, Y. Lu, and M. Du, "Development and application of a portable 3-axis transient magnetic field measuring system based on AMR sensor," in *IEEE International Conference on Smart Instrumentation, Measurement and Applications (ICSIMA)*, Nov. 2013, pp. 1–9.

# 5

# Magnetic Field Measurement for Power Distribution Systems

## 5.1 Introduction

The modern electric power system has evolved into a highly interconnected, complex, and interactive network. The growing demand for energy independency, modernization of old infrastructure with new technologies, integration of various renewable energy and storage solutions, more efficient utilization of energy and assets, and self-healing and resilience to attack and natural disasters, together with pressure for sustainable development, has motivated the development of the smart grid in the 21st century [1]. The smart grid concept has become more consolidated and gained more and more attention since 2009 after many countries or economic union announced their plans for smart grids [59]. The smart grid is generally envisioned as the platform for the implementation of the strategic development of power grids and optimized allocations of energy and resources. It is not only a revolution of the electric power industry, but also the catalyst to create or breed new industries.

Power distribution networks, which are undergoing revolutionary change by smart grid initiatives, play an important part in power systems. The power distribution network receives electrical power from the transmission network and delivers electric power to the end customers. With the development of smart grids, higher requirements have been proposed for electrical equipment in modern distribution systems to provide more accurate, real-time information about the parameters in the distribution network to obtain better information, which is essential for better decisions.

Current measurement is one of the fundamental tasks in power system instrumentation, serving functionalities such as metering, control, protection, and monitoring. Traditionally, current measurement in a power system is accomplished using magnetic core current transformers (CTs). However, with the rapid development of modern power systems, CTs, because of their intrinsic disadvantages (e.g. nonlinearity of the CT magnetic core characteristic, do not work for DC measurement, narrow bandwidth, etc.) are inadequate for either quantitative or qualitative evaluation of current [2; 3]. For example, in modern power transmission and distribution networks, high-voltage direct current (HVDC), known to be very energy efficient and economical, is popularly used for transmitting long-distance power. The current measurement in HVDC has to rely on expensive fiber optic sensors. The disadvantages and deficiencies of CTs are further magnified when they are used in modern power quality measurement and monitoring, which are crucial as suitable theoretical approaches and indices in power quality are properly defined [4]. On the one hand, information about both the steady and transient

*Magnetic Field Measurement with Applications to Modern Power Grids*, First Edition.
Qi Huang, Arsalan Habib Khawaja, Yafeng Chen and Jian Li.
© 2020 John Wiley & Sons Ltd. Published 2020 by John Wiley & Sons Ltd.

power quality is getting more attention. In modern power systems, utility companies and end users are concerned not only with steady power quality issues such as voltage fluctuation, flicker, frequency fluctuation, and harmonics, but also with transient power quality issues such as voltage swell, voltage sag, and voltage interruption. As most power electronics devices work at over 20 kHz, current measurement capable of fast response and broadband is required. For example, current sensors are required to capture fast varying (millisecond or second) transient waveforms, which cannot be done by traditional CTs. On the other hand, with the development of smart home systems, users have become more concerned about the safety of the magnetic field in their living environment, as a result of which intelligent devices to measure magnetic field are required by end users.

Electricity is an electromagnetic phenomenon in nature. Its generation, transmission, and utilization all rely on the physics of electromagnetics. Magnetism, an interaction among moving charges, is one of the oldest branches of science under constant active study, with great implications for energy and environment. Voltage and current are the fundamental parameters in an electric power system. It is known that the magnetic field is always associated with current. Therefore, magnetics-related phenomena will play an important role in power systems [1]. Against this background, measurement based on magnetic theory may have an advantage because magnetic field is generally a vector and is distributed in space. Due to the structure of power distribution networks, which consist of an overhead network, electric cables, poles and towers, distribution transformers, isolating switches, and reactive power compensation controllers, the magnetic environment is complex. Devices are therefore being investigated that measure the required quantities without physical contact, i.e. they are non-contact. This characteristic is very important in distribution networks to reduce the cost of measurement devices.

## 5.2 Magnetic Field Measurement Based Non-invasive Detection

In recent years, both industrial and domestic users have come up with higher requirements for monitoring energy use [5]. For industry, for instance, the government needs to make sure that systems of pollution treatment occur under normal running conditions to prevent companies from not disposing of pollutants in order to save these energy costs. In light of this situation, monitoring an industry's pollution treatment system is an effective approach. For home users, having more information about their home energy use is helpful to them in making decisions about how to use energy efficiently. On the other hand, utility companies can gain a more comprehensive understanding of residential loads and make better decisions about how to distribute energy to end users when constructing smart grids. However, novel monitoring devices should be installed without affecting the existing equipment. Against this background, magnetic field measurement based methods demonstrate an ability to monitoring energy use due to their non-invasive characteristics. In terms of the relationship between the current and the magnetic field, methods based on magnetic sensors are one potential approach to realize non-invasive detection.

Magnetic sensors have long been used for applications such as current sensing, encoders, gear tooth sensing, linear and rotary position sensing, rotational speed sensing, and motion sensing. Solid-state magnetic field sensors, which are generally employed for this purpose, have an inherent advantage in size and power consumption compared with search coil, fluxgate, and more complicated low-field sensing techniques, such as superconducting quantum interference detectors (SQUIDs) and spin resonance magnetometers. This holds true even for high-frequency applications. These sensors work on the principle of converting the magnetic field into a voltage or resistance. The sensing can be easily done in an extremely small area, further reducing size and power requirements.

One advanced solid-state magnetic field sensor is the magnetoresistance (MR) sensor. It is based on the MR effect, i.e. the change of the resistivity of a material due to a magnetic field. In recent years, novel types of MR materials with much higher sensitivity to small changes in magnetic fields have been developed through theoretical understanding and experimental study, discovery of new magnetic materials, fabrication of next-generation electronics, and applications in various aspects, giving magnetism a key role in consumer electronics, power grids, energy, and the environment. MR is the ability of a material to change the value of its electrical resistance when an external magnetic field is applied to it. The first generation of application is anisotropic magnetoresistance (AMR) [6]. The AMR phenomenon, in which the resistance of a piece of iron increases when the current is in the same direction as the magnetic force and decreases when the current is at 90° to the magnetic force, was first discovered by William Thomson (Lord Kelvin) in 1851. The AMR effect is used in a wide array of sensors for measurement of the Earth's magnetic field (electronic compass), for electric current measuring (by measuring the magnetic field created around the conductor), for traffic detection, and for linear position and angle sensing. An AMR sensor is made by depositing a very thin film of permalloy. When a magnetic field is applied, the magnetic domains "swing" round and the electrical resistance changes by around 2–3%. Many AMR products are now available in market.

The AMR effect is usually quite small, just a few percent at room temperature, and is a consequence of bulk scattering. Moreover, in thin films it typically decreases with decreasing film thickness as scattering from the surfaces of the film becomes more important. This limits the applications of AMR and has also motivated the development of novel types of MR sensors. Giant magnetoresistance (GMR), tunneling magnetoresistance (TMR), and colossal magnetoresistance (CMR) sensors are a relatively new generation of solid state magnetic sensors. The development of these new types of sensors has been accelerated by the development of spintronics, i.e. a new paradigm of electronics based on the spin degree of freedom of electrons. In the past decade, the field of spintronics has blossomed with the development and application of magnetically engineered thin-film spintronic magnetic field sensors.

Spintronics integrating magnetism and electronics has enormous applications in novel electronic devices whereby the magnetic field applied on a spintronic device interacts with the spin of the electrons determining the electrical current and thus controlling the resistance of the device [7]. It provides a promising route for smaller, faster, and cheaper devices to record and convey information. In addition, it also largely accelerates the advance of magnetometers [8; 9]. Spintronic sensors, including GMR and TMR sensors, are vector sensors for magnetic fields [10].

These new sensors can be made very sensitive, e.g. TMR sensors make use of the quantum nature of electrons. They consist of two magnetic layers separated by an insulating oxide barrier. The tunneling current between the two magnetic layers changes as the angle between the magnetization vectors in the two layers changes in an external applied field. According to the Julliere's model, TMR materials exhibit a dramatic change of tunneling current in magnetic tunnel junctions (MTJs) when relative magnetizations of the two ferromagnetic layers change their alignment [11]. Electrons of one spin state from the first ferromagnetic film are accepted by unfilled states of the same spin in the second ferromagnetic film. If the two ferromagnetic films are magnetized parallel, the minority spins tunnel to the minority states and the majority spins tunnel to the majority states, resulting in a low-resistance state. On the other hand, when the two ferromagnetic films are magnetized antiparallel, the identity of the majority- and minority-spin electrons is inverted. Thus, the majority spins of the first film tunnel to the minority states in the second film and vice versa, resulting in a high-resistance state. The high MR (the latest report is 604% at room temperature [12]) and the extremely low magnetic coupling between the layers result in extremely sensitive sensors. Potentially, TMR sensors can reach a sensitivity of 1 picoTesla [9] and some studies have reported the possibility of achieving 1 femtoTesla with other MR sensors [13].

These sensitive MR sensors made with spintronic materials can be used to measure the magnetic field generated by current in high-voltage transmission lines. The magnetic signal can then be converted into an electrical signal for state estimation and analysis. This measurement technique is more direct than CT measurement because CT measurement relies on the measurement of the secondary current induced by the magnetic field generated by the primary current whereas MR sensors can measure primary current directly from the emanated magnetic field without the need for a secondary induced current. Because of all kinds of factors related to CT measurement accuracy, such as burden, rating factor, saturation of magnetic cores, external electromagnetic field, temperature, and physical configuration, CTs are not the preferred devices for 21st century smart grids. In terms of safety, MR sensor measurement is a non-contact method which gives convenient installation and safe operation. It is also an economical method because it does not require power cut-off when installation or maintenance are carried out. MR sensor technology can also enable current measurement on a distribution level at low cost. In addition, MR sensors have the advantages of high sensitivity, small size (sensor chip size around 3 × 3 mm), good temperature tolerance, low power consumption (around 10 mW), and wide operating frequency from DC to several megahertz. This innovative MR sensor current measurement technique will be very useful in power systems that are rapidly expansion and upgrading.

In summary, magnetic field measurement at the power distribution level can be used for current measurement and magnetic field investigation. Based on these two fundamental functions, various applications can be developed. It should be noted that millions of points need to be monitored for utilities, industrial applications, and household applications at power distribution level.

## 5.3 Magnetic Sensors for HEMSs

With developing countries becoming more and more advanced, the standard of living is rising worldwide and, at the same time, the demand for energy is increasing. Electrical energy, easy to use and move from one location to another, is the most convenient form of energy for most human uses. It can be used to power computers and most appliances, home heating, and even transportation. Electricity makes up 18% of end use energy in industry, households, and businesses worldwide. With the development of civilization, new demands are being made in the production of electricity.

The increasing demand for electricity and the emergence of smart grids have presented new opportunities for home energy management systems (HEMSs) in demand response markets. The common objectives of different demand side management (DSM) and demand–response (DR) strategies in smart grids are reductions in electricity cost and minimization of energy consumption in peak hours. A typical HEMS is shown in Figure 5.1 [14].

Smart grids make the integration of renewable energy systems and distributed generation practicable, and involves residential and commercial users in DSM and DR activities. DSM is the modification of consumer demand for energy through various methods such as financial incentives and behavioral change through education. Usually, the goal of DSM is to encourage consumers to use less energy during peak hours and move the time of energy use to off-peak hours. DSM techniques are used to optimize the energy consumption pattern, to efficiently utilize limited energy resources, and to enhance the overall efficiency of the power system. The term DR is used for programs designed to encourage end users to make short-term reductions in energy demand in response to a price signal from the electricity hourly market or a trigger initiated by the electricity grid operator. It is a change in the power consumption of an electric utility consumer to

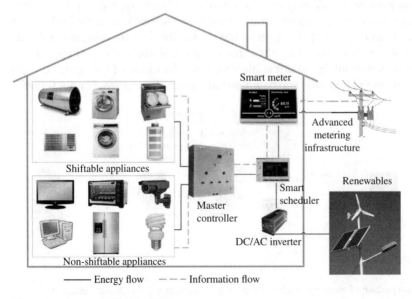

**Figure 5.1** A typical HEMS for DSM.

better match the demand for power with the supply. DR seeks to adjust the demand for power instead of adjusting the supply. However, it is totally impractical to ask consumers to schedule their energy usage by compromising their comfort level. Therefore, an automatic HEMS is required, but a little awareness of consumers is needed to establish the benefits of various scheduling schemes.

The key factors that make smart grids superior to traditional grids are two-way communication, advanced metering infrastructure (AMI), and information management units. These factors introduce intelligence, automation and real-time control to power system. Two-way communication not only keeps end users well informed about varying electricity prices, distribution network maintenance schedules, and events/failures that arise from equipment failure or natural disasters, but also enables the operator to monitor and analyze the real-time data of energy consumption and make real-time decisions about operation activities and standby generators.

Here, a non-invasive HEMS will be introduced that monitors the energy consumption of home equipment/appliances non-intrusively, down to individual appliances.

### 5.3.1  Magnetic Sensors Enable Non-intrusive Monitoring for HEMSs

Traditionally, to obtain information about electricity usage in a home it was necessary to deploy a load monitoring system, which is a common metering system that measures the energy consumption of an appliance by connecting a power meter to it. Each appliance in the household has to have its own power meter so this requires entering the house, thus the system is referred to as intrusive. Such a system provides accurate results, but also imposes high costs and is a complex installation that usually requires wiring and data storage units for the households concerned. Because of their advantages, described in previous section, magnetic sensors have been applied to provide accurate and real-time information for HEMS. Figure 5.2 compares an invasive monitoring system and a non-invasive monitoring one. In a traditional (invasive) monitoring system, sensors have to be installed indoors, while in the non-invasive monitoring scheme, a sensor or sensor array is installed outside of the house. The data are collected and analyzed by the correct algorithm to log the starting up and shutting down of appliances. The installation is non-intrusive and straightforward. This kind of configuration can find application both in domestic and commercial/industrial scenarios. For households,

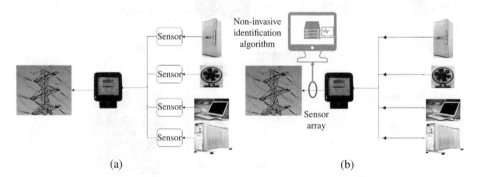

Figure 5.2  A comparison of monitoring schemes (a) invasive monitoring and (b) non-invasive monitoring.

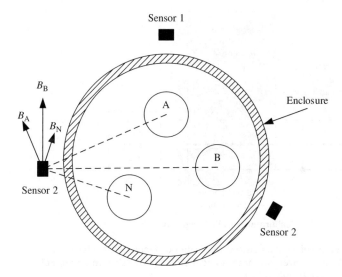

**Figure 5.3** Magnetic sensor array for home energy use monitoring.

cost-effective benefits will be enjoyed, such as ease of monitoring, reduced power consumption, and green credentials, while for commercial/industrial users these devices can replace up to five sub-meters and provide ease of data collection, report generation and upgrades, flexible configuration, lower electricity consumption, lower electricity bills, and a greener environment.

To achieve these goals, research has been carried out into novel techniques. For instance, in [15] an innovative measurement technique with a magnetic sensor array (see Figure 5.3) is proposed to measure the currents flowing in three conductors. The relationship between the sensor output and the relevant current is confirmed by a specially designed calibration scheme. From the results of field tests, this technology can realize long-time monitoring with an energy estimation error below 3%.

Such a system can especially be useful for the end user to identify their energy use pattern and coordinate with the provider to optimize their energy bill by shifting energy use time. In addition, this type of system can easily be extended to large industrial or commercial situations to provide the following advantages:

**Sub-metering**: (a) The power of infrastructure equipment can be monitored, e.g. the amount of power consumed by infrastructure equipment such as air-conditioners, lifts, and lighting can be recorded and stored in servers, (b) power consumption can be allocated to multiple tenants, e.g. allocation of costs for each tenant/department for power consumption made easier by having the actual usage record, and (c) the data can be used for energy audit requirements.

**Early warning system**: (a) An early warning system, initiating pre-emptive maintenance before equipment fails, e.g. setting an alert level for maintenance based on usage of equipment, can be set to generate a trigger when the set level is reached, e.g. in the maintenance of lifts, which may lead to longer asset life and (b) equipment usage comparisons can be made using historical equipment data.

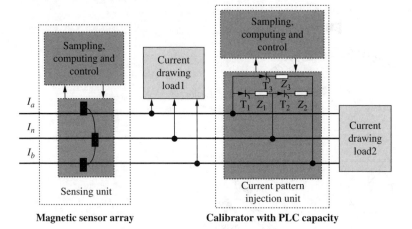

**Figure 5.4** The measurement system of a sensor array with acalibrator. (*Source:* Reprinted with permission from P. Gao, S. Lin, W. Xu, A novel current sensor for home energy use monitoring, *IEEE Transactions on Smart Grid*, 5, 2021–2028, July 2014.)

Two important components constitute the proposed measurement technique: the set of magnetic sensors surrounding the conductors under measurement and the calibration scheme needed to determine the relationship between the current of interest and the sensor output.

The measurement of such currents is a typical inverse problem, therefore a matrix coefficient $[K]$ should be determined. The authors in [15] designed a magnetic field sensor array based calibration system, as shown in Figure 5.4. In this scheme, a calibrator is connected to the conductors at a home outlet downstream of the sensor array to create three distinct momentary current patterns. To send the current information to the sensor array, a power line communication (PLC) is designed.

### 5.3.2 Detection Method for Edge Identification

Non-invasive load identification has two functionalities. For power distribution, the utility company will have more detailed information about the composition of end-user load, hence efficiency can be improved through optimized power distribution and enhanced power transmission management. The utility company can guide users to use their energy at certain times to reduce the peak valley difference and regulation loss. If there is an outage, the cause can be identified faster and more efficiently. For customers, a non-invasive load identification system can provide timely and accurate data, helping users to understand information about load composition and the operation conditions of the assets. Energy bills can be minimized by reducing unnecessary energy use, hence economic and environmental optimization can be achieved.

Non-invasive load identification typically involves edge detection, in which the change in the operation state of a certain appliance will be identified by character extraction, in which the characteristics concerning a certain operation state of an appliance are extracted, and character matching, in which the detected measurements are compared with those stored in a database. Edge detection is the basis for fast and accurate identification of load.

### 5.3.2.1 Introduction

Besides collecting power energy information, another important function of the magnetic sensor based non-invasive method for HEMSs is to analyze the start–stop time and working mode of each user's electrical equipment. Through the acquisition of energy load information, this helps the electric power company to understand the composition of the user load, analyze the usage rules of the equipment, evaluate the working mode of the equipment, and then strengthen the load side management, guiding the user to reasonably arrange the use of the load and improve the efficiency of their electric energy use. Finally, the purpose of adjusting peak-to-valley differences and reducing network loss is achieved [16].

When the working conditions of the user-side electrical equipment change (such as starting and turning off equipment, adjustment of controls on a microwave oven or a heater, etc.), the load of the electrical equipment connected to the user will change. Consequently, the transient load characteristics such as current, active power, and reactive power will change, and these transient load superpositions will cause corresponding changes in the load on the bus, resulting in user load fluctuations in the home bus. Traditional load detection obtains the power consumption information for equipment in the home and then analyzes the energy usage behavior of the device by installing a sensor for each piece of equipment. This method increases the cost and operational complexity for the user [17]. The non-invasive identification method identifies each piece of equipment through a central detector on the user's home bus to measure information such as the bus high-frequency voltage and current. This method eliminates the need to enter the user's home to install specific equipment, which effectively reduces the cost and maximizes the usage. Non-invasive load identification methods include edge detection, feature extraction, feature matching, etc. Edge detection is the most important basis for realizing fast and accurate load identification [18]. Much research [19; 20; 21; 22; 23] has shown that the main factors affecting edge detection accuracy and recognition time are interference in the detection environment and the complex waveform changes of different modal loads when transforming.

To solve these problems, a novel edge detection method based on Gaussian filtering and the cumulative sum (CUSUM) algorithm for industrial detection has been designed. First, due to the ability of the Gaussian filter to effectively de-noise, an adaptive Gaussian filtering algorithm is designed to solve the problem that the modal noise of the electrical equipment is so dense and complicated that it interferes with the detection. At the same time, the problem of the traditional filtering method weakens the abrupt signal is solved; in view of the complex transient load change during the mode conversion of different electrical equipment, this method realizes the sharp capture of information for the positive–negative edge fluctuation by combining with the sensitivity of the improved CUSUM detection algorithm to the fluctuation. Thereby, the accuracy of the working mode detection of the electrical equipment is greatly improved.

In the non-invasive identification process for electrical equipment load, to effectively distinguish the different working modes of each piece of electrical equipment, it is necessary to extract the characteristic quantity for each equipment. The types and differences in device features will directly affect the speed and accuracy of load identification [24].

Since each electrical device has different electrical characteristics, when its working state changes, the active power, reactive power, current harmonics, etc. on the monitoring bus will change accordingly. Based on these changes, the time and type of load

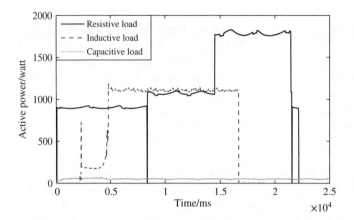

**Figure 5.5** Working state change of different types of load active power. (*Source:* Reprinted with permission from H. Qu et al., Study of fast identification edge detection method for electrical equipment, *Proceedings of the CSEE*, April 2018.)

condition for each device can be identified. There are three characteristic parameters that contain the characteristics of all electrical equipment [25]:

- Active power
  Active power ($P$) is the most important indicator for detecting the working state of the load. It changes with the change in voltage and current values of the detected electrical equipment. In the transient process of the state change of electrical equipment, the instantaneous power value of different types of load varies greatly. For instance, Figure 5.5 shows the characteristic curves of three types of electrical equipment. Active power can be expressed as:

$$P = UI \cos \phi \tag{5.1}$$

where $U$ and $I$ are the instantaneous voltage and current values of the ports of the powered device, respectively. $\phi$ is the phase angle of the voltage and the current.

- Reactive power
  Reactive power ($Q$) is the extra power required to keep power equipment running normally. It is often used as an important indicator to distinguish different types of loads from edge detection. The instantaneous value of each load is shown in Figure 5.6 when the load changes.

$$Q = UI \sin \phi \tag{5.2}$$

- Current harmonic
  The current harmonic is a detailed description of the characteristics of different electrical equipment and the characteristics of the same electrical equipment in different working modes. By Fourier transform, the sampled current data can be decomposed into

$$I_a(t) = I_{a1} \cos(\omega t + \theta_{a1}) + I_{a3} \cos(3\omega t + \theta_{a3}) + \cdots + I_{ak} \cos(k\omega t + \theta_{ak}) \tag{5.3}$$

where $I_t$ is the total current under operating conditions of electricity load, $I_{a1}$ is the fundamental amplitude of the current under the operating conditions of electricity

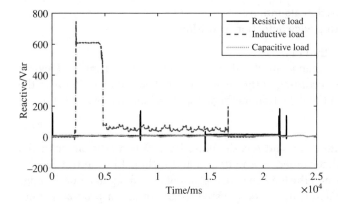

**Figure 5.6** Working state change of te reactive power of different types of load. (*Source:* Reprinted with permission from H. Qu et al., Study of fast identification edge detection method for electrical equipment, *Proceedings of the CSEE*, April 2018.)

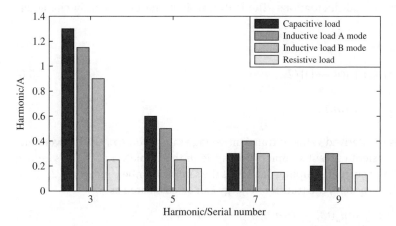

**Figure 5.7** Harmonics of different types of loads. (*Source:* Reprinted with permission from H. Qu et al., Study of fast identification edge detection method for electrical equipment, *Proceedings of the CSEE*, April 2018.)

load, $\omega$ is the angular frequency of the fundamental component in the operating current of the electrical load, $\theta_{a1}$ is the initial phase angle of the fundamental component in the operating current of the electrical load, and $k$ is odd, representing the harmonic number. When $k$ is 1, this represents the fundamental component. $I_{ak}$ is the amplitude of the $k$ th harmonic component in the steady-state operating current of the electrical load and $\theta_{ak}$ is the initial phase angle of the $k$ th harmonic component in the steady-state operating current of the electrical load.

Figure 5.7 shows the current data of Fourier transform decomposition values of multiple electrical equipment. It can be seen that the difference between the first and seventh harmonic values of different devices and different working modes of the same device is large.

### 5.3.2.2 Traditional Edge Detection Algorithms

There are four types of existing edge detection algorithm:

- Threshold detection algorithm (TDA)

  The TDA [26] detects a change in the data information by setting a dynamic threshold, which changes according to the change in the steady state to improve the anti-interference ability of the detection method. The threshold is represented as:

$$H = \bar{q} + \beta \tag{5.4}$$

  where $\bar{q}$ is the steady-state mean and $\beta$ is the noise level. When $H$ is larger than a certain value, it will be determined as a mutation and included in the test results. This method is effective when the load change is prominent and regular, but the edge mutation information, which is more complicated for the transient process, cannot be quickly and effectively identified, and is susceptible to interference and results in error detection.

- Transient energy based algorithm (TEA)

  The core idea of the TEA [27; 28] is to detect the load state with the typical physical characteristics of load fluctuations. The load transient energy is mainly characterized as

$$V(m) = v(m) - v(m-1) \tag{5.5}$$

$$I(m) = [i(m) + i(m-1)]/2 \tag{5.6}$$

$$U_T = \sum_{m=0}^{m} V(m)I(m) \tag{5.7}$$

  where $V(m)$ is the derived value of transient voltage sample $m$, $I(m)$ is the averaging value of the transient current sample, $v(m)$ is the $m$ th sample value of the sampled voltage, $i(m)$ is the current sampling data, $M$ is the sample number, and $m = 1, 2, \ldots, M$. For a three-phase system, the transient energy can be calculated as

$$U_T = \sum [V_a(m)I_a(m) + V_b(m)I_b(m) + V_c(m)I_c(m)] \tag{5.8}$$

  where $V_a$, $V_b$, and $V_c$ are the derived values of the three-phase voltage. The key idea of this algorithm is to accumulate the absolute value of the load, to accelerate the edge detection. The physical characteristics of the load considered are single dimension, leading to poor anti-interference ability and low accuracy.

- Differential operator algorithm (MDA)

  The MDA [29] is a commonly used edge detection method, the core idea of which is to obtain a fitting function based on the sampled data. When the fitting function changes, the first derivative extreme value and the second derivative zero crossing point are used to judge the edge of the detected signal. The method is sensitive to noise, and when there are many noise sources false edge information will be detected, resulting in lower edge detection accuracy.

- Data fitting method

  The data fitting method [30] is based on the least squares method. It segments the load data waveform and describes it with a linear combination of a set of basis functions. In practice, low-order polynomials are often used and the edge position is obtained according to the fitting parameters. The advantage of this algorithm is strong noise immunity. However, it adopts the segmented signal fitting method, leading to a heavy computation burden and a long edge detection time.

### 5.3.2.3 An Improved Load Detection Method

In order to solve the problems of the existing edge detection algorithms, and effectively improve edge detection speed and accuracy, an adaptive Gaussian filter is used to effectively filter the noise on the basis of retaining the information of the sudden change point to ensure the accuracy of the edge detection. The industrial detection CUSUM algorithm can sensitively detect changes in device status and working mode, and improve the performance of edge detection.

#### A. Adaptive Gaussian Filtering Algorithm

When the electrical equipment is working, some disturbance or noise will be generated and the load detection signal acquisition unit will also bring some noise, so the noise level will fluctuate with the change in load power [31]. These interferences or noises can affect the load edge detection and even generate some edgeless information, reducing the accuracy of the identification. Adding a Gaussian filter to the load detection filters out noise in the acquired signal. The basic principle of Gaussian filtering is as follows [32].

Let the one-dimensional Gaussian function be

$$f(t, \sigma) = \frac{1}{\sqrt{2\pi}\sigma} \exp\left(-\frac{t^2}{2\sigma^2}\right) \tag{5.9}$$

Its first derivative is

$$f'(t, \sigma) = \frac{-t}{\sqrt{2\pi}\sigma^3} \exp\left(-\frac{t^2}{2\sigma^2}\right) \tag{5.10}$$

where $f'(t, \sigma)$ is called the Gaussian filter. The results $N(t, \sigma)$, filtered by the Gaussian filter of the original signal of $f_0(t)$, become

$$N(t, \sigma) = f_0(t) * f'(t, \sigma) \tag{5.11}$$

where $*$ is the convolution operator and $\sigma$ is the standard deviation of the Gaussian function. The smoothness of the filtered function changes as the value of $\sigma$ changes.

The standard Gaussian filter is often applied to the signal function whose noise is normally distributed, and all the mutations in the function are smoothed as it weakens the abrupt signal when de-noising. Therefore, the Gaussian filtering algorithm can retain load breakpoint information while filtering out the noise.

Set the load sampling waveform signal to $2T$, where $t$ is the sampling time, $g(t, \sigma)$ is Gaussian filtering weight, and $f_{k+1}(t)$ is the filter output after the $k+1$th iteration. Let the length of the filtering window be $2T$, where $g_k(t + i, \sigma)$ is the weight of each point in the window after the $k$ th iteration and $f_k(t + i, \sigma)$ is the filter output of each point in the window after the $k$ th iteration, therefore

$$f_{k+1}(t) = \frac{\sum_{-T}^{t} f_k(t + i, \sigma) g_k(t + i, \sigma)}{\sum_{-T}^{t} g_k(t + i, \sigma)} \tag{5.12}$$

From (5.12) it can be seen that when $g_k(t + i, \sigma) = 1$,

$$f_{k+1}(t) = \frac{1}{N} \sum_{i=-T}^{T} f_k(t + i) \tag{5.13}$$

That is, in the $k+1$th iteration, the values of all points in the $k$ th iteration are smoothed and filtered, and the mutation signals are not effectively distinguished. If

$$g_k(t + i, \sigma) = \begin{cases} 0 & i \neq 0 \\ 1 & i = 0 \end{cases} \tag{5.14}$$

then when the mutation occurs,

$$f_{k+1}(t) = f_k(t) \tag{5.15}$$

From (5.15) it can be seen that the load mutation can be effectively retained.

Since the time and amplitude of the load change are unknown, the weight cannot be directly set so an adaptive method is introduced. Let

$$g_k(t, \sigma) \exp[-\frac{|f'_k(t)|^2}{2\sigma^2}] \tag{5.16}$$

where $\sigma$ is a constant parameter, $k$ represents the iteration number, and $f_k(t)$ is the first derivative of $f'_k(t)$:

$$f'_k(t) = \frac{1}{2}[f_k(t + 1) - f_k(t - 1)] \tag{5.17}$$

It can be seen from (5.12) to (5.17) that when $\sigma$ is set to a larger value, the de-noising effect is better. Nevertheless, a sudden change point cannot be retained. When $\sigma$ is set to a smaller value, the effect due to the noise after iteration on the experimental results will be strengthened. When $f_k)\cdot(t) \approx \sigma$, the sudden change point can be retained after several convolutional iterations.

### B. CUSUM Algorithm Based Double-sided Detection Method

In the complex transient state when the load state changes during the actual detection process, the load transient characteristics are prone to missed or false detection [33]. A CUSUM algorithm based double-sided detection method is proposed.

As a commonly used monitoring algorithm, the CUSUM algorithm is generally used for industrial quality monitoring, network fault monitoring, and economic and financial monitoring [34]. The idea is to accumulate the small fluctuations generated in the process of monitoring the object and record the cumulative sum to improve the sensitivity of the system and better identify the detection object. The deviation accumulation of the detection sequence and its effect is used to realize the effective detection of any change in the system [35].

Let $q_1, q_2$, and $q_t$ be independent and identically distributed, obeying the normal distribution $N(0, 1)$ and $q_{t+1}, q_{t+2}, \ldots$ be independent and identically distributed, obeying the normal distribution $N(\sigma, 1)$.

$t$ is the detected point, and let $t = a(a < n)$. When $t = \infty$, the likelihood ratio statistic [36] is

$$
\begin{aligned}
W_{n,a} &= \frac{\prod_{i=1}^{a} \theta(q-i) \prod_{i=a+1}^{a} \theta(q-i-\sigma)}{\prod_{i=1}^{n} \theta(q-i)} \\
&= \frac{\prod_{i=a+1}^{a} \theta(q-i-\sigma)}{\prod_{i=a+1}^{n} \theta(q-i)} \\
&= \exp\{\sigma \sum_{i=a+1}^{n} (q_i - \frac{\sigma}{2})\}
\end{aligned}
\tag{5.18}
$$

Since $\prod_{i=n+1}^{n} \theta(q_i) = 1$ and $\sum_{i=n+1}^{n} q_i = 0$, after taking logarithms (5.18) becomes

$$
A_{n,a} = \ln W_{n,a} = \sigma \sum_{i=a+1}^{n} (q_i - \frac{\sigma}{2})
\tag{5.19}
$$

If variants of $q_1, q_2, q_t$ and $q_{t+1}, q_{t+2}$ drift, the likelihood statistics of (5.18) after taking logarithms is

$$
E_n = \max_{1 \le a \le n} E_{n,a} = \max\{\sigma \sum_{i=a+1}^{n} (q_i - \frac{\sigma}{2})\}
\tag{5.20}
$$

If the monitoring object is positive fluctuation, i.e. $\sigma > 0$, it can be derived that

$$
E_n = M_n = \max_{1 \le a \le n} \sum_{i=a+1}^{n} (q_i - \frac{\sigma}{2})
\tag{5.21}
$$

If the $n-1$ detected values do not drift, $M_i \le h$, and $i = 1, 2, \ldots, h$ are threshold values. If at some moment $q_t - \frac{\sigma}{2} \ge h$ or $q_t + q_{t-1} + q_{t-2} - \frac{3\sigma}{2} \ge h$ or $q_t + q_{t-1} + \cdots + q_1 - \frac{n\sigma}{2} \ge h$, then it is determined that the mean shift of the observed sequence occurs.

When $\tilde{q}_i = q_i - \frac{\sigma}{2}$ and $\tilde{q}_0 = 0$, $\tilde{S}_k = \sum_{i=0}^{k} \tilde{x}$ and $\tilde{S}_0 = 0$, it can be derived that

$$
E_n - E_{n-1} = \max\{\tilde{q}_n, -E_{n-1}\}
\tag{5.22}
$$

where $n = 1, 2, \ldots, n$.

If the threshold of $h > 0$, $n$ exists and $E_n > h, (E_i \le h, i = 1, 2, \ldots, n-1)$, then the system is detecting the abnormal state.

The traditional CUSUM algorithm uses a sequential probability ratio test to establish a parametric model of a random sequence, and detects the observed sequence by a probability density function to solve the practical problem [37]. However, the load of the electrical equipment is constantly changing, and there are fluctuations in both positive and negative directions. It is often difficult to obtain a random probability model of each parameter of the load.

A non-parametric based double-sided CUSUM edge detection algorithm is proposed for the case where the mathematical model cannot be given by the load of the electrical equipment. The power changing values of $E_n$ are recorded according to the changes in the equipment or load at the operating state. The problem of low detection accuracy for load transients in the edge detection process is therefore solved.

In the edge detection algorithm,

$$
E_n = \begin{cases} E_n^+ = \max(0, E_{n-1}^+ + q_{n-1} - \overline{q}_{n-1}) \ q_n > q_{n-1} \\ E_n^- = \max(0, E_{n-1}^- - q_{n-1} + \overline{q}_{n-1}) \ q_n < q_{n-1} \end{cases}
\tag{5.23}
$$

where $\bar{q}_{n-1}$ is the averaging load value before the mutation. Supposing that the previous detected sudden change point is $q_t$, then

$$\bar{q}_{n-1} = \frac{\sum_{i=t}^{i=n-1} q_i}{n-t} \tag{5.24}$$

The initial values are set as $E_n = E_n^+ = E_n^- = 0$.

When $E_n = E_a > 0$, point a is recorded. Then when $E_n = E_c = 0$, point c is also recorded. If during the time range of [a, c] all the value of $E_n$ is less than $H$, no transient events occur. The values of a and c are set to zero. If during the time range of [a, c] the value of $E_n = E_b > H$ exists, mutation occurs during this detected sequence. The time span of the transient events is [a, c] and the variants of the characteristics are stored.

The steps of the improved CUSUM algorithm based edge detection method are as follows:

1) When there is a load detection request on the device side, the statistics are initialized, the initial value of $E_n$ is 0, and the load point of the power device is detected in turn. When the state of the power device is steady, $E_n$ is a random variable fluctuating around 0.
2) When the state of the electrical equipment changes, that is, there is an abrupt change in the detection sequence, if the load increases, the average value of the statistics will increase to $\bar{q}_t > \bar{q}_{t-1}$. This positive increasing deviation will continue to accumulate in $E_n^+$, resulting in the increase in $E_n^+$. If the load decreases, downshift or shutdown of the state of the powered device occurs, there will be negative drift of the statistics, and deviation will be accumulated into $E_n^-$. When $E_n$ is larger than $H$, the system will return information to the user. The sudden change point is detected.
3) When the sudden change point is detected, if the value of $E_n^+$ is not zero, then it represents an increase in the load of the electrical equipment, otherwise the load decreases.

The flow chart of the non-invasive identification edge detection method is shown in Figure 5.8.

### C. Simulation

A portable power failure analysis device, ZH-102 from Zhongyuan Huadian, is utilized to sample the electrical signal from devices. The sampled signal is then imported into Matlab as the input signal of $f_{k+1}(t)$, which is transmitted to the module of edge detection. The edge detection module will output triggering results when detecting the change in load events. Taking six types of electrical equipment as the tested objects in the commercial power environment, the working state of the experimental objects is changed in a certain period of time and ten experiments are performed. The improved CUSUM, TDA, TEA and MDA algorithms are used to analyze the signal and the averages of the experimental results is compared.

Figure 5.9 shows the comparison of the detection and recognition time and the occurrence time of the actual load event. A good test result is obtained in which the variance of the ten sets of data is 0.11–3.12 ms. As can be seen from Figure 5.9, the horizontal coordinate represents the number of random combinations of devices at each test, and the ordinate represents the difference between the detected time and the real time of the powered device when its state changes. The difference is the time required for the algorithm to detect the change in load. The detection and recognition times of several

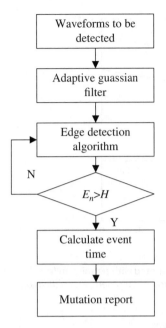

**Figure 5.8** Edge detection flow chart diagram. (*Source:* Reprinted with permission from H. Qu et al., Study of fast identification edge detection method for electrical equipment, *Proceedings of the CSEE*, April 2018.)

traditional detection methods are not much different, but the improved CUSUM has the shortest detection and recognition time.

From the design principle of the algorithm, the TDA, TEA and MDA detection methods cannot effectively extract useful information, eliminate interference noise, and make full use of edge information, resulting in worse edge detection effects. The new algorithm is a comprehensive detection algorithm. As a result of the Gaussian filtering process, the effective information of the edge is improved. By using the idea of accumulation in CUSUM, the detection speed and accuracy of the edge detection are improved.

### D. Experimental Verification

Using the RT-LAB real-time simulation platform, the cross-core AC voltage transmitter (JLT4I), the primary current transformer (JLBH25), and the electrical equipment to be tested, a non-intrusive identification experiment platform, as shown in Figure 5.10, was constructed. Experimental verification of edge detection and device identification was developed. The voltage measurement uses a cross-core AC voltage transmitter module (JLT4I) to convert AC220V to AC0-5/10V. To sample the current, a primary current transformer (JLBH25) is utilized. The sampled voltage and current signals pass through the No. 1 and No. 3 analog signal interfaces of the RT-LAB PCB board, thereby realizing the real-time input of electrical signals. The signal acquisition equipment adopts precision constant current technology and linear temperature compensation technology to isolate the signal into a standard signal output to ensure the safety of the experimental equipment.

A Simulink analog circuit was built in the RT-LAB system and imported into the RT-LAB project. At the same time, splitting the circuit model into multiple cores makes

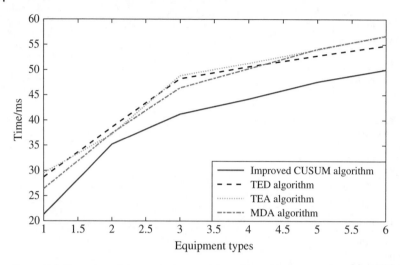

**Figure 5.9** Detection of change and recognition of time. (*Source:* Reprinted with permission from H. Qu et al., Study of fast identification edge detection method for electrical equipment, *Proceedings of the CSEE*, April 2018.)

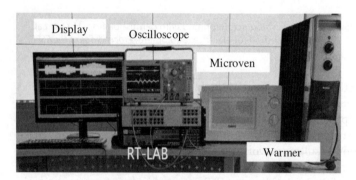

**Figure 5.10** Experimental platform for non-invasive identification of electric equipment. (*Source:* Reprinted with permission from H. Qu et al., Study of fast identification edge detection method for electrical equipment, *Proceedings of the CSEE*, April 2018.)

calculation more efficient. The sub-systems are divided into console subsystems (SC) and main subsystems (SM). Finally, the OpComm module was built and the communication information configured so that the input signal of the analog signal interface is used as the Simulink circuit signal generator.

After the circuit was built, the performance of the adaptive Gaussian filter algorithm and the load detection method based on the improved CUSUM algorithm were tested. Finally, the test results were output to the analog circuit display and compared with the oscilloscope results to verify the effectiveness of the algorithm.

(1) Random noise with mean value 0 and range [−4, 4] units were added to the original signal. The median filtering algorithm, Gaussian filtering algorithm, and adaptive Gaussian filtering algorithm were utilized to filter. The reliability of the method was verified by comparing the retention ratio, the de-noising ratio, and the signal-to-noise ratio of the sudden change point.

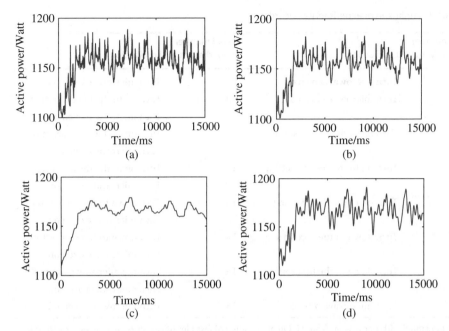

**Figure 5.11** Comparison of different de-noising effects: (a) original waveform, (b) median filtering, (c) Gaussian filtering, and (d) adaptive Gaussian filtering. (*Source:* Reprinted with permission from H. Qu et al., Study of fast identification edge detection method for electrical equipment, *Proceedings of the CSEE*, April 2018.)

The signal-to-noise ratio is

$$R = 10 \lg \frac{\sum_{i=0}^{T} m_2(i)}{\sum_{i=0}^{T} [x(n) - m(n)]^2} \tag{5.25}$$

In (5.25), $T$ is the length of the sampled load, i.e. the discrete number of the sampled periodic signal, $m(n)$ is the original value of the $n$ th point, and $x(n)$ is the de-noising value.

The filtered results are shown in Figure 5.11.

From Table 5.1 and Figure 5.11, it can be seen that the signal-to-noise ratio is 16.5 dB with noise added. Although the sudden change point is well retained with the median filter technique, the noise is well retained and the signal-to-noise ratio decreases. With Gaussian filtering, all the mutations in the function are smoothed, and the transient signal is weakened when de-noising the function. Nevertheless, the adaptive Gaussian filtering algorithm effectively preserves the transient signal and filters out the random noise better.

(2) The detection process of the improved CUSUM algorithm is briefly described by taking the collection data of a user's electricity information as an example. The edge detection sudden change point is shown in Figure 5.12.

During the detection process, if the electric equipment is turned on or the load is increased to make $E_n$ reach the threshold of $H$, it is judged that there is a load state change event at the moment, and the algorithm evolves as shown in Figure 5.13.

**Table 5.1** Comparison of experimental records

| Serial number of load event | Load action | Serial number of load event | Load action |
| --- | --- | --- | --- |
| 1 | Turn microwave oven on | 9 | Turn microwave oven on |
| 2 | Turn microwave oven off | 10 | Turn on to the lowest setting |
|   |   |   | Turn microwave oven on |
| 3 | Turn on to the lowest setting | 11 | Turn on to the second setting |
|   |   |   | Turn microwave oven on |
| 4 | Turn on to the second setting | 12 | Turn on to the third setting |
|   |   |   | Turn microwave oven on |
| 5 | Turn on to the third setting | 13 | Turn down to the second setting |
|   |   |   | Turn microwave oven on |
| 6 | Turn down to the second setting | 14 | Turn down to the first setting |
|   |   |   | Turn microwave oven on |
| 7 | Turn down to the first setting | 15 | Turn the microwave oven off |
|   |   |   | Turn microwave oven on |
| 8 | Turn microwave oven off | 16 | Turn microwave oven off |

*Source:* Reprinted with permission from H. Qu et al., Study of fast identification edge detection method for electrical equipment, *Proceedings of the CSEE*, April 2018.

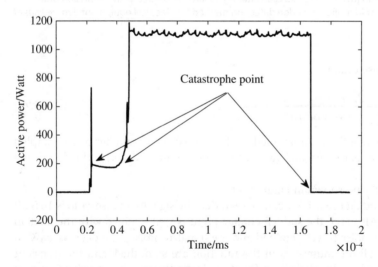

**Figure 5.12** Edge detection of mutation point. (*Source:* Reprinted with permission from H. Qu et al., Study of fast identification edge detection method for electrical equipment, *Proceedings of the CSEE*, April 2018.)

If the load change $E_n$ does not reach the threshold $H$ during the detection process, it is not sufficient to reach the threshold value of the change and the detection result does not change, as shown in Figure 5.14.

(3) The load mode is changed by using the switch on the microwave oven and the different settings of the heater, and the load changing time and the load event number are

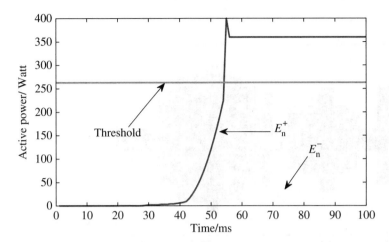

**Figure 5.13** The case of $E_n$ passing over $H$. (*Source:* Reprinted with permission from H. Qu et al., Study of fast identification edge detection method for electrical equipment, *Proceedings of the CSEE*, April 2018.)

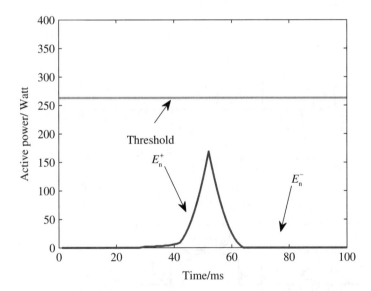

**Figure 5.14** The case where $E_n$ does not reach $H$. (*Source:* Reprinted with permission from H. Qu et al., Study of fast identification edge detection method for electrical equipment, *Proceedings of the CSEE*, April 2018.)

recorded, thereby verifying the non-invasive load identification method based on the improved algorithm. The original waveform is shown in Figure 5.15 and the experimental results are recorded in Table 5.1.

Through the proposed method analysis, the experimental algorithm results are displayed in RT-LAB as shown in Figure 5.16.

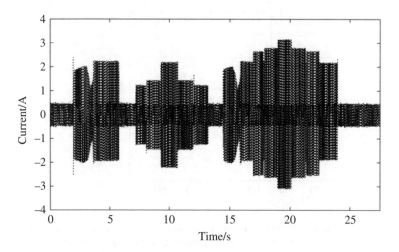

**Figure 5.15** Raw data waveform. (*Source:* Reprinted with permission from H. Qu et al., Study of fast identification edge detection method for electrical equipment, *Proceedings of the CSEE*, April 2018.)

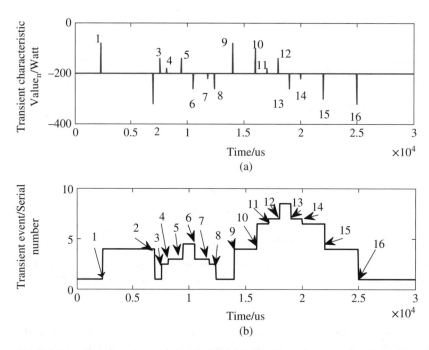

**Figure 5.16** Detected edge results. (*Source:* Reprinted with permission from H. Qu et al., Study of fast identification edge detection method for electrical equipment, *Proceedings of the CSEE*, April 2018.)

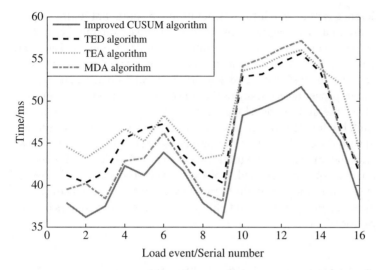

**Figure 5.17** Comparison of recognition time in edge detection. (*Source:* Reprinted with permission from H. Qu et al., Study of fast identification edge detection method for electrical equipment, *Proceedings of the CSEE*, April 2018.)

In Figure 5.16, the $x$ axis represents time and Figure 5.16a shows the amplitude map of the transient eigenvalue over time. The raw current waveform data are processed and the power changing value is calculated by analyzing the change in the running equipment or the load. When a transient event occurs, the greater the magnitude of the deviation from the horizontal axis, the greater the $E_n$. The transient event above the horizontal axis represents the open or upshift, and the transient event below the horizontal axis represents the shutdown or downshift. The coordinate in Figure 5.16b represents the serial number of the transient event, and the situation represented by each transient event corresponds to a unique serial number value. The numbers 1–16 correspond to the transient events represented by the serial numbers in Table 5.1.

Figure 5.17 is a comparison of the experimental performance of the improved CUSUM algorithm with several other algorithms. The horizontal coordinate indicates the transient event sequence number, corresponding to events 1–16 in Table 5.1, and the ordinate represents the consuming time for the algorithm to identify the corresponding event. Experiments show that the algorithm can quickly and accurately detect and identify different power equipment conditions.

### 5.3.3 Discussion

Until now, using magnetic sensors for monitoring the energy usage has still had some problems. One key component of the technique in [15] is how to confirm the relationship between the sensor output and the interested current. The authors utilized a calibrator connected to the targeted conductors at a home outlet. Strictly speaking, as part of the monitoring technique, the calibration scheme is not non-invasive. In addition, magnetic field interference is not taken into consideration. The potential solutions proposed by the authors are to utilize magnetic shielding and increase the number of sensors. To confirm the relationship between the sensor output and the interested current, other solutions are based on the geometric information of the conductors and sensors. However, displacement of the conductors and sensors exists when manufacturing the devices. Algorithms to improve the accuracy of the sensor measurement are proposed with calibration. In [38], the authors developed a control-theoretical calibration, but this solution is not suitable for a fixed sensor array arrangement. In [39], the authors introduced a method with the condition that the ambient sensors have correlated measurements, which is not the same as the magnetic sensor array. In [40], a method based on the least mean square algorithm was used. However, this method still needs to connect a high precision sensor to the conductor to measure the total current as the reference magnitude. This is still an open research field as the use of energy requires further attention and policy reform.

## 5.4 Magnetic Field Measurement Based Fault Location and Identification

### 5.4.1 Introduction

One current solution to magnetic field measurement based fault location and identification is the transformation of the existing power grid into a smart grid with cutting edge information and communication technologies (ICTs) [41]. These advanced ICTs not only enable smart grids to incorporate the distributed generation and renewable energy systems, but also enhance the stability and reliability of power system. The European technology platform [42] defines a smart grid as "an electricity network that can intelligently integrate the actions of all users connected to it – generators, consumers, and those that do both" in order to efficiently deliver sustainable, economic, and secure electricity supplies. Smart grids have different kinds of operational and energy measures, such as smart meters, smart appliances, renewable energy, and electric energy storage resources. The vital aspect of smart grids is the control of power production, transmission, and distribution through advanced ICTs. These ICTs enable smart grids to send control commands within the time limits defined by numerous international standards, e.g. IEEE standard 1547 (the standards defined for the control and management of distributed energy resources) [42]. Moreover, smart grids makes access to the power system operator and end users possible.

Electricity consumers expect the energy supply to be highly reliable, but unfortunately, for various reasons, power quality is not always as high as the customer might want it to be. Aside from the possibility of a complete power failure, it is essential, when assessing power quality, that factors such as voltage and frequency fluctuations, overtone distortion, etc. are correctly assessed, and are within the limits specified by regulations and standards.

When investing in electrical devices such as computers, communications systems, computer-controlled managers, and other equipment that might be sensitive to disruption, it is essential to consider the quality of the power supply available in a given location in order to avoid unnecessary cost at a later date.

### 5.4.2  MR Based Non-invasive Identification Technique

#### 5.4.2.1  Current Fault Identification Techniques

Fault indicators and distribution power automation terminals are two methods of distribution network fault identification. Fault indicators are usually applied in high-voltage and medium-voltage distribution networks. They use the positive mutation of the current and line outage to indicate the path of the fault current. When the fault current develops to a certain level, depending on the active value of the indicator, the display window of the fault indicator will be illuminated. A fault indicator with a communication mode can report the pulse information representing the fault current to the master station. The master station will intelligently locate the fault according to the wiring topology of the distribution network and the reported information. In [43], the authors built an intelligent electronic device (IED) module of a digital fault line selection based on the IEC 61850 standard. They used a field programmable gate array system to handle the captured IED sampling values of the fault line selection and applied the manufacturing message standard (MMS) to report the selection results to the background management system. Such fault indicator systems face some shortcomings:

- Because the ground fault indicator can only indicate the fault locally and the maintenance personnel need to find the fault location, this kind of indicator is only suitable for areas where the wires are short and the topology and the geographic environment are simple, otherwise the maintenance efficiency is extremely low.
- The working power is supplied by coils, which may fail when the line load is small.
- The remote fault location system relies on the master station for communication. If the node information is lost, the complete information for the fault path cannot be created.
- The communication protocol between the master station and the remote fault indicator is based on IEC 61870-5-101/104[44], which requires that every device should be configured according to this standard. However, the plug and play function of the device cannot be realized and devices for different manufactories cannot be interchanged.

To overcome these disadvantages, a novel smart fault identification and location technique can be established for distribution networks based on non-contact MR sensors.

#### 5.4.2.2  Novel MR Based Identification Techniques

Smart grids have proposed a requirement of automation for future distribution networks [45]. Conventionally, intrusive current transformers are employed for fault identification in distribution lines. The sensing units are mainly based on the coils of Faraday's law of electromagnetic induction. These sensing units need to connect the feeders directly by cross-core or clamp-shape forms. Thus, the sensing units operate at high potential. The work [46] by Kansai Electric Co. points out that fault identification can be carried out from the relay protection system in substations. However, the relay protection system sometimes needs a long time to track and detect the fault points. In this context, three-phase sensors can locate faults quickly and accurately. As a result of the advantages of MR sensors, non-invasive methods can be used to identify faults that occur on the overhead distribution lines, which improves safety for maintenance personnel. The sensing device can be installed easily at a distance from the electrical

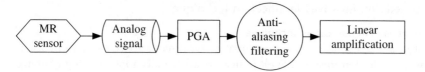

**Figure 5.18** Hardware design of MR sensor based fault location and identification system. PGA, programmable gain amplifier.

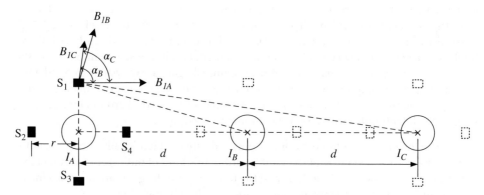

**Figure 5.19** A mathematic model of the sum-averaging algorithm.

parts and dismantled conveniently. It is also free from complex insulation problems. The design of MR based fault identification devices includes of signal conditioning and an analog-digital conversion circuit, a fault identification algorithm, communication interfaces, and power modules. A hardware design of such a device is shown in Figure 5.18.

The principle of MR sensor based fault identification is based on the relationship between the current and the magnetic field. According to this relationship, current information can be reconstructed by algorithms such as the simple sum-averaging algorithm in which the average results of detected magnetic fields are utilized to estimate the current information based on the Biot–Savart law. As shown in Figure 5.19, the current is measured while two other parallel currents interfere with the detection of the sensors. Supposing that $I_A(t)$ is $100\sin(100\pi t - 2\pi/3)$, $I_B(t)$ is $100\sin(100\pi t)$ and $I_C(t)$ is $100\sin(100\pi t + 2\pi/3)$. The distance $r$ from each sensor $(S_1, S_2, S_3, S_4)$ to the conductor is 1.5 cm. The distance $d$ between two conductors is 30 cm. The magnetic fields detected by each sensor are represented as $B_1, B_2, B_3$, and $B_4$. Based on the Biot–Savart law, the current under measurement can be estimated by

$$I = \frac{2\pi r(B_1 + B_2 + B_3 + B_4)}{\mu_0} \tag{5.26}$$

From numerical simulations, the relative errors between the peak value of the reference current and the estimated current are 0.0783% for $I_A$, 0.1603% for $I_B$, and 0.0783% for $I_C$. It can be demonstrated that this method can be used to monitor the current information and provide useful information for identification and the location module.

### 5.4.3 Distributed Sensor Network Based Fault Location and Identification

In remote, inconvenient, and dangerous areas, maintenance of the distribution network is a huge problem and as a result the outage time is long when a fault occurs, which reduces the reliability of the power supply. It is therefore necessary to use novel fault identification and location techniques in less populated areas. The Vattenfall Power Company in Sweden and the Iberdrola Power Company in Spain have proposed a system consisting of intelligent electronic devices, fault passage indicators, and a communication service with a SCADA user interface control center to identify and assess the grounding faults in a medium-voltage compensation network [47].

Compared with developed countries, fault identification and location technique based on sensors in China is still in the early stages. In [48], a fault location system consisting of an earthed fault signal source and fault indicator is proposed. This paper also points out that a fault location system can help to reduce the operation time for tracking and locating the fault points. Reference [49] discusses the lack of research on medium-voltage distribution networks in China and proposes a fault location strategy based on a wireless sensor network. Compared with traditional fault location systems, this novel system has advantages such as higher accuracy, smaller blind area, simpler routing, and better synchronization. In [50], the authors designed a fault-indicating device made up of a module with a three-phase current sensor and a panel indicating module. This device increases the accuracy and reliability of the detection and has the advantage of a self-power-supply module. However, this device is complicated to incorporate when building the network.

Here, a distributed sensor network based on MR sensors is described, which has advantages over the traditional fault indicator limitation that the indicator relies heavily on topology of distribution lines. The arrangement of te sensor network does not rely on the existing distribution topology. Each sensor node can coordinate its action based on layered protocols and layered algorithms, so an independent network can be formed quickly and automatically. In addition, it is convenient to increase or decrease the sensor nodes.

Since each sensor node in a distributed sensor network can describe and publish its location, the geographic information maps for the distributed sensor network and the distribution lines can be associated based on GIS and the visual location of the fault points can be realized on the map. Meanwhile, the fault points of the sensor itself can be located using the self-description information of the sensor nodes for the sake of convenient operation and maintenance.

#### 5.4.3.1 Key Technologies

The complete system architecture of the non-invasive fault identification and location distribution network is shown in Figure 5.20. From previous discussion, there are five key technologies that need to be studied.

- **The detection of zero sequence current by non-invasive magnetic field measurement**. Since whether or not a grounding fault occurs can be judged with the variance of zero sequence current, detecting the zero current is an effective approach to identifying the fault section.
- **Fault phase identification by non-invasive magnetic field measurement**. A single MR sensor fails to identify that the fault phase for the detected magnetic field is the

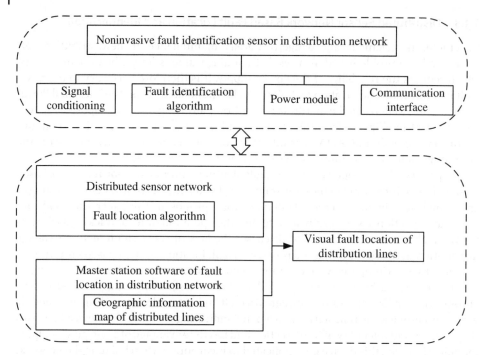

**Figure 5.20** Non-invasive fault identification sensor in a distribution network.

vector sum of three phase currents. Nevertheless, using several sensors can compensate for this situation because the detected results of each sensor at different positions are different. Hence, how to arrange sensor positions needs to be studied.

- **Power supply for the sensors**. Since the fault identification device is intended to be installed in less populated areas, the device needs a reliable power supply. However, since this device is non-invasive, finding a novel method to provide power poses a challenge.
- **Fault identification and location based on the distributed sensor network**. How to arrange the sensor nodes, how to synchronize the node information and which method should be utilized to transmit the information are problems that need to be solved in a distributed sensor network.
- **Visual fault identification technique based on GIS in depopulated areas**. The visual fault identification method is based on the GIS technique. Researchers therefore need to know how to use a geographic information map to describe the distribution lines in the depopulated areas and how to link this information to the fault location system.

The research results will effectively improve the automation level of the distribution network, improve the reliability of the power supply, reduce the blackout time, eliminate the power bottleneck caused by the failure of distribution lines, and provide a stable and reliable power supply for the majority of people in remote areas. On the other hand, the research will also affect the operation and maintenance of the distribution network in remote areas, providing an advanced and reliable technical methods for maintenance personnel and improving work efficiency. The application of the non-invasive fault identification and location technique can effectively reduce the blackout time, thereby reducing the economic losses caused by power outages. It can also greatly improve the efficiency of maintenance personnel in carrying out maintenance work, which reduces operation and maintenance costs and brings considerable indirect economic benefits. Overall, this technique will enable the development of the distribution network to be more intelligent.

## 5.5 Magnetic Sensors for Survey of EMF Exposure

### 5.5.1 Magnetic Fields and Health

As a natural physical energy, a magnetic field spreads all over the Earth. For hundreds of millions of years, this magnetic field has supported the activities of life on Earth. In the same way as air, sunlight, and water, the magnetic field is an indispensable basic environmental element for the survival and development of life on Earth. It can interact with various intrinsic biological magnetic fields formed in the living body to maintain the magnetic balance between the living body and the environment, affecting the various physiological and biochemical life processes of the organism's molecules, cells, organs, and even the whole body.

When the human body is in an electric field environment, its conductivity causes current to flow into the Earth through the skin, and the magnetic field may affect the iron molecules in the blood when it passes through. When the human body is in a strong magnetic field environment, various magnetic substances in the body will be subjected to magnetic attraction. Due to the magnetic evoked action (the object forming the magnetic field magnetizing other objects), the magnetization phenomenon occurs, that is, magnetic objects, such as iron (Fe) in the red blood cells in the body, are attracted or magnetized, thereby affecting the remaining magnetic substances. Obviously, this is harmful to human health. If magnetic substances are present in the blood or cells, magnetic induction will impede their normal activities. Heavy metal will accumulate in the body, making people easy to be magnetically induced to the fatal body blow. However, the influence of electromagnetic waves on the human body is difficult to examine methodically and scientifically. Because various environmental factors are integrated into the human body, it is difficult to conduct long-term follow-up investigations. The effects of electromagnetic waves on the human body have been verified by epidemiological investigations and animal experiments.

In 1979, Professors N. Wertheimer and E. Leeper conducted an epidemiological investigation of the link between high-voltage transmission lines and childhood cancer [51]. The results showed that the incidence of childhood leukemia in children living in proximity to a strong electromagnetic field was more than three times higher than that in other children [52]. The so-called epidemiological investigation is research that aims to find the relationship between the factors causing the cancer and the rate of increase in cancer patients. The most famous epidemiological studies were on smoking and lung cancer, in a bid to determine the relationship between them. The study of the relationship between electromagnetic waves and childhood cancers used the same methods. The resulting report has had a great impact not only in the USA, but also in European countries, and people are now actively studying the effects of electromagnetic waves on the human body. At present, through various animal experiments, experts and scholars in the USA, Europe, Japan, and other countries believe that strong electromagnetic waves are harmful to the human body. In November 1995, the Swedish and Danish joint research organization published the results of the study in the *European Journal of Cancer*, reporting that the incidence of childhood leukemia in environments with a magnetic field above 5 mG was five times higher than that found in normal environments [53].

Various animal experiments have shown that electromagnetic waves have a great impact on health [54]: (1) they cause changes in neurotransmitters, (2) they cause changes in the calcium content of the cells and surfaces of chickens, pigs, and mice, leading to malformed fetuses and causing malignant lymphoid tumors, and (3) they reduce the ability of mice to respond to stimulation, and reduce testicular weight, change brain chemical composition, and reduce the body growth rate. The effects of electromagnetic waves on the human body can be roughly divided into thermal, stimulating, and non-thermal effects.

Whether long-term low-frequency micro-electromagnetic waves affect the health of the body has become the focus of debate on the effects of electromagnetic waves. The harmful effects of strong electromagnetic waves on the human body have been scientifically verified, so countries around the world have defined maximum time limits for exposure to electromagnetic fields.

Children under the age of 15 who are exposed to an average magnetic induction intensity greater than 0.2 $\mu$T are 2.7 times more likely to develop leukemia than those exposed to normal environments. When the magnetic induction is greater than 0.3 $\mu$ T, the rate increases to 3.8 times. Electromagnetic fields can, to a certain extent, cause an increased risk of childhood and adult leukemia, adult malignant brain tumors, amyotrophic lateral sclerosis, abortion, etc. [55; 56].

Currently, authoritative scientific groups believe that the data are not sufficient to support the conclusion that extremely low frequency electric or magnetic fields cause cancer, or lead to abnormalities or learning and behavior problems. However, some guidelines on exposure have been established. For example, CENELEC-EN 50413, Basic standard on measurement and calculation procedures for human exposure to electric, magnetic and electromagnetic fields (0 Hz to 300 GHz), deals with quantities that can be measured or calculated in free space, notably electric and magnetic field strength or power density, and includes the measurement and calculation of quantities inside the body that forms the basis for protection guidelines. Generally accepted guidelines have been established for safe public and occupational exposure to power-frequency electromagnetic forces. According to the International Commission on Non-Ionizing Radiation Protection (ICNIRP: Guidelines for Limiting Exposure to Time-Varying Electric, Magnetic, and Electromagnetic Fields (up to 300 GHz), 2014) the reference levels for general public exposure are:

- for electric field strength, $E < 5$ kV/m
- for magnetic field strength, $H < 80$ A/m
- for magnetic flux density, $B < 100$ $\mu$T.

The levels for safe occupational exposure are:

- for electric field strength, $E < 10$ kV/m
- for magnetic field strength, $H < 400$ A/m
- for magnetic flux density, $B < 500$ $\mu$T.

| Average magnetic fields of some appliances (µT) | Distance from source | | |
|---|---|---|---|
| | 15 cm | 30 cm | 1.2 m |
| Iron | 0.8 | 0.1 | – |
| Dishwasher | 2 | 1 | – |
| Electric stove element | 3 | 0.8 | – |
| Straight-tube fluorescent light | 4 | 0.6 | – |
| Electric mixer | 10 | 1 | – |
| Microwave oven | 20 | 1 | 0.2 |
| Hair dryer | 30 | 0.1 | – |
| Vacuum cleaner | 30 | 6 | 0.1 |
| Electric can-opener | 60 | 15 | 0.2 |
| Photocopier | 90 | 20 | 1 |

Figure 5.21 Magnetic field in a typical household.

Many factors contribute to the magnetic field in our homes: use of electrical appliances (see Figure 5.21), the amount of current circulating in the ground wire of your electrical distribution panel, power consumption in your neighborhood, the distance between your house and the next, and between your house and the power distribution system, etc. The magnetic field is invisible and you cannot touch it, but it exists. A transient magnetic three-dimensional field monitoring device for the home is therefore needed to continuously monitor the living environment of the family.

Such a device could also track the ambient temperature, humidity, environmental noise, and ambient illumination. Devices would be small and portable, so could integrate functions such as detecting and measuring formaldehyde and PM2.5 as multifunctional home environment safety monitoring equipment.

### 5.5.2 Magnetic Environment Monitoring Systems

Three-dimensional magnetic induction sensors (HALL, AMR, GMR, TMR) are used in environmental magnetic field sensing devices, along with a precision amplifier for detecting weak signals, an anti-aliasing filter, and a high-precision analog to digital converter (ADC) to accurately collect sensor signals. An ARM-based STM32 forms the core of the system to process the acquired spatial magnetic field signal and calculate the spatial magnetic field strength below 1 MHz by fast Fourier transform (FFT) conversion processing. According to the GB8702-2014 Electromagnetic Environment Control Limits standard, differences in the limits of magnetic field intensity are set at different frequencies. A liquid crystal display (LCD) screen is used to display the exact value of the determined magnetic field intensity and whether the spatial magnetic field at the current position meets safety standards.

At the same time, the system also integrates functionalities such as monitoring the temperature, humidity, illuminance, formaldehyde, total volatile organic compounds (TVOCs), PM2.5, etc. of the current environment to realize multi-functional home environment monitoring.

The hardware schematic of the monitoring system is shown in Figure 5.22. The environmental information collection module includes temperature, humidity, ambient light, formaldehyde, TVOC, and PM2.5 sensors. It uses a digital interface to directly

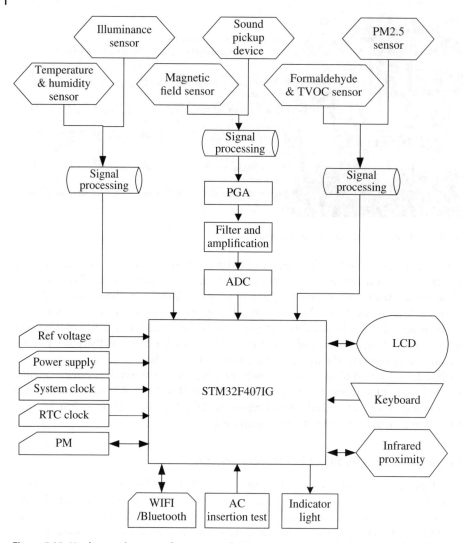

**Figure 5.22** Hardware schematic of a magnetic field environment monitoring terminal model.

connect to the I/O port of the microcontroller (MCU). The noise acquisition circuit outputs an analog signal that is directly connected to the MCU's internal ADC. TVOC and PM2.5 sensors require active measurement to ensure the equipment responds accurately and quickly.

The three-dimensional magnetic field strength acquisition module is composed of a sensor and its peripheral circuits. The three-dimensional magnetic field sensor outputs an analog signal, but since it needs to acquire the signal at high sampling rate, an external high-speed ADC is required as a signal converter, and then the signal is input to the digital signal processing core inside the STM32 for FFT conversion to obtain the amplitude–frequency characteristics of the signal.

The data transmission analysis processing module includes MCU, real-time clock management, power management, Wi-Fi/Bluetooth, and AC intrusion detection.

The module is mainly responsible for data collection, processing, judgment, and transmission. In order to enable users to view device information remotely, it is necessary to use Wi-Fi to transmit data. The user interaction module includes an LCD for displaying information such as the working state, the state of charge, and the state of the environment detection, several buttons, where the user can change the working conditions, and a reflective modulation infrared proximity sensor.

The analog signal front end is amplified using a programmable gain amplifier. By using the amplifier, the system can intelligently adjust the amplification of the analog signal amplitude, realize a wider range of signal intensity acquisition, and effectively improve the acquisition precision of the system.

### 5.5.3 Selection of Sensors

The front end of the three-dimensional magnetic field strength module uses a three-dimensional magnetic field sensor to convert the magnetic field signal into a voltage signal. Commonly used sensors are Hall, AMR, GMR, and TMR. The characteristics of the four sensors are shown in Table 5.2.

According to the standard GB8702-2014 Electromagnetic Environment Control Limits, the public exposure limit of magnetic induction intensity in the frequency range of 1 Hz to 3 MHz is 0.12 $\mu$T (1.2 mOe). To achieve effective detection of this value, high-precision sensors and hardware circuits are required to collect the ambient magnetic field strength, and related software algorithms (such as the Kalman filter) are needed to reduce the error generated during the acquisition process. How to achieve high-precision data acquisition and data restoration is a vital but difficult part of this system design. Based on the comparison of the four sensor resolutions in Table 5.2, AMR and TMR sensors are most suitable for this function.

From Table 5.2 and Figure 5.23, it can be seen that the sensitivity of TMR sensors is much higher than that of AMR sensors. This means that within a certain range, the TMR sensor will have a much higher voltage output change caused by the same amplitude of the magnetic field than the AMR sensor, which may reduce the dependence of the sensor on the high-precision amplifier circuit and the ADC collector, and is beneficial in improving magnetic detection. Figure 5.24 shows that the TMR sensor has better temperature characteristics than the AMR sensor.

As an example, the commonly used AMR sensor HMC105X/HMC104X has a sensitivity of 1 mV/V/Oe and an accuracy of 0.12 $\mu$ T, which does not meet the design requirements. The TMR23XX series TMR sensor has a maximum sensitivity

**Table 5.2** Magnetic Sensor Performance Parameters.

| Item | Current (mA) | Size (mm) | Sensitivity (mV/V/Oe) | Work scope (Oe) | Resolution (mOe) | Temperature (°C) |
|------|------|------|------|------|------|------|
| Hall | 5–20 | 1 × 1 | 0.05 | 1–1000 | 500 | <150 |
| AMR | 1–10 | 1 × 1 | 1 | 0.001–10 | 0.1 | <150 |
| GMR | 1–10 | 2 × 2 | 3 | 0.1–30 | 2 | <150 |
| TMR | 0.001–0.01 | 0.5 × 0.5 | 20 | 0.001–200 | 0.1 | <200 |

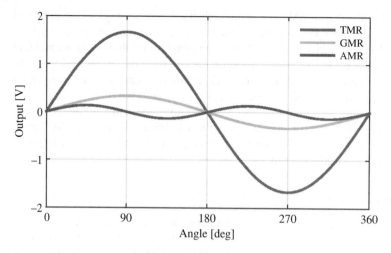

**Figure 5.23** Comparison of TMR, GMR, and AMR outputs.

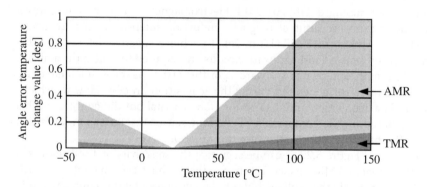

**Figure 5.24** Comparison of the temperature characteristics of TMR and AMR sensors.

of 100 mV/V/Oe, which fully meets the design requirements. The TMR23XX series product-related parameters are shown in Table 5.3.

Considering all aspects of performance, the TMR sensor is used as the linear magnetic detection device sensor in this design. Both TMR2309 and TMR2305 can be used but TMR2305 may be cheaper for mass production.

Due to the extremely high sensitivity of the TMR sensor, circuit shielding needs to be considered in the circuit design. The circuits and components other than the TMR are effectively placed in the shield to reduce the interference of the circuit and the components themselves, and improve the measurement accuracy.

### 5.5.4 System Design

#### 5.5.4.1 Hardware

The hardware of the designed magnetic environment monitoring system is as shown in Figure 5.25. STM32F4 chose as MCU the master chip. TMR2305 is utilized to measure the magnetic environment.

**Table 5.3** TMR23XX series product list.

| Product type | Sensitivity (mV/V/Oe) | Resistance (kOhm) | Dynamic range (Oe) | Noise (nT/rtHz @1Hz) |
|---|---|---|---|---|
| TMR2301 | 1 | 15 | +/−500 | 100 |
| TMR2303 | 3 | 30 | +/−150 | 30 |
| TMR2305 | 25 | 9 | +/−10 | 2 |
| TMR2307 | 8 | 1.5 | +/−30 | 1 |
| TMR2309 | 100 | 15 | +/−8 | 0.15 |

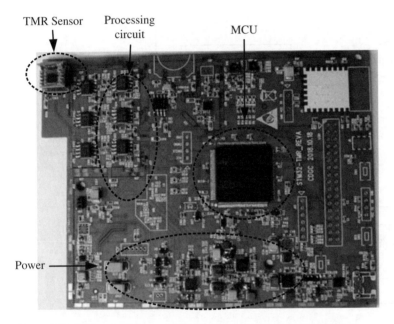

**Figure 5.25** Hardware of the magnetic environment monitoring system.

### 5.5.4.2 Software

The software design of this system has two parts: the main program, which includes data collecting and results display, and the magnetic field monitoring module. The flow chart for the magnetic field monitoring module is shown in Figure 5.26. It performs functions such as data storage, external data interaction, data communication, alarm etc.

### 5.5.4.3 Technical Parameters

The environmental monitoring system has the following technical parameters:

- magnetic induction frequency range: 1 Hz −1 MHz, sensitivity 0.01 $\mu$ T
- temperature measurement: −5 − 60°C, accuracy ±1°C
- humidity measurement: 0−100%. accuracy ±10%
- ambient light measurement: 1−50000 lux, accuracy ±0.1 lux

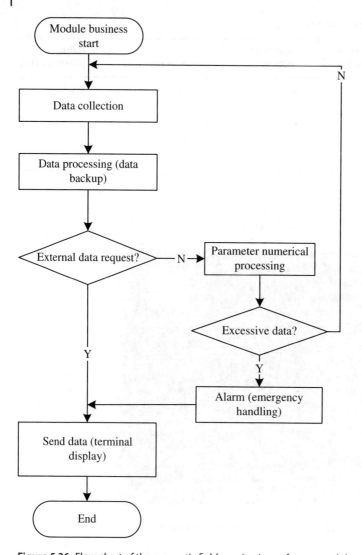

**Figure 5.26** Flow chart of the magnetic field monitoring software module.

- noise measurement: frequency response 20 Hz −8 kHz, range 30−120 dB, accuracy ±1.5 dB
- external power supply: 5V/2A (USB type C)
- operating temperature range: −10−55°C
- operating humidity range: 0−90% RH
- physical size: ≤60 × 60 × 150 mm
- portable requirements: powered by a rechargeable lithium battery pack, battery powered for up to 5 days of continuous operation
- system signal transmission method: wireless (wifi)
- mobile phone app (Android and IOS).

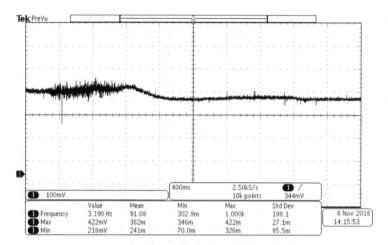

**Figure 5.27** Home transient magnetic field with mobile phone approaching.

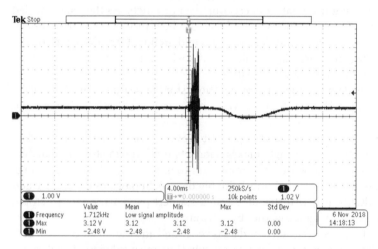

**Figure 5.28** Home transient magnetic field with audio equipment turning on and off.

With the designed system, the tested measurement results are shown on an oscilloscope. Figures 5.27 and 5.28 show the home transient magnetic field with an electromagnetic device nearby. Figure 5.27 shows that when a mobile phone is close to the magnetic sensor, the transient magnetic field will change. In Figure 5.28, it can bee seen that when audio equipment turns on or off, the environment magnetic field fluctuates. This demonstrates that the designed system is useful for monitoring the magnetic field.

## 5.6 Collection of Energy Big Data

### 5.6.1 Concept of Big Data

As the world becomes increasingly digital, new approaches to aggregating and analyzing data will bring huge benefits to fields as diverse as health care, astrophysics, genetics, business, and public policy. The development of data science and data intensive applications has lead to an expanding new field, big data [57]. Big data is a buzzword, or catch-phrase, used to describe a collection of data sets (both structured and unstructured) so large and complex that it becomes difficult to process using hands-on database management tools or traditional data processing applications. In most enterprise scenarios the data are too big or moves too fast or exceeds current processing capacity. While the term may seem to refer to the volume of data, that is not its full meaning. The term big data, especially when used by vendors, may refer to the technology (including tools and processes) that an organization requires to handle the large amounts of data and storage facilities.

In the science field, big data is a result of the development of data science, which is the study of the generalizable extraction of knowledge from data. Data science incorporates varying elements and builds on techniques and theories from many fields, including signal processing, mathematics, probability models, machine learning, computer programming, statistics, data engineering, pattern recognition and learning, visualization, uncertainty modeling, data warehousing, and high performance computing, with the goal of extracting meaning from data and creating data products. As the data is scaled up, big data becomes an important aspect of data science. In this case, big data can be understood as a new generation of technologies and architectures designed to economically extract value from very large volumes of a wide variety of data by enabling high-velocity capture, discovery, and/or analysis.

The challenges of big data include capture, curation, storage, search, sharing, transfer, analysis, and visualization, the 5Vs: Volume, Velocity, Variety, Verification/Veracity, and Value.

**Volume** refers to the vast amounts of data generated every second. The amount of data continues to explode. This increasingly makes data sets too large to store and analyse using traditional database technology. It is important to improve archiving and storage to accommodate the fast-growing volume. With big data technology one can store and use these data sets with the help of distributed systems, where parts of the data are stored in different locations and brought together by software.

**Velocity** refers to the speed at which new data is generated and the speed at which data moves around. As data is generated and moves quickly, the time needed for data designs, performance tuning, and especially maintenance will be compressed. Automated processing and data management should be developed. Big data technology allows us now to analyse the data while it is being generated, without ever putting it into databases.

**Variety** refers to the different types of data for use. In the past we focused on structured data that neatly fits into tables or relational databases. With all the storage capabilities available, the amount of structured and unstructured data and its diverse sources will continue to explode. It is necessary to develop new data management strategies for integrating these diverse structured and unstructured data types quickly into useful information. With big data technology we can harness different types of data (structured and unstructured), including messages, social media conversations, photos, sensor data, video or voice recordings, and bring them together with more traditional, structured data.

**Verification/Veracity** refers to the quality or trustworthiness of the data. With many forms of big data, quality and accuracy are less controllable. It is necessary to develop automated processes and tools that will automatically verify quality and compliance issues. Big data and analytics technology allow this type of data to be used because the volumes often make up for the lack of quality or accuracy.

**Value** refers to the costs and benefits of the data. It is all well and good having access to big data but unless one can obtain value from it, it is useless. It is necessary to have data management support and extract the value of the data, taking advantage of the insights and benefits arising from the increased amount of data. Statistical, hypothetical, and correlation approaches should be developed to accomplish this task.

Big data analytics can be carried out with the software tools commonly used as part of advanced analytics disciplines such as predictive analytics and data mining. However, the unstructured data sources used for big data analytics may not fit into traditional data warehouses. Furthermore, traditional data warehouses may not be able to handle the processing demands posed by big data. As a result, a new class of big data technology has emerged and is being used in many big data analytics environments. The technologies associated with big data analytics include NoSQL databases, Hadoop, and MapReduce. These technologies form the core of an open source software framework that supports the processing of large data sets across clustered systems.

Big data is a phrase that echoes across all corners of business. It is the biggest game-changing opportunity for marketing and sales since the Internet went mainstream almost 20 years ago, particularly because of the unprecedented array of insights into customer needs and behaviors it makes possible. It is hoped that big data will have significant effects on various industries, including smart grids.

### 5.6.2 Energy Big Data

Data is essential for grid management and is a critical element in solving key business problems for utility companies. Big data is everywhere in today's energy industry and is getting bigger. To most people, energy data is about consumption metrics, but data is changing the market in all parts of the value chain. Generation companies use market data to optimize their dispatch, ramping flexible assets up and down in response to real and near real-time supply and demand forecasts. To do this they combine data from their own assets relating to plant operational performance, availability, and technical parameters (which can be variable, e.g. ambient temperature) with information from the market on the availability of other assets and prices. Energy traders can balance and optimize their portfolios internationally, trading in the wholesale markets and on interconnectors. The more sophisticated players have invested in fast and flexible IT systems allowing them to exploit short-lived arbitrages wherever they arise. Transmission system operators (TSOs) rely on demand and generation forecasts in order to ensure the grid is balanced at all times and frequency is stabilized. Historically, the system was simpler and much of the data was based on models and forecasts due to the limited deployment of half-hourly metering. As the electricity system becomes more decentralized and more complex, and as technology develops to increase opportunities for load management, the associated data is multiplying. Smart meters are expected to provide both users and suppliers with granular consumption data, and with the development of the Internet of Things (IoT) and a wide range of networked appliances expected to be available in the future, huge data transfers are also expected.

The explosion of data in the energy markets presents a range of challenges, or opportunities, depending on the perspective around data handling, data security, and standardization. The collections of data or documents so large and complex that they are difficult to process using hands-on database management tools or traditional data processing applications. Data analytics and data mining techniques have been developed to analyze large amounts of data in order to find patterns that may not otherwise be obvious. This can provide significant insights into a range of relevant areas from technical generation and grid operations to organizational and consumer behaviors.

As the information age continues to bring more sophisticated data innovations with each passing year, the energy industry is starting to catch on in a big way. Utilities are gradually introducing intelligent data analytics to improve their knowledge of core aspects of their business and lay the foundation for the next generation of smarter energy generation, distribution, and consumption.

Smart grid technologies, coupled with big data analytics, are going to be transformative to utilities. They will disrupt the status quo of operations, organizational management, customer relations, and regulatory interactions. This will require that utility managers embrace data in new ways. The utility industry will need to go from being reactive to being proactive in using data.

The data sources in a modern power system include:

- transmission level monitoring data, such as phasor measurement unit output or other state variable measurements (e.g. power flow, bus voltage, and transmission line current)
- distribution level monitoring data, such as AMI or smart meter output
- generation level monitoring data, especially those for integration of renewable energy
- environmental data such as weather condition/prediction, temperature, humidity etc.
- customer and power market data.

In the smart grid era, most applications are data intensive, a term that describes applications that are I/O bound or have a need to process large volumes of data. The following novel applications from the smart grid concept are especially data intensive:

- proactive load management
- demand response
- accurate billing
- revenue protection
- advanced outage management.

Utilities' spending on data analytics is increasing every year, transforming this service into one of the industry's biggest growth markets. According to a 2012 report by GTM Research, utility spending on data analytics is set to reach $3.8 billion by 2020, with US utilities expected to spend as much as US$100 per home on grid operations and consumer analytics in that period [58].

As the transmission and distribution power grids are modernized and additional data are collected at various points in the system, utilities are finding that big data is touching nearly all aspects of the industry. Utilities must manage and process increasing amounts of data to create actionable information. For example, distribution previously fell in the domain of customer or field operations, with a focus on trouble-call management. But with new sensors, SCADA, and programs for measuring demand response and time-of-use rates, distribution operations are becoming a more formal set of operational functions that require a sophisticated data management system.

Utilities need to innovate business processes through analytics-driven operational excellence to increase agility and responsiveness, reduce operational costs, and improve asset reliability. Data analytics software will allow utilities to track, visualize, and predict everything from grid operations to electricity consumption, e.g. integrate alternative energies, expand situational awareness across the system, and deepen their relationships with customers, while continuing to do what they have always done, deliver reliable, safe, and affordable energy to everyone. Organizations that want to expand their business are adopting analytics to increase agility and responsiveness, reduce operational costs, and improve asset reliability.

### 5.6.3  Non-invasive Collection of Energy Big Data

The development of industry unavoidably carries a risk of harmful effects on the natural environment. In most countries, new enterprises are required to install proper waste treatment equipment to treat waste water or waste gas before releasing. However, some unscrupulous enterprises may release the waste directly to save money. The general practice in most countries is that the government has a environmental monitoring program that monitors the quality of the environment by sampling at specific locations and then performs environmental impact assessments. Once a environmental pollution event occurs, an analysis will be performed to locate the organization that released the waste.

Here, by extending the magnetic sensor based non-invasive measurement technology described above, a novel monitoring scheme is described. This scheme, unlike traditional environmental monitoring programs, which take time and effort to locate the unscrupulous organization, can directly monitor the operation of the waste treatment equipment without intrusive wiring.

The system is shown in Figure 5.29. The magnetic sensor array is installed on the inlet power line at the power distribution system (outside of the factory). By measuring the

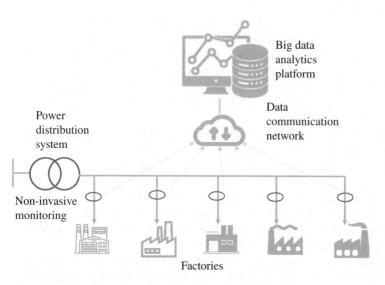

**Figure 5.29** A non-invasive energy big data collection system.

current and associated power of the factory, and with the power equipment identification strategies described above, it can monitor the use of the treatment equipment. All the factory data are integrated in a big data analytics platform for various applications.

To allow government agencies to supervise the operation of the factory, the energy use can be correlated to the operation of the waste treatment equipment, and any unscrupulous organization can then be identified. Since there is a strong correlation between economics and energy use, a by-product of the system is analysis of the economic prosperity of a specific region. It is clear that a higher gross domestic product is connected with a larger electricity supply.

## Bibliography

1 Q. Huang, Y. Song, X. Sun, L. Jiang, and P. Pong, "Magnetics in smart grid," *IEEE Transactions on Magnetics*, vol. 50, no. 7, pp. 1–7, 2014.

2 A. Cataliotti, D. D. Cara, A. Emanuel, and S. Nuccio, "Characterization of current transformers in the presence of harmonic distortion," in *Proceedings of the IEEE Instrumentation and Measurement Technology Conference (IMTC)*, 2008, pp. 2074– 2078.

3 ——, "Frequency response of measurement current transformers," in *Proceedings of the IEEE Instrumentation and Measurement Technology Conference (IMTC)*, 2008, pp. 1254–1258.

4 M. F. McGranaghan and S. Santoso, "Challenges and trends in analyses of electric power quality measurement data," *EURASIP Journal on Advances in Signal Processing*, vol. 2007, pp. 171–176, 2007.

5 T. J. Lui., W. Stirling, and H. O. Marcy, "Get smart," *IEEE Power and Energy Magazine*, vol. 8, no. 3, pp. 66–78, May 2010.

6 V. V. Amelichev, I. A. Gamarts, and S. Polomoshnov, "Anisotropic magnetoresistive sensor of magnetic field and current," in *International Siberian Conference on Control and Communications*, 2009, pp. 222–226.

7 S. A. Wolf, D. D. Awschalom, R. A. Buhrman, J. M. Daughton, S. von Molnr, M. L. Roukes, A. Y. Chtchelkanova, and D. M. Treger, "Spintronics: A spin-based electronics vision for the future," *Science*, vol. 294, no. 5546, pp. 1488–1495, Nov 2001.

8 A. Edelstein, J. Burnette, G. Fischer, S. F. Cheng, W. Egelhoff, P. W. T. Pong, R. McMichael, and E. Nowak, "Advances in magnetometry through miniaturization," *Journal of Vacuum Science Technology A: Vacuum, Surfaces, and Films*, vol. 26, no. 4, pp. 757–762, 2008.

9 W. E. Jr., P. Pong, J. Unguris, R. McMichael, E. Nowak, A. Edelstein, J. Burnette, and G. Fischer, "Critical challenges for picoTesla magnetic-tunnel-junction sensors," *Sensors and Actuators A: Physical*, vol. 155, no. 2, pp. 217–225, 2009.

10 P. P. Freitas, R. Ferreira, S. Cardoso, and F. Cardoso, "Magnetoresistive sensors," *Journal of Physics: Condensed Matter*, vol. 19, no. 16, pp. 165–221, 2007.

11 J. Qiu, G. Han, W. K. Yeo, P. Luo, Z. Guo, and T. Osipowicz, "Structural and magnetoresistive properties of magnetic tunnel junctions with half-metallic $Co_2MnAl$," *Journal of Applied Physics*, vol. 103, no. 7, p. 07A903, 2008.

12 S. Ikeda, J. Hayakawa, Y. Ashizawa, Y. M. Lee, K. Miura, H. Hasegawa, M. Tsunoda, F. Matsukura, and H. Ohno, "Tunnel magnetoresistance of 604% at 300K by suppression

of Ta diffusion in CoFeB/MgO/CoFeB pseudo-spin-valves annealed at high temperature," *Applied Physics Letters*, vol. 93, no. 8, 2008.

13 M. Pannetier, C. Fermon, G. L. Goff, J. Simola, and E. Kerr, "Femtotesla magnetic field measurement with magnetoresistive sensors," *Science*, vol. 304, no. 5677, pp. 1648–1650, 2004.

14 A. Ahmad et al., "An optimized home energy management system with integrated renewable energy and storage resources," *Energies*, vol. 10, no. 4, p. 549, 2017.

15 P. Gao, S. Lin, and W. Xu, "A novel current sensor for home energy use monitoring," *IEEE Transactions on Smart Grid*, vol. 5, no. 4, pp. 2021–2028, July 2014.

16 M. Dong, P. C. M. Meira, W. Xu, and C. Y. Chung, "Non-intrusive signature extraction for major residential loads," *IEEE Transactions on Smart Grid*, vol. 4, no. 3, pp. 1421–1430, 2013.

17 J. A. Mueller and J. W. Kimball, "Accurate energy use estimation for nonintrusive load monitoring in systems of known devices," *IEEE Transactions on Smart Grid*, vol. 9, pp. 2797–2808, 2016.

18 K. Basu, V. Debusschere, S. Bacha, U. Maulik, and S. Bondyopadhyay, "Nonintrusive load monitoring: A temporal multilabel classification approach," *IEEE Transactions on Industrial Informatics*, vol. 11, no. 1, pp. 262–270, 2015.

19 X. Wang, R. Li, D. Zhou, H. Zhou, and W. Hu, "Non-intrusive power load disaggregation method based on decision fusion and its applications," *Power System Protection and Control Press*, vol. 44, pp. 115–121, 2016.

20 B. Liu, "*Non-intrusive power load monitoring and disaggregation technique*," Tianjin University, 2013.

21 Abdullah-Al-Nahid , Y. Kong, and M. N. Hasan, "Performance analysis of Canny's edge detection method for modified threshold algorithms," in *International Conference on Electrical & Electronic Engineering*, 2016, pp. 93–96.

22 R. Li, M. Huang, D. Zhou, H. Zhou, and W. Hu, "Optimized nonintrusive load disaggregation method using particle swarm optimization algorithm," *Power System Protection & Control*, vol. 44, pp. 30–36, 2016.

23 W. Xu, M. Dong, P. Meira, and W. Freitas, "An event window based load monitoring technique for smart meters," in *IEEE PES General Meeting*, July 2014, pp. 1–1.

24 H. H. Chang, K. L. Chen, Y. P. Tsai, and W. J. Lee, "A new measurement method for power signatures of nonintrusive demand monitoring and load identification," *IEEE Transactions on Industry Applications*, vol. 48, no. 2, pp. 764–771, 2012.

25 P. Anghelescu, V. G. Iliescu, C. Mara, and M. B. Gavriloaia, "Automatic thresholding method for edge detection algorithms," in *International Conference on Electronics, Computers and Artificial Intelligence*, 2017, pp. 1–4.

26 J. M. Bruno, B. L. Mark, and Z. Tian, "An edge detection approach to wideband temporal spectrum sensing," in *IEEE Global Communications Conference (GLOBECOM)*, 2016, pp. 1–6.

27 S. Welikala, C. Dinesh, R. I. Godaliyadda, M. P. B. Ekanayake, and J. Ekanayake, "Robust non-intrusive load monitoring (NILM) with unknown loads," in *IEEE International Conference on Information and Automation for Sustainability*, 2017, pp. 1–6.

28 H. Chang, C. Lin, and J. Lee, "Load identification in nonintrusive load monitoring using steady-state and turn-on transient energy algorithms," in *14th International Conference on Computer Supported Cooperative Work in Design*, 2010, pp. 27–32.

**29** E. Holmegaard and M. B. Kjaergaard, "NILM in an industrial setting: A load characterization and algorithm evaluation," in *IEEE International Conference on Smart Computing*, 2016, pp. 1–8.

**30** S. Makonin, "Investigating the switch continuity principle assumed in non-intrusive load monitoring (NILM)," in *2016 IEEE Canadian Conference on Electrical and Computer Engineering (CCECE)*. Vancouver, BC, Canada: IEEE, 2016, pp. 1–4.

**31** P. Lindahl, S. Leeb, J. Donnal, and G. Bredariol, "Noncontact sensors and nonintrusive load monitoring (NILM) aboard the USCGC spencer," in *2016 IEEE AUTOTESTCON, Anaheim, CA*. IEEE, 2016, pp. 1–4.

**32** A. Galletti and G. Giunta, "On the construction of a second-order Gaussian recursive filter," in *International Conference on Signal-Image Technology & Internet-Based Systems*, 2017, pp. 705–712.

**33** J. B. Xu, S. Wang, J. L. Nie, and X. H. Xu, "A kind of robust processing for Gaussian filtering mean line of surface profile," in *International Forum on Strategic Technology*, 2017, pp. 311–313.

**34** W. Jiang, "New edge detection model based on fractional differential and sobel operator," *Computer Engineering & Applications*, vol. 48, no. 4, pp. 182–185, 2012.

**35** J. U. Ping, W. Liu, X. Li, Y. U. Yiping, M. Ding, and C. Qian, "Automatic postdisturbance simulation based method for power system load modeling," *Automation of Electric Power Systems*, vol. 37, pp. 60–64, 2013.

**36** Y. Liu and X. R. Li, "Average run length function of CUSUM test with independent but non-stationary log-likelihood ratios," in *American Control Conference*, 2013, pp. 2803–2808.

**37** N. Amirach, B. Xerri, B. Borloz, and C. Jauffret, "A new approach for event detection and feature extraction for NILM," in *IEEE International Conference on Electronics, Circuits and Systems*, 2015, pp. 287–290.

**38** R. Tan, G. Xing, X. Liu, J. Yao, and Z. Yuan, "Adaptive calibration for fusion-based wireless sensor networks," in *Proceedings IEEE Infocom*, March 2010, pp. 1–9.

**39** M. Takruri, S. Challa, and R. Yunis, "Data fusion techniques for auto calibration in wireless sensor networks," in *12th International Conference on Information Fusion*, July 2009, pp. 132–139.

**40** G. B. Samson, M. A. Levasseur, F. Gagnon, and G. Gagnon, "Auto-calibration of Hall effect sensors for home energy consumption monitoring," *Electronics Letters*, vol. 50, no. 5, pp. 403–405, Feb 2014.

**41** IEEE, "IEEE standard for interconnecting distributed resources with electric power systems," *Application Guide for Standard 1547(TM)*, pp. 1–217, 2009.

**42** M. Zhang, X. Xiao, and Y. Li, "Speed and flux linkage observer for permanent magnet synchronous motor based on EKF," *Proceedings of the CSEE*, no. 36, pp. 36–40, 4 2007.

**43** Q. Jia, L. Shi, J. Tian, H. Dong, and B. Fan, "Fault line selection IED model for digital substation," *Automation of Electric Power Systems*, pp. 43–46, 60, 2011.

**44** G. Clarke and D. Reynders, *Practical modern SCADA Protocols: DNP3, IEC 60870.5 and related systems*. Boston: Newnes, 2004.

**45** T. Balachandran, V. Aravinthan, and T. Thiruvaran, "Local detection of distribution level faults in a distributed sensor monitoring network using HMM," in *Electrical Engineering Conference*, 2016, pp. 25–30.

**46** T. Ito, K. Abe, D. Dodo, T. Koike, and M. Inai, "Evaluation of detecting the breaking of wires on medium-voltage system by three-phase sensors," in *International Symposium on Smart Electric Distribution Systems and Technologies*, 2015, pp. 302–306.

**47** F. Carlsson, N. Etherden, A. K. Johansson, D. Wall, A. Fogelberg, and E. Lidstrm, "Advanced fault location in compensated distribution networks," in *CIRED Workshop*, Helsinki, Finland, 2017, pp. 1–4.

**48** K. Zhou and H. Huang, "Application of fault location system for fault treatment in distribution network," *Shaanxi Electric Power*, vol. 38, pp. 71–73, 2010.

**49** S. Miao, X. Chen, P. Liu, and X. Huang, "A distribution lines fault location scheme based on the wireless sensor network," *Automation of Electric Power Systems*, vol. 82, pp. 61–66, 2008.

**50** X. Fan, X. Zhang, K. Chen, J. Fan, C. Liang, and W. Shi, "A smart fault indicator system and its application research for distribution," *Electrical Measurement and Instrumentation*, vol. 60, pp. 43–46, 2012.

**51** P. K. Verkasalo, E. Pukkala, J. Kaprio, K. V. Heikkil, and M. Koskenvuo, "Magnetic fields of high voltage power lines and risk of cancer in finnish adults: Nationwide cohort study," *British Medical Journal*, vol. 313, no. 7064, pp. 1047–1051, Oct 1996.

**52** G. Draper, T. Vincent, M. E. Kroll, and J. Swanson, "Childhood cancer in relation to distance from high voltage power lines in england and wales: a case-control study," *British Medical Journal*, vol. 330, no. 7503, p. 1290, 2005. [Online]. Available: https://www.bmj.com/content/330/7503/1290

**53** (1997) A news report on the health effects of clectromagnetic energy. [Online]. Available: https://www.emfacts.com/download/Forum_3.pdf

**54** "Potential health effects of exposure to electromagnetic fields (EMF)," Scientific Committees, European Commission, Tech. Rep., 2015.

**55** A. Morgan and K. Martin. (2016) Do electric and magnetic fields cause childhood leukaemia? A review of the scientific evidence. [Online]. Available: https://www.childrenwithcancer.org.uk/wp-content/uploads/2016/12/Taking_forward_results_-_Magnetic_fields_report.pdf

**56** J. D. Brain, R. Kavet, D. L. McCormick, C. Poole, S. L. B., T. J. Smith, P. A. Valberg, R. A. Van Etten, and J. C. Weaver, "Childhood leukemia: electric and magnetic fields as possible risk factors," *Environmental Health Perspectives*, vol. 111, no. 7, pp. 962– 970, 2003.

**57** A. Katal, M. Wazid, and R. Goudar, "Big data: issues, challenges, tools and good practices," in *6th International Conference on Contemporary Computing (IC3)*, 2013, pp. 404–409.

**58** "The soft grid 2013-2020: Big data & utility analytics for smart grid," GTM Research, Tech. Rep., 2012.

**59** Europa EU (2019) Smart Grid European Commission. [Online]. Available: https://ec.europa.eu/research/energy/pdf/smartgrids_en.pdf.

# 6

# Innovative Magnetic Field Measurement for Power Generation Systems

## 6.1 Introduction

The modern power system is a highly complex network. Security/reliability, efficiency/economy, sustainability, controllability/observability, and user interaction are the five fundamental objectives of smart grids, as shown in Figure 6.1. The first two objectives are short-term goals to satisfy the immediate needs of existing power systems on system reliability, security, and cost-effectiveness. The third objective intends to achieve sustainability by implementing renewable energy and carbon reduction as long-term goals of smart grids. The last two objectives are the technical approaches for realizing a smart grid. They stress the observability of the power system and the interaction with users, which rely heavily on sensing and communication technologies [1].

Challenges such as pollution, technical innovation, and energy sources follow with the realization of the five objectives of smart grids. These challenges also guide the developing direction of power generation systems: to be smarter, cleaner, and more efficient. To satisfy the requirement that modern power generation systems should be clean, environmental friendly, and require low investment, new generation techniques have been proposed, such as distributed generation. Nevertheless, with new generation techniques being applied to power grids, new problems that threaten the safety and reliability of power systems appear. To address these problems, key information about power generation system needs to be real-time and accurate, including information related to the magnetic field supports potential monitoring, and measurement approaches in power generation systems.

The magnetics-related phenomena and technologies relevant to smart grids are categorized into three types (see Figure 6.1). Electromagnetic interaction is the fundamental principle for power generation and transmission. Many electrical events can produce electromagnetic disturbances composed of magnetic fields and electric fields. Electromagnetic compatibility (EMC) is very important as it is critical for the proper operation of the essential facilities in the smart grid. With the rapid development of magnetic field sensors, non-contact monitoring by magnetic field measurement can provide an advanced sensing technology for the smart transmission and distribution network. Magnetic fields can store energy and act as the media for transferring energy. Magnetic field based energy storage/conversion may become one of the solutions for energy generation in smart grids.

*Magnetic Field Measurement with Applications to Modern Power Grids*, First Edition.
Qi Huang, Arsalan Habib Khawaja, Yafeng Chen and Jian Li.
© 2020 John Wiley & Sons Ltd. Published 2020 by John Wiley & Sons Ltd.

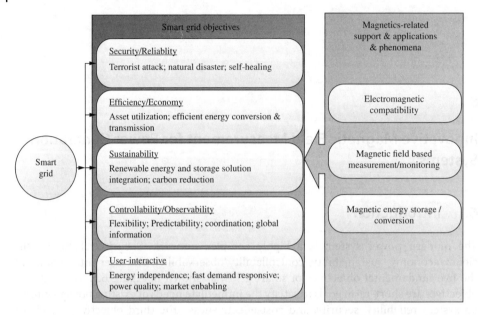

**Figure 6.1** The five fundamental objectives of smart grids. (*Source*: Reprinted with permission from Q. Huang et al., Magnetics in smart grid, *IEEE Transaction on Magnetics*, 50, 1–7, July, 2014.)

Measurement in power generation systems may include following aspects:

- measurement of voltages, including terminal voltage or voltage across some point in the generator
- measurement of currents, including the terminal current or other leakage current in the generator
- measurement of field, including magnetic field and electric field
- measurement of frequency
- power quality monitoring
- measurement of resistance and susceptance
- measurement for fault detection
- health monitoring of the equipment
- other electrical characteristics measurements, such as partial discharge etc.
- other chemical or mechanical measurement, e.g. rotation speed etc.

Magnetic field based measurement has many applications in these areas.

## 6.2 Condition Monitoring of Synchronous Machines

### 6.2.1 Introduction

In the end consumption of energy, electrical energy comprises 20% and is still increasing due to its ease of transmission and use, especially cleanness. In the future, much of the increase in electrical energy will come from the electrification of transport, including transition of cars (internal combustion engines) from petroleum to electric power.

But where does the electricity come from? Electrical energy is obtained from other types of energy stored in nature, such as coil, gas or petroleum, and water. To generate electricity, one needs to have some (mechanical) mechanism to turn a crank that rotates a loop of wire between stationary magnets. The faster this crank is turned, the more current can be generated. Popular methods of turning the crank include water falling onto it (hydro power), directing steam at it (coal- or nuclear-powered steam plant), or the wind turning it (windmill). Electrical energy is all basically generated the same way regardless of the starting form of the energy.

It can be safely stated that the electrical energy of the world is generated through utilization of magnetic fields. Some power generation systems do not generate power through magnetic fields, such as photovoltaic (PV) systems and fuel cells, but they make up a very small share of the total energy consumed. For example, according to the US Energy Information Administration (EIA), in 2016 the USA generated about 4 trillion kilowatt-hours of electricity at utility-scale facilities. About 65% of the electricity generated was from fossil fuels (coal 30%, natural gas 34%, and petroleum 1%). Other sources included nuclear 20%, hydropower 6.5%, biomass 1.5%, geo-thermal 0.4%, wind 5.6%, solar 0.9%, and other gases <1%.

Traditionally, generator maintenance scheduling has been implemented using highly conservative maintenance policies based on manufacturing specifications and engineering expertise on the type of generators. However, as instrumentation and information systems tend to become cheaper and more reliable, recent advances in sensor technology, signal processing, and embedded online diagnosis have provided more unit-specific information on the degradation characteristics of the generators. This helps the transition of time-based maintenance (TBM), also known as periodic maintenance, to condition-based maintenance (CBM). Nowadays, CBM is an important tool for running a plant or factory in an optimal manner, leading to improved system reliability, lower production cost, and less use of resources [2].

### 6.2.2 Speed Monitoring

Generators are one of the most important components of a power system and is it vital to safeguard them. Synchronous machines are commonly used in power systems. These are able to provide power during electrical power system outages and allow the distributed resource owner to control the power factor at the facility [3]. Synchronous machines can be categorized into synchronous generators and synchronous motors. In modern power generation systems, synchronous generators are prominently employed. Nearly 99% of electrical power is generated by synchronous machines, therefore the dynamic performance of synchronous machines has a huge effect on the operation of power systems. In contrast to induction machines, the speed of which varies with load changing, the speed of the motor in synchronous machines is constant. It has to be driven at a speed corresponding to the pole numbers of the machine and the frequency of the electric power system [3]. Thus, in steady-state operation, the rotor speed is related to the frequency of the power grid:

$$n = n_s = 60\frac{f}{p} \tag{6.1}$$

where $n_s$ is the synchronous speed, $f$ is the frequency of the power gird, and $p$ is the number of the pole pairs. When the power system stably operates, the speed of the synchronous machine is a constant number.

To ensure the normal operation of the synchronous machine, methods to monitor the synchronous machine speeds are under research. In [4], the authors proposed a speed and flux linkage observer for a permanent magnet synchronous motor based on the extended Kalman filter. In their work, a mathematic model of a non-salient-pole permanent magnet synchronous motor was built with the stator flux as the state variable in static coordinates:

$$u_\alpha = R_s i_\alpha + L \frac{d(L i_\alpha + \psi_r \omega \cos\theta)}{dt} \tag{6.2}$$

$$u_\beta = R_s i_\beta + L \frac{d(L i_\beta + \psi_r \omega \cos\theta)}{dt} \tag{6.3}$$

where $u_\alpha$, $u_\beta$, $i_\alpha$, and $i_\beta$ are voltages and currents in $\alpha$–$\beta$ coordinates, respectively. $R_s$, $L$, and $\psi_r$ are stator resistance, stator inductance, and rotor permanent magnet flux. $\omega$ and $\theta$ are angular velocity and the position of the rotor. Algorithms based on this model have been applied in their work, but detailed discussion about how to obtain the stator flux is not included.

Currently, the Hall effect component is employed to measure the magnetic field of the permanent magnet synchronous machine. The magnetic flux density can be confirmed by measuring the Hall voltage when the input controlling current is constant.

In recent years, magnetoresistance (MR) sensors have been applied to magnetic field measurement. Compared with Hall effect sensor, an MR sensor has higher sensitivity. Therefore, MR sensors can response to the speed change of the synchronous machine accurately and quickly. The operation status of the synchronous machine can be monitored by measuring the magnetic field of the synchronous machine.

### 6.2.3 Vibration Monitoring

To prevent sudden damage to the generator or unplanned downtime for the power plant, it is important to continuously monitor the condition of the generator. Vibration monitoring has been an important research field since motors started to be used in industy. Vibration is inevitable due to the operation of the rotor inside the synchronous machine. On the one hand, vibration can cause loss during the operation of the machine. Vibration analysis provides fundamental information on the vibration loss to help reduce the cost. On the other hand, an effective monitoring method for the vibration is necessary due to the vibration isolation and reduction techniques applied in the machine design. Vibration monitoring enables the mechanical structure and its behaviour during operation to be inspected. Thereby, vibration analysis is needed to handle the problems contributed by vibration reduction and isolation. Commonly used methods to detect vibrations are the measurement of acceleration, velocity, and displacement.

Sources of noise in electrical machines can be mechanical, aero-dynamical, or electromagnetic. Mechanical noises usually come from the bearings, thus they occur for most machines, but they are not the main source of noise, especially in the case of low rotational speed machines. It is suspected that magnetic unbalance may be the cause of many rotating machine vibrations [5]. Radial magnetic forces are among the main sources of vibration in electrical machines. These forces, produced by flux density waves, act on the stator bore and cause vibrations and noise. Electromagnetic noises are even more important to consider in low-speed machines as the frequency of those noises can be close to the natural frequency of the machine's structure, which would dramatically amplify the vibration and the noise. Currently there is no simple method available to check the integrity of the rotor. Measurement of the rotor shaft current may look attractive to detect the fault in rotor coils, but it is hard to install a current sensor inside the rotor coil and hard to locate the defective pole. The measurement of the magnetic field might be more straightforward. In [6] and [7] vibrations are detected by measuring the magnetic field of the machines.

In [6], the authors propose a solution with three giant magnetoresistance (GMR) sensors [8; 9] based vibration detection by measurement of magnetic field variations. The magnetic field variations are used to analyze the vibration of the machines.

According to their analysis, the magnetic perturbations, which are generated by the magnetic variations translated by vibration, can be converted to resistance variations inside the MR sensors. Therefore, by measuring the variations in sensor resistance, magnetic variations can be confirmed [10; 11; 12]. In this work, two GMR resistors in a half-Wheatstone bridge configuration are used to produce the differential output voltage signal based on the phenomenon that one resistor will decrease while the other will increase when a time-varying magnetic field is presented [13; 14; 15; 16]. Three sensors are used to measure the magnetic field along the $x$, $y$, and $z$ axis independently. In the detecting process, the output signal of the sensors is amplified and transmitted to an acquisition system. With LabView, the vibrations along the three axes can be visualized.

In this design, three GMR sensors form an array, with the direction of each sensor orthogonal to the axes. To reduce the power consumption, a generalized impedance converter supplies a 1 mA current for each sensor [17; 18]. Traditionally, current up to 5 mA is supplied to each sensor for vibration measurement. Therefore, the design in [6] is low cost and has low power consumption. Weak vibrations translated into small perturbations over the Earth's magnetic field are hard for the sensor to detect. To amplify the signal with a gain larger than 60 dB, two amplification stages have been designed. The gain of the first stage is higher than that of the second stage [19]. Since the gain is high, a differential high-pass filter is utilized to filter the DC noise of the signal and narrow the bandwidth [20]. Then a differential instrumentation amplifier (INA 128) amplifies the voltage by 60 dB. Next, the signal is filtered by a second-order active low-pass filter (LPF) and the signal is amplified by 12 dB.

In the work of [7], investigation of a GMR sensor applied to detect the vibration of machines is presented. Different from the work of [6], where the vibration occurred in the drilling machine, the faulty bearing parts of a machine were measured in [7] because faulty bearings generate higher frequencies than those of the rotation frequency [21; 22]. In addition, only one sensor is utilized. The GMR sensor is fixed on a machinery fault simulator. The measurement results and the faulty bearings are recorded in the simulator. The frequency of the machine is initially set to 50 Hz. A large spectral peak occurs at 50 Hz with 23 dB over the noise floor. Then, as the bearing changes to a faulty condition, new peaks occur at 100, 150, and 200 Hz. From the results, it can be seen that GMR sensors demonstrate feasibility in monitoring vibration.

These kinds of scenarios may provide an in-depth application field for magnetic field measurement. In the future, more sensors, magnetic or fiber optic, will be permanently installed on machines and more specific tests will be tried for monitoring purposes and early detection of upcoming generator damage.

### 6.2.4 Crack Detection

Electrical generators are the core of every kind of thermal power plant, wind turbines, hydroelectric dams, and more recently, wave power technologies. In generation systems, accidents due to cracks in the electrical equipment can cause huge damage, therefore actions need to be taken to monitor and prevent this. Most of the crack identification techniques and models in the available literature are based on vibration-based methods. The main idea of vibration-based methods is that the presence of a crack in a rotor induces a change in its mass, damping, and stiffness, and consequently detectable changes appear in the modal properties (natural frequencies, modal damping, and mode shapes). A non-invasive magnetic field can be realized based on the relationship between the magnetic field and the electrical current. Currently, magnetic flux leakage (MFL) and eddy current testing are useful approaches to detect cracks.

In [23], the authors tried to solve the problem of corner cracks in places such as aircraft lap joints or splice plates, which are difficult to detect based on eddy current. Compared with previous work in [24; 25; 26; 27; 28; 29; 30; 31; 32], this work overcame the problems of the ultrasonic approach whereby an ultrasonic system needs the sealant between layers to be inspected. However, although eddy current testing is free from the problem the sealant, low frequencies are required by eddy current testing. When low frequencies are used, a coil usually performs with low sensitivity and poor spatial resolution. In this context, magnetic sensors like GMR or Hall effect sensors show potential for the detection of the magnetic field generated by the eddy current. The authors investigated three probe configurations to demonstrate the feasibility of GMR sensors to detect corner cracks. In the probe, a coil and GMR sensor are combined with a lock-in-amplifier. A lock-in amplifier is used to excite the coil with reference to the eddy current. The amplifier outputs a DC component that is proportional to the signal magnitude.

In one of the configurations, the GMR sensor was placed inside the coil that was used to generate eddy current. The sensitive direction of the GMR sensor and the coil axis were perpendicular to the inspected part. In these experiments, the results from using the GMR sensor were compared with those obtained using the coil. Due to the better sensitivity of the GMR sensor, the curve of the phase variation with the GMR sensor was several times larger than that obtained without use of a GMR sensor.

In another configuration, the GMR was located near the coil. Using only a coil, a small difference in phase could be used to detect a small notch in a hole. In the configuration using a GMR sensor, the detection of the notch was enhanced.

### 6.2.5 Electrical Machine Signature Identification

From the discussion of magnetic sensors for home energy management systems in Chapter 5, it can be seen that such techniques are feasible for homes and small businesses. As for medium to large industrial facilities, techniques with better performance are required to handle the challenges of high rates of event generation, load balancing, and power factor correction. In [33] the authors review some advanced techniques by monitoring higher harmonics, transient detection, and disaggregating continuously variable loads.

Because of the inherent physical characteristics or the presence of power electronics, many loads draw distorted, non-sinusoidal currents. By using the higher harmonics, it is possible to distinguish the loads. The field monitoring system proposed by the authors uses a phase-locked short-time Fourier transform of current waveforms to calculate the spectral envelops that summarize the time-varying harmonic content. The current waveforms are collected at samples rates of 8000 Hz or higher. The load monitor in [33], which depicts the distinguishing individual loads, is able to examine harmonic content up to the seventh harmonic. With appropriate adjustment, the system can examine higher harmonics.

Since the transient behavior of each load varies from one to the other due to the physical task of the load, a transient detection based monitoring system is able to recognize individual loads by distinguishing the transient shapes of the load behaviors. Previous defined transient signatures and exemplars can be used to identify the behavior of the electrical equipment. On the other hand, the transient shapes of most loads have repeatable profiles, which support the basis for the pre-training approach of the loads. To be specific, each section of the exemplar can be shifted in time and offset to match incoming transient data. A better fit can be realized with an overall gain applied to all sections of the exemplar. If the fit is good enough, the event is classified as a match to the exemplar. Otherwise, it is left unclassified. Additionally, transient analysis can provide diagnostic information based on the relationship between the electrical transient and load physics.

The third advantage of such load monitoring systems is their ability to monitor buildings with continuously variable loads. Examining the spectral envelopes that roughly track the shape of the real power trace makes it possible to disaggregate the variable load when the non-intrusive load monitoring (NILM) recognizes a continuously variable load like variable speed drive (VSD). The NILM can recognize that a VSD is active but cannot provide a means for continuously tracking the power consumption of the variable speed drive. The method proposed overcomes this shortcoming. In addition, the monitoring system is able to catch the failure mode that occurs when the closed controlling loop is poorly tuned. With this information, energy waste can be saved.

These analytical enhancements have been implemented for non-intrusive monitoring on a computer platform for application in the field. A suite of NILM tools has been developed under the Linux operating system environment. The tools include sophisticated model-based diagnostic algorithms that track (and could in principle trend) model parameters to determine the health of critical loads. All of these software tools run on a Pentium-class personal computer (300 MHz clock or higher) with a PCI-bus data-acquisition card. The platform of this monitoring system is easy to build. Any personal computer, laptop, or embedded system, such as a PC104 chassis, could be used. The success of these platforms with in monitoring buildings and transportation systems demonstrates the potential of the system for monitoring transient events.

### 6.2.6 Magnetic Field Measurement for Condition Monitoring of Synchronous Generators

Generators are complex machines that generally work under atrocious environments. All rotating electrical machines are subjected to electrical stress. In addition, the internal losses in such machines cause an increased temperature that might lead to thermal stress. Machine rotation and attractive forces between machine parts can produce mechanical stresses and many machines today are exposed to dirt and contamination, resulting in environmental or chemical stress. These stresses can occur in different parts of a rotating electrical machine, where they are likely to cause failure.

The faults in a generator may be classified into stator-related faults, rotor-related faults, bearing faults, and electricity-related faults. The reasons faults occur include corrosion, contamination, improper lubrication or improper installation, and the effect on the generator may take the form of vibration, distorted current, cracking, fatigue, insulation degradation or failure.

Magnetic field measurement can find plenty of applications in condition monitoring of synchronous machines. The measurement of the magnetic field is often the final verification of the complex design and fabrication process of a magnetic system. In several cases, when seeking high accuracy, the measurement technique and its realization can result in considerable effort.

Condition monitoring of large synchronous generators plays a critical role in power generation systems. Many approaches have been developed for effective condition monitoring, such as artificial neural networks, fuzzy logic, etc, while model-based methods are gaining a special interest. The best-known method for synchronous machine health monitoring includes creating a model of the synchronous machine and simulating synchronous machine operation. However, synchronous machine models are created on the assumption that the spatial distribution of coils inside the machine is sinusoidal. The coils of a synchronous machine with an inter-turn fault no longer have a sinusoidal distribution. In other words, the model may not take into account the inherent asymmetry in the winding distribution under fault conditions. This may lead to erroneous fault detection. Magnetic field based measurement can directly detect the non-symmetry of the structure, together with its non-invasive character, hence will find more applications in future.

## 6.3 Magnetic Field and Renewable Energy

Faced with new challenges from the exhaustion of energy resources, air pollution, and innovative of technology, power grids have been confronted with unprecedented opportunities and challenges. The time is past where simply adding larger and larger generators was the best way to satisfy growing demand for electric energy. To handle these challenges, the power grid needs to introduce low-carbon technology, develop energy-saving devices, and exploit new energy or renewable resources actively. Renewable energy sources, needed to create independence from fossil fuels and reducing greenhouse gas emissions, will show much growth in the coming decades. A new energy revolution has begun to build a smarter, cleaner, and more efficient power system. Distributed generation, which is cleaner, more environmentally friendly, and requires low investment, is an ideal renewable energy resource and has been developed rapidly. However, when distributed generation systems are connected to power grids, the drawbacks of intermittent and plug-in characteristics are increasingly serious, which shakes the foundation of interdependent protection systems, power quality, and power system stability.

## 6.3.1 Commonly Used Renewable Energy Sources

The reserves of fossil fuels are limited and, more importantly, it is impossible to make use of all of those reserves. At the current rate reserves will not be exhausted for a few decades, but they will become a more and more marginal source of energy. Even prior to that, the price of using fossil fuels will substantially increase because of the higher costs needed to acquire them and the carbon taxes levied on burning them. Recent advances in the renewable resources of solar energy and wind energy are developing fast.

### 6.3.1.1 Solar Energy

Solar energy, radiant light, and heat from the sun are being harnessed using a range of ever-evolving technologies such as solar heating, solar PV, solar thermal electricity, solar architecture, and artificial photosynthesis [34]. The solar energy marketplace is one of the fastest growing renewable energy markets in the world.

Solar technologies are widely depicted as passive or active solar relying on the pattern of solar energy capture, conversion, and distribution. Passive solar techniques include choosing materials with good thermal mass or light-dispersing properties to build eco-system in a building. Active solar techniques include harnessing energy by using PV panels and solar collectors.

According to Table 6.1 [35], the amount of solar energy reaching the surface of the planet is so vast that in one year it is about twice as much as will ever be obtained from all of the Earth's non-renewable resources of coal, oil, natural gas, and mined uranium combined. The total solar energy absorbed by the Earth's atmosphere, oceans, and land masses is approximately 3 850 000 EJ per year. Photosynthesis captures approximately 3000 EJ per year in biomass.

### 6.3.1.2 Wind Energy

Wind power is the conversion of wind energy into a useful form of energy, e.g. by using wind turbines to make electrical power, windmills for mechanical power, wind pumps for water pumping or drainage, or sails to propel ships.

Large-scale wind farms consist of hundreds of wind turbines connected to the power transmission network. Recently, onshore wind has been shown to be a much cheaper source of electricity compared with fossil fuel plants in many places around the world [36]. In some situations small wind farms can provide basic energy to people, especially in isolated locations or mountain areas. Small domestic wind turbines can produce electricity for communal facilities such as street lights.

**Table 6.1** Renewable energy sources and yearly human energy consumption (in exajoules, EJ)

| | |
|---|---|
| Solar | 3 850 000 EJ |
| Wind | 2250 EJ |
| Biomass | 100–300 EJ |
| Primary energy use (2010) | 539 EJ |
| Electricity (2010) | 66.5 EJ |

From the latest research it would appear that solar, wind, and biomass could be sufficient to supply all of our energy needs, but the increased use of biomass has had a negative effect on global warming and dramatically increased food prices by diverting forests and crops into biofuel production. As intermittent resources, solar and wind raise other issues.

### 6.3.2 Potential Applications

Renewable energy systems are diverse in terms of resources and conversion technologies. Measurement solutions should be the first tool developed to handle various aspects of the impacts of renewable energy development.

Renewable energy generation systems are costly and must be monitored properly as the operation of wind generation or solar PV systems is closely related to environment. It is important to collect data and accumulate operational experience for future renewable energy development. For example, each year new wind power plant sites are commissioned worldwide. However, it is difficult to accurately predict how well a turbine will perform in different atmospheric conditions.

The intermittency of renewable energy causes great challenges to the integration of renewable energy in the power grid. The requirements for instrumentation and measurement for renewable power plant energy systems are generally high. Those functions should be coordinated to manage the utility connection, such as assessing reactive power supply, voltage control, fault ride-through, and power quality monitoring.

In summary, in renewable energy technology, sensors can help to ensure that systems run smoothly. Measurements are fundamental to any renewable energy project, both during the development stage and throughout the operational life. The data gathered by the measurement system has been of fundamental importance in the development of the renewables industry.

#### 6.3.2.1 Power Quality Monitoring

The impact of wind turbines and PV systems on network operation and power quality (harmonics and voltage fluctuations) is very important. The capability of the power system to absorb this perturbation is dependent on the fault level at the point of common coupling. Basic power quality requirements must be met for harmonics, voltage, frequency, etc. for interconnecting any equipment to the grid at the interface between the power grid and renewable power generation. As power electronics are used to convert energy at the interface, the power quality problem is serious. Renewable energy generators with their associated power electronics generate harmonics and have electrical characteristics under voltage and frequency excursions that may make it difficult to meet these requirements [3]. Large-scale wind farms and large-scale PV systems present a spectrum of technical challenges arising mostly from the expanding application of power electronic devices at high power ratings.

When incorporating renewable energy sources within power distribution networks, it is very important to provide power quality analysis at many nodes in the network, for example voltage fluctuations and the presence of harmonics in the network, due to wind gusts and due to non-natural wind oscillations caused by the presence of the tower, stability problems caused by faults in the system such as short circuits, lightning surges, manoeuvres, etc., or due to the great variability of the wind [37]. In general, power quality disturbances include those from short to long duration variations and flicker, and degrade product quality, increase downtime, and reduce customer satisfaction. International standard IEC 61400-21 has been developed to define and specify the measurement to quantify the power quality of a grid-connected turbine [38].

It is necessary to develop a method that is suitable for efficient monitoring of power qualities in renewable energy systems. The emphasis should be on low computational power required to perform the necessary calculations, as well as the possibility of detecting as many categories of PQ disturbances as possible (P, active power; Q, reactive power). Magnetic field measurement technology may therefore help in various high performance current detection systems [39].

### 6.3.2.2 Magnetic Bearings

The proportion of renewable energy is increasing worldwide and it is important to maintain this trend because of the increased demand for electricity in the developing countries, e.g. wind power has a growth rate of about 20%. Wind turbines, with the advantages of endless resources, low cost, and convenient use, have become one of the most promising new energy technologies and are now a mature technology implemented on a large scale. Traditional mainstream wind turbines are supported by mechanical bearings, which have many shortcomings, such as high starting speed, large starting torque, high maintenance cost, low utilization rate of wind energy, short life of rolling bearing, relatively low precision, need for lubrication, lubrication oil pollution etc. [40]. These shortcomings have seriously restricted the development of wind turbine technology. For years, wind turbine manufacturers have been searching for ways to make direct-drive turbines competitive with gearbox turbines. Direct-drive technology has been praised for its design, which is less complex than gearbox technology, leading to easier operation and maintenance [41]. This appeal has made direct drive especially coveted for use in offshore wind developments.

Direct-drive permanent magnet (PM) generators are gaining popularity in the wind power industry, particularly for large offshore wind turbines with high power ratings. In this configuration, the vulnerable spot (the gearbox) is eliminated. Elimination of the gearbox, which has a relatively high failure rate, improves the reliability of the conversion system and reduces the maintenance work considerably. This is a distinct advantage, particularly in the growing number of offshore wind farms where maintenance operations are difficult, time-consuming, and expensive.

Direct-drive machines are very heavy and expensive for large wind turbines (normally larger and heavier than generators in geared systems) [42], causing difficulties in production, transportation, and installation, as well as high usage of permanent magnets and greatly increased cost. The use of magnetic bearings has the potential to reduce the weight of the direct-drive machine, this may be an effective solution. Magnetic bearings are high-performance bearings that use magnetic force to suspend the rotor in space without mechanical contact between the rotors and the stators. At present, the most widely used magnetic bearings are one of three types: passive magnetic bearings, active magnetic bearings, and hybrid magnetic bearings [43]. Just like mechanical gears, magnetic gears transform rotational power between different speeds and torques, but instead of physically interlocking teeth, they use magnetic fields. By using magnets to transmit torque between the input and output shafts of the gear, they avoid mechanical contact. This provides a number of advantages such as reduced maintenance, reduced acoustic noise and vibration, and improved overall reliability. This system increases reliability substantially and reduces maintenance costs. With maintenance time reduced, production time is increased, which provides improved returns.

Along with the enlargement of wind turbine capacity and the development of offshore wind energy applications, the reliability, volume, cost, and lifetime of wind turbines are of increasing concern. As wind power generation is one of the main contributions to the growth of the bearings business, magnetic bearings will play an important role in future wind power and wave power generation. Hence, magnetic field related measurement will find applications in the evaluation or detection of the magnetic field in such systems.

### 6.3.2.3 Wide-area Monitoring

Inverters connected with renewable energy sources, nonlinear customer loads, and power electronics devices in a renewable energy penetrated power system introduce harmonics to the distribution network, causing overheating of transformers, tripping of circuit breakers, and reduction in the life of connected equipment [44]. Harmonics is therefore one of the most dominant attributes that needs to be kept at a minimum level to ensure the power quality of the power network. The estimation of harmonics has become very important for design, analysis, tariff, control, and monitoring purposes. In a renewable energy system, since the harmonics are generated everywhere in the system and interact with each other, the monitoring of harmonics phenomena should be systematic.

In addition, recent years have witnessed emerging oscillatory stability issues in power systems with high-penetration renewable energies [45]. Power systems are integrating more and more with renewable power generation, and are generally interfaced with such systems via power electronic converters. As a result, the interaction between converters and the AC/DC grid has significantly changed the system dynamics and caused new types of oscillation issues, e.g. high-frequency harmonic oscillations in microgrids, sub/supersynchronous oscillations at wind farms, and low-frequency oscillations induced by constant power loads. These emerging oscillations can impair the stable operation of the system and the efficient accommodation of renewable power. A new type of sub- and supersynchronous interaction (broadband oscillation) between STATCOMs and the weak AC/DC grid have been reported [46]. Those phenomena are fundamentally wide area, hence wide-area monitoring is needed.

With recognition that synchrophasor technology, with high-speed, wide-area, time-synchronized grid monitoring, and sophisticated analysis, could become a foundational element of grid modernization for transmission systems, the utilities have continued to expand their investment and industry partnership in the areas of synchrophasor measurement devices, communications, applications, measurements, and technical interoperability standards. Traditional synchrophasors rely on current transformers and potential transformers physically connected to transmission lines or buses to acquire input signals for phasor measurement. Some authors have presented two innovative designs for non-contact synchronized measurement devices (NCSMDs), including an electric field sensor based non-contact SMD (E-NCSMD) and a magnetic field sensor based non-contact SMD (M-NCSMD) [47].

Traditionally small-scale and limited to transmission systems, phasor measurement unit (PMU) deployments have been hindered by system complexity and other limitations associated with network communication, performance, and data management issues. However, with recent breakthroughs in smart grid technologies, advanced PMUs are now being developed for deployment worldwide and for integration into power distribution networks.

### 6.3.2.4 Monitoring for Demand-side Management and Optimized Operations

Consumer-side controlled demand can balance the load on the transmission and distribution grid, and postpone investment into new transmission assets. Demand-side management (DSM) is normally associated with encouraging or forcing a reduction in maximum demand to enable a constrained generation capacity to match demand [48]. Residential homes usually use DSM programs to manage their consumption patterns according to varying electricity prices over time. Subscribers can participate in DSM in different ways, such as reducing their load demand, shifting their load demand to off-peak hours, and relying on renewable energy to limit their dependence on grid energy [49].

One of the challenges is that renewable energy continues to grow in most power systems. A power system with high penetration of renewable energy will be not easy to dispatch, and hence poses a challenge to the balancing of the power system. In low or moderate penetrations, the renewable energy power generation may help to supply some of the local load, resulting in reduced energy imports from higher levels of distribution or transmission networks. But when the energy production exceeds load and the distribution network starts to export power, the renewable energy power generation system may introduce problems, since these distribution systems were not traditionally designed to deal with energy production. Moreover, other problems, such as voltage rise or increased power losses, are intensified by the intermittent production of renewable energy, which further decreases power quality and puts additional strain on distribution system equipment. One of the key problems associated with renewable energy generation is that the time period in which renewable energy systems produce energy often does not coincide with the period when the energy is demanded. This will present new challenges for the optimal operation of power systems [50].

Measuring and managing concepts go hand in hand: one cannot manage what one cannot measure. DSM is based on real-time information of transmission and distribution devices from a wide-area measurement system. Technically, DSM can be an artificial power plant that balances deviations within energy contracts and avoids spot market expenses for reserve power. Together with the two-way communication infrastructure built along with smart grids, various data should be collected by measurement and monitoring systems to implement the optimal operation of power networks containing a significant amount of renewable energy.

### 6.3.2.5 Measurement for Planning and Integration

The impact of renewable energy requires an active contribution from each renewable energy system to ensure power quality and grid stability, and therefore a system approach is necessary. Although renewable energy is used extensively worldwide, the benefits of it are overshadowed by its intermittent nature. A high penetration level of intermittent renewable power leads to a series of technical and economic impacts on power systems. These impacts reduce the operators' motivation to integrate renewables into power systems. It is therefore necessary to study intermittency measurements and mitigation solutions right from the planning stage.

Reliable, robust, and validated data are critical for informed planning, policy development, and investment in the clean energy sector. Data-driven decisions enable ambitious, cost-effective, and achievable outcomes for renewable energy deployment. Data support a wide variety of renewable energy analyses and decisions, including technical and economic potential assessment, renewable energy zone analysis, grid integration, risk and resiliency identification, electrification, and distributed solar PV potential.

Generally, the key data sets to assist renewable energy decisions include [51]:

- renewable energy resource data, e.g. wind (wind speeds, power density, ground measurements), solar (irradiance, ground measurements), biomass (crop or forestry residue), and other renewable resources, etc.
- administrative, e.g. population density or distribution and load locations, etc.
- environmental, e.g. topographic limitations, environmental attributes, and land-use constraints, designated protected areas, elevation, water bodies (e.g. rivers, lakes), etc.
- infrastructure, e.g. roads, etc.
- power grids, e.g. electricity transmission and/or distribution lines, substations, power plants (e.g. location, type, operational status), other energy infrastructure (e.g. natural gas pipelines), etc.
- development, e.g. electrification rates, poverty rates, other data sets that can help inform the identification of tradeoffs and synergies between renewable energy development and other national development goals, etc.
- natural hazards, e.g. extreme weather events and other natural hazards, which can affect achievable energy generation and feed into resilience planning and decisions, fault lines, landslide, earthquake, drought, flood, tornado risk, etc.
- energy demand and costs, e.g. electricity and/or heating demand and price, critical loads, other cost and policies, etc.

Some of these data may be collected from historic recording or statistical investigation, while some should be measured and may be monitored for certain period of time. Magnetic field based measurement can contribute in a lot of applications to radiation detection and electrical current monitoring.

This measurement/monitoring can be used for some new renewable energy development, e.g. marine renewable energy technology, which is currently increasing across the world. Offshore wind has led the way, but wave and tidal technologies are being extensively tested and will most likely be operational within the next few years. With interest increasing, understanding of the interactions between marine renewable energy and the environment is poor. There is an increasing number of marine renewable energy devices and their associated subsea cables, which will connect to existing medium voltage (MV) or low voltage (LV) power distribution networks in the future. However, very few studies have considered the environmental impacts connected with their installation and operation [52].

### 6.3.2.6 Other Protection and Control Problems

The increasing penetration rate of renewable energy in power systems is raising technical problems, as voltage regulation, network protection coordination, loss of mains detection, and renewable energy operation following disturbances on the distribution network. These problems must be quickly solved in order to fully exploit the opportunities and benefits offered by renewable energy. The intermittent nature of renewable energy sources, in particular solar and wind energy, has an impact on system operations, including voltage and frequency, harmonics, and power quality in general, and influences the overall performance of the power network as well as the distribution network. Adding protection applications will make sure the power flow is only interrupted when absolutely necessary, assuring the highest reliability and quality of service for customers. With an increasing amount of renewable energy resources in the grid, the need for accurate selectivity has never been higher.

As described in the previous chapter, lightning monitoring is also very important in renewable energy. Winter lightning damages many wind turbines in Japan every year [53]. A non-contact measurement method would be very helpful in the attempt to measure a large lightning current, where the direct measurement of equipment is threatened by the risk of burnout.

The current measurement system (CMS) has a significant role in renewable energy development [54]. The CMS measures currents directly in the final circuit and can therefore ensure the error-free operation of a system: continual measurement in the final circuit detects potential hazards like failures, power drops or other abnormal behavior before major damage is caused. Constant measurement of current flow in overcurrent protection devices can monitor whether or not a cable is loaded beyond its nominal current range. Tripping can be prevented before it is too late. CMS also can offer an uncomplicated yet highly effective solution to use energy more efficiently and hence reduce the cost. It measures the current in the individual final circuits, allowing the end users to precisely trace energy flows and triangulate where electricity consumption is too high.

As the demand for electricity that is reliable, affordable, sustainable, and resilient continues to grow, new and more sophisticated tools, practices, and analytical methods are being developed to economically and reliably plan and operate the power systems of the future. Other than a simple overload warning system with CMS, more advanced systems, e.g. to increase transparency, degree of customer awareness of product options, and fair market prices for a given amount of specific energy, will be developed and supported by CMS.

### 6.3.3 Challenges

The electric power sector around the world is undergoing long-term technical, economic, and market transformations. Part of these transformations is the challenge of integrating high shares of renewable energy, particularly variable wind and solar. This presents great challenges to the flexibility of a power system and is key in terms of balancing these variable sources while keeping the system in its normal operating state. Change needs to take place in the areas of technology, economics, planning, operations, business, and policy, while technological supports, especially real-time accurate information, are fundamental to tackling the challenge.

Distributed generation changes the unidirectional flow structure of a distribution system so that the magnitude and direction of flow cannot be predicted. Due to the installation location of distributed generation, the flow of feeder line sections may increase or reduce [55]. Hence the protection theory and configuration will be changed.

Renewable energy such as solar energy and wind energy possess the characteristics of uncertainty and fluctuation. Distributed generators starting or stopping frequently may cause voltage fluctuation, voltage flicker, frequency fluctuation, and other power quality problems. Different types of distributed generation (inverter and traditional rotary) will cause different degrees of harmonic distortion and in some cases distributed generation will cause the rising or falling of system voltage [56]. The measurement of line voltage and current signals for power quality analysis demands much higher accuracy than is needed for protection purposes. Traditional voltage and current transducers used in the primary circuits of power stations cannot meet the requirements of increased accuracy and wide measurement bandwidth. New types of voltage and current transducers are needed with frequency measurement ranges equal to at least the 40th harmonic of fundamental frequency, high dynamic range, and very good linearity [57].

Increasing number of renewable energy sources and distributed generators requires new strategies for the operation and management of the electricity grid in order to maintain or even to improve power supply reliability and quality. In addition, liberalization of the grids leads to new management structures in which trading of energy and power is becoming increasingly important [58]. The integration as well as future peer-to-peer transactions of energy may bring new challenges for grid operators in terms of maintaining stability, and quality and adequacy of supply [59]. As a first step to the seamless integration of renewables, monitoring is essential. An automation solution offers a modular way of providing monitoring, control, measurement, and protection to enable rapid detection of complex situations ongoing in the grid and an effective response to the consequences of decentralized energy feed-in. Innovative measurement technology will definitely facilitate the integration in a cost-effective way in the future.

# Bibliography

1 Q. Huang, Y. Song, X. Sun, L. Jiang, and P. Pong, "Magnetics in smart grid," *IEEE Transactions on Magnetics*, vol. 50, no. 7, pp. 1–7, 2014.

2 M. Yildirim, X. A. Sun, and N. Z. Gebraeel, "Sensor-driven condition-based generator maintenance scheduling. Part I: Maintenance problem," *IEEE Transactions on Power Systems*, vol. 31, no. 6, pp. 56–63, 2016.

3 IEEE, "Interconnecting distributed resources with electric power systems," *IEEE Standard 1547.2-2008*, pp. 1–217, April 2009.

4 M. Zhang, X. Xiao, and Y. Li, "Speed and flux linkage observer for permanent magnet synchronous motor based on EKF," *Proceedings of the CSEE*, no. 36, pp. 36–40, 4 2007.

5 F. Lalonde, *Magnetic Field Measurement*. Rotterdam: Balkema, 1992.

6 J. P. Sebastia, J. A. Lluch, J. R. L. Vizcaino, and J. S. Bellon, "Vibration detector based on GMR sensors," *IEEE Transactions on Instrumentation and Measurement*, vol. 58, no. 3, pp. 707–712, March 2009.

7 K. M. Goh, H. L. Chan, S. H. Ong, W. P. Moh, D. Tws, and K. V. Ling, "Wireless GMR sensor node for vibration monitoring," in *5th IEEE Conference on Industrial Electronics and Applications*, June 2010, pp. 23–28.

8 D. Ramirez and J. Pelegri-Sebastia, "GMR sensors manage batteries," *EDN Network*, vol. 44, 1999.

9 J. Pelegrí, J. B. Ejea, D. Ramírez, and P. P. Freitas, "Spin-valve current sensor for industrial applications," *Sensors & Actuators A: Physical*, vol. 105, no. 2, pp. 132– 136, 2003.

**10** R. L. White, "Giant magnetoresistance: a primer," *IEEE Transactions on Magnetics*, vol. 28, no. 5, pp. 2482–2487, 2002.

**11** J. M. Daughton, P. A. Bade, M. L. Jenson, and M. M. M. Rahmati, "Giant magnetoresistance in narrow stripes," *IEEE Transactions on Magnetics*, vol. 28, no. 5, pp. 2488–2493, 1992.

**12** J. L. Brown, "High sensitivity magnetic field sensor using GMR materials with integrated electronics," *Proceedings of the IEEE International Symposium on Circuits and Systems*, vol. 3, pp. 1864–1867 vol.3, 1995.

**13** P. B. Aronhime and F. W. Stephenson, *Analog Signal Processing*. Springer US, 1994.

**14** J. L. Schmalzel and D. A. Rauth, "Sensors and signal conditioning," *IEEE Instrumentation & Measurement Magazine*, vol. 8, no. 2, pp. 48–53, 2005.

**15** J. Fraden and L. G. Rubin, "AIP handbook of modern sensors," *Physics Today*, vol. 47, no. 6, pp. 74–75, 1994.

**16** J. Heremans, "Solid state magnetic field sensors and applications," *Journal of Physics D: Applied Physics*, vol. 26, no. 8, p. 1149, 1999.

**17** D. Ramirez, S. Casans, C. Reig, and E. Al, "Build a precise DC floating-current source," *EDN Network*, vol. 50, no. 8, pp. 83–84, 2005.

**18** D. R. Munoz, S. C. Berga, and C. R. Escriva, "Current loop generated from a generalized impedance converter: A new sensor signal conditioning circuit," *Review of Scientific Instruments*, vol. 76, no. 6, pp. 517–409, 2005.

**19** D. K. Cheng, *Fundamentals of Engineering Electromagnetics*, ser. Addison-Wesley Series in Electrical Engineering. Addison-Wesley, 1993.

**20** O. Casas and R. Pallas-Areny, "Basics of analog differential filters," *IEEE Transaction on Instrument and Measurement*, vol. 45, no. 1, pp. 275–279, 1996.

**21** V. Dubickas and H. Edin, "High-frequency model of the Rogowski coil with a small number of turns," *IEEE Transactions on Instrumentation and Measurement*, vol. 56, no. 6, pp. 2284–2288, 2007.

**22** T. A. Harris, *Rolling Bearing Analysis*. Wiley, 1984.

**23** R. T. Ko, S. Sathish, J. S. Knopp, and M. P. Blodgett, *Hidden Crack Detection with GMR Sensing of Magnetic Fields from Eddy Currents*. College Park, MD: AIP Conference Proceedings, 2007.

**24** R. E. Beissner, G. L. Burkhardt, E. A. Creek, and J. L. Fisher, *Eddy Current Probe Design for Second-Later Cracks under Installed Fasteners*. Springer US, 1993.

**25** F. Thollon, B. Lebrun, N. Burais, and Y. Jayet, "Numerical and experimental study of eddy current probes in NDT of structures with deep flaws," *NDT & E International*, vol. 28, no. 2, pp. 97–102, 1995.

**26** M. Gibbs and J. Campbell, "Pulsed eddy current inspection of cracks under installed fasteners," *Materials Evaluation*, vol. 49, pp. 51–59, 1991.

**27** R. A. Smith and G. R. Hugo, "Deep corrosion and crack detection in aging aircraft using transient eddy-current NDE," *Proceedings of the 5th Joint NASA/FAA/DoD Aging Aircraft Conference, Orlando, Florida*, 2001.

**28** W. W. W. Iii and J. C. Moulder, *Low Frequency, Pulsed Eddy Currents for Deep Penetration*. Springer US, 1998.

**29** Y. S. Sun, "Electromagnetic-field-focusing remote-field eddy-current probe system and method for inspecting anomalies in conducting plates," Patent, 1999, US6002251A.

**30** W. F. Avrin, *Magnetoresistive Eddy-Current Sensor for Detecting Deeply Buried Flaws*. Springer US, 1996.

**31** B. Wincheski, J. Simpson, P. Dan, E. Scales, and R. Louie, *Development of Giant Magnetoresistive inspection system for detection of deep fatigue cracks under airframe fasteners.* College Park, MD: AIP Conference Proceedings, 2002.

**32** T. Dogaru, C. H. Smith, R. W. Schneider, and S. T. Smith, "Deep crack detection around fastener holes in airplane multi-layered structures using GMR-based eddy current probes," in *AIP Conference*, 2004, pp. 398–405.

**33** C. Laughman, K. Lee, R. Cox, S. Shaw, S. Leeb, L. Norford, and P. Armstrong, "Power signature analysis," *IEEE Power and Energy Magazine*, vol. 99, no. 2, pp. 56–63, March 2003.

**34** "Solar energy perspectives," International Energy Agency, Tech. Rep., Dec. 2011. [Online]. Available: https://www.iea.org/publications/freepublications/publication/Solar_Energy_Perspectives2011.pdf

**35** Wikipedia. (2014) Solar energy. [Online]. Available: http://en.wikipedia.org/wiki/Solar_energy

**36** H. Holttinen, B. Lemstrom, P. Meibom, H. Bindner, A. Orths, F. Van Hulle, C. Ensslin, A. Tiedemann, L. Hofmann, W. Winter et al., *Design and operation of power systems with large amounts of wind power: state of the art report.* Helsinki: Julkaisija-Utgivare, 2007.

**37** M. Sanz, A. Llombart, A. A. Bayod, and J. Mur, "Power quality measurements and analysis for wind turbines," in *Proceedings of the 17th IEEE Instrumentation and Measurement Technology Conference*, vol. 3, 2000, pp. 1167–1172 vol.3.

**38** *Wind Turbine Generator Systems. Part 21: Measurement and assessment of power quality characteristics of grid connected wind turbines.* Geneva: International Electrotechnical Commission, 2008.

**39** S. H. Laskar, S. Khan, and Mohibullah, "Power quality monitoring in sustainable energy systems," in *IEEE International Symposium on Sustainable Systems and Technology (ISSST)*, May 2012, pp. 1–6.

**40** C. Kumbernuss, J. and J. W. Jian, "A novel magnetic levitated bearing system for vertical axis wind turbines (VAWT)," vol. 90, no. 1, pp. 148–153, 2012.

**41** H. Li and Z. Chen, "Optimal direct-drive permanent magnet wind generator systems for different rated wind speeds," in *European Conference on Power Electronics and Applications*, 2007, pp. 1–10.

**42** H. Polinder, F. F. A. van der Pijl, G. de Vilder, and P. J. Tavner, "Comparison of direct-drive and geared generator concepts for wind turbines," *IEEE Transactions on Energy Conversion*, vol. 21, no. 3, pp. 725–733, 2006.

**43** J. Sun, Z. Ju, C. Peng, Y. Le, and H. Ren, "A novel 4-DOF hybrid magnetic bearing for DGMSCMG," *IEEE Transactions on Industrial Electronics*, vol. 64, no. 3, pp. 2196–2204, 2017.

**44** G. M. Shafiullah and A. M. T. Oo, "Analysis of harmonics with renewable energy integration into the distribution network," in *IEEE Innovative Smart Grid Technologies – Asia (ISGT ASIA)*, 2015, pp. 1–6.

**45** C. S. Lai, Y. Jia, L. L. Lai, Z. Xu, M. D. McCulloch, and K. P. Wong, "A comprehensive review on large-scale photovoltaic system with applications of electrical energy storage," *Renewable and Sustainable Energy Reviews*, vol. 78, pp. 439 – 451, 2017.

**46** D. Shu, X. Xie, H. Rao, X. Gao, Q. Jiang, and Y. Huang, "Sub- and supersynchronous interactions between STATCOMs and weak AC/DC transmissions with series compensations," *IEEE Transactions on Power Electronics*, vol. 33, no. 9, pp. 7424–7437, Sept 2018.

47 W. Yao, Y. Zhang, Y. Liu, M. J. Till, and Y. Liu, "Pioneer design of non-contact synchronized measurement devices using electric and magnetic field sensors," *IEEE Transactions on Smart Grid*, vol. 9, no. 6, pp. 5622–5630, Nov 2018.

48 Energy demand management. [Online]. Available: https://en.wikipedia.org/wiki/Energy_demand_management

49 C. Eid, E. Koliou, M. Valles, J. Reneses, and R. Hakvoort, "Time-based pricing and electricity demand response: Existing barriers and next steps," *Utilities Policy*, vol. 40, pp. 15–25, 2016.

50 P. S. Moura and A. T. de Almeida, "Multi-objective optimization of a mixed renewable system with demand-side management," *Renewable and Sustainable Energy Reviews*, vol. 14, no. 5, pp. 1461–1468, 2010. [Online]. Available: http://www.sciencedirect.com/science/article/pii/S1364032110000055

51 S. Cox, A. Lopez, A. Watson, N. Grue, and J. E. Leisch, "Renewable energy data, analysis, and decisions: A guide for practitioners," National Renewable Energy Laboratory, Colorado, Tech. Rep., 2018.

52 A. G. Borthwick, "Marine renewable energy seascape," *Engineering*, vol. 2, no. 1, pp. 69 – 78, 2016.

53 J. Montanya, F. Fabro, and O. van der Velde, "Global distribution of winter lightning: a threat to wind turbines and aircraft," *Natural Hazards and Earth System Science*, vol. 16, pp. 1465–1472, 2016.

54 H. Kirkham, "Current measurement methods for the smart grid," in *IEEE Power Energy Society General Meeting*, 2009, pp. 1–7.

55 P. P. Barker and R. W. de Mello, "Determining the impact of distributed generation on power systems: I. Radial distribution systems," in *IEEE Power Engineering Society Summer Meeting*, vol. 3, 2000, pp. 1645–1656.

56 M. T. Doyle, "Reviewing the impacts of distributed generation on distribution system protection," in *IEEE Power Engineering Society Summer Meeting*, vol. 1, 2002, pp. 103–105.

57 A. Nowakowski, A. Lisowiec, and Z. Kolodziejczyk, *Power Quality: Power Quality Monitoring in a System with Distributed and Renewable Energy Sources*. IntechOpen, 2011.

58 J. M. Carrasco, L. G. Franquelo, J. T. Bialasiewicz, E. Galvan, R. C. PortilloGuisado, M. A. M. Prats, J. I. Leon, and N. Moreno-Alfonso, "Power-electronic systems for the grid integration of renewable energy sources: A survey," *IEEE Transactions on Industrial Electronics*, vol. 53, no. 4, pp. 1002–1016, 2006.

59 J. Guerrero, A. C. Chapman, and G. Verbi, "Decentralized P2P energy trading under network constraints in a low-voltage network," *IEEE Transactions on Smart Grid*, pp. 1–1, 2018.

# 7

# Future Vision

Measurement and sensors are vital to the development of many industrial processes. One of the major trends driving and driven by technological advancement is the increasing use of sensors and instruments in industrial processes. Researchers and engineers continually looking for innovative methods to improve efficiency, increase reliability, minimize risk, and reduce cost.

One of the main incentives in electrical measurement is the evolution of smart grids. In this evolution, new measurement tools, strategies, and bidirectional communication systems are deployed to implement not only simple functionalities (such as the disconnection of a distributed generator in the case of unwanted islanding), but also more complex tasks such as the management of islanding in the presence of faults in other parts of the network (i.e. safe islanding), reconfiguration of the network topology, remote control of distributed generators (to allow their participation in the regulation of voltage and frequency), power quality analyses, and other automation functions (ability to adapt and self-heal).

## 7.1 Magnetic Field Based Instrumentation and Measurement in Smart Grids

Smart grids are the enablers for the future vision of a clean, renewable, locally generated, secure, open, observable, and controllable modern power grid that is required to meet society's energy challenges. The smart grid revolution involves in every aspect of power systems, including power generation, transmission, and distribution as well as energy utilization and transactions.

### 7.1.1 Transmission Systems

Transmission systems must be kept highly reliable to prevent blackouts and ensure robust energy markets. Technical advances have occurred throughout the history of transmission, with advances in monitoring, protection, analysis, and control, accompanied by periodic breakthroughs in transmission capacity. Recent developments in power transmission systems include capacity improvement in high-voltage alternating current (HVAC, 765 kV and above) and high-voltage direct current (HVDC, ±800 kV

*Magnetic Field Measurement with Applications to Modern Power Grids*, First Edition.
Qi Huang, Arsalan Habib Khawaja, Yafeng Chen and Jian Li.
© 2020 John Wiley & Sons Ltd. Published 2020 by John Wiley & Sons Ltd.

and above). Power electronics have contributed to the development of transmission systems with advanced flexible AC transmission systems, which have greatly improved the controllability of transmission systems. Other technologies, such as optimized transmission dispatch, high capacity conductors, advanced storage, etc., are being adopted in transmission systems as part of the development of smart grid initiatives. To achieve long distance transmission of renewable energy, some light HVDC transmission systems are being used, flexibly transmitting the intermittent renewable energy.

Sensing and measurement are the key to smart power transmission, recognizing that one can only manage what one measures. A smart grid is never smarter than the quality of its measurements. Hence there is a clear need for more and better measurement tools. Conventional centralized grids can be thought of as passive one-way bulk energy systems whereas smart grid systems are active systems consisting of multiple bi-directional energy clients. These distributed systems are highly complex, difficult to optimize, and vulnerable to instability. This leads to a paradigm shift in the instrumentation and control requirements for smart grids so that a stable high-quality electricity supply can be assured.

Typical innovative measurement/monitoring tools in smart power transmission systems include following aspects to make them smarter:

- Dynamic line and equipment rating, which can provide the actual current-carrying capacity of overhead lines based on real-time operating conditions, and hence greatly improve the utilization of transmission assets and therefore efficiency.
- Synchrophasor monitoring, which is one of main smart grid enablers that provides the instantaneous measurement of electrical magnitudes and angles to reveal emerging instability. System-wide deployment of a synchrophasor measurement system will greatly improve reliability and security.
- Reliability assessment, which provides sophisticated tools that reveal issues and offers mitigating solutions by identifying long-term emerging issues and trends.
- Advanced metering infrastructure (AMI), which provides not only billing information, but also other valuable information to the regional transmission operator about system problems as well as voltage and power quality measurements at customer sites and substations.

Non-contact magnetic field based measurement will find plenty of applications in smart power transmission systems. In the future, based on various applications already established, such as current measurement and operation parameters (such as sag, vibration, galloping etc.), further applications (such as lightning strike monitoring, measurement under faulty transient, bias magnetization of transformers, etc.) will be developed. Another aspect will be the non-contact measurement system itself, for which more integrated compact system will be developed, allowing it to be used in a more convenient, more robust, and more effective manner.

### 7.1.2 Distribution Systems

The entire distribution system has seen very little technological change over most of its existence, but there is now a huge need and opportunity for improvement, made possible by today's digital communications and control tools.

Modern power distribution systems are characterized by the penetration of renewable energy. This trend has been further developed recently, especially in China. A new practice, the so-called energy internet, or internet + smart energy system, or smart grid 2.0, is flourishing in China and is quickly spreading around the world [1]. In this system, the various types of energy sources, interconnected by an electric power network, can exchange flexibly and hence can be controlled to be mutually complementary with each other. Hence the various energy sources can be utilized comprehensively to achieve maximum comprehensive efficiency.

This development trend, together with ongoing smart grid initiatives, requires ubiquitous monitoring in the energy network. In such an energy network, monitoring of power offers benefits on numerous levels: in addition to cost savings through optimized consumption, the monitoring of power networks and power quality in infrastructures and industrial plants ensures greater reliability. At the same time, systematic power monitoring raises awareness about actual power consumption and is therefore also an important prerequisite for higher energy efficiency.

The reliable operation of smart grid depends upon the accurate data collected by different measurement devices and effective design of AMI. The fundamental changes happening in distribution grids, including the deployment of volatile and distributed generation, the increase in prosumer behavior, and new demand-side services, are accompanied by increasing system complexity, dynamics, and uncertainty. Situation awareness is of essential importance to enable reliable and efficient control and operation applications [2]. Advanced control and operation systems for distribution grids can be facilitated by the use of emerging measurement technologies such as smart metering, synchronized phasor measurement, and other novel monitoring devices, providing more accurate and extensive measurement information to enable real-time monitoring, state estimation, and system analysis. Magnetic field based measurement can provide various novel solutions for current measurement with much better performance for most special applications, such as broadband monitoring under faulty conditions, fast and accurate fault location and identification, harmonics tracking, and high-confidence state estimation and parameter identification, etc., to support more advanced applications.

### 7.1.3 Generation Systems

From the power system view point, the development trend of generation systems includes two aspects: more penetration of renewable energy and smart power generation systems.

The increased use of renewable energy is vital to meet emission reduction targets and secure the energy supply in the world. Unless the energy flows can be accurately measured and controlled, the intermittency of renewable power generation will result in costly power quality degradation, ultimately leading to widespread blackouts. Intermittency has been the major obstacle for renewables in the past, but seems likely that this issue will be solved in near future. As many important companies have already set their eyes on large-scale battery storage, there is reason for the players in renewable power generation to be optimistic [3]. New tools are needed by network operators to measure the quality and stability of the electricity supply and enable its steady operation. Decentralization of the electricity system coupled with smarter technology should translate into better asset utilization, more security, and enhanced reliability.

In parallel with smart grids, there is also a smart power generation (or smart power plant) initiative, which is a little less developed and emerging at a slower pace than smart grids. Power generation is a complex process, in which various pieces of information has to be collected and then, based on the collected information, optimized decision are made automatically or with human intervention.

Smart power plants, based on digital power plants, aim to be highly secure, energy efficient, and environmentally friendly by incorporating information, communication, control, and sensor technologies. Smart power plants can significantly improve information management and service, master the production process, increase the controllability of the production process, lower human intervention, and produce an optimized production plan with full timely information about the production process.

The fundamental characteristics of a smart power plant include the following [4]:

- Digitization is the basis of smart power plants. Through various advanced sensor and network communication technologies, the production and management of power plants are presented in digital form and shared with each other, providing information for smart control and decision making.
- Self-adaptation: Modern control theory and technology such as data mining, self-adaptive control, predictive control, fuzzy logic control, artificial neural network self learning, process optimization, and self decision making are extensively used to make the power plant operate optimally, securely, and in an environment friendly way. A self-adapted system automatically changes the control strategy and management modes according to changes in environmental conditions, the environmental protection index, and the fuel state.
- Interaction: Production and management activities are planned to make the electricity produced satisfy the users' requirement for security and fast-responsiveness, by interacting and sharing data with smart grids, energy internets, and bulk power user information systems, using real-time analysis and prediction of the supply–demand in power market.

Magnetic field based measurement, as wellas providing current measurement solutions in smart power plants, can provide solutions for the monitoring of generators, and due to its non-contact characteristics it will contribute to the evolution of generation equipment to smart power generation equipment, which can be characterized by digital measurement, networked control, visualized state monitoring, integrated functionality, and interactive information.

## 7.2 Integration with Existing Power Systems

Smart grid development in power systems is an evolving process and the pace of deployment depends on many factors. Governments, regulators, industry organizations, and end users have proposed smart grids to enhance customer options, support climate change initiatives, and enhance the reliability of the electrical energy system. The evolving integration of smart grids will require significant changes in power system planning, design, and operation practices. The entire electricity infrastructure and associated social-technical system, including transmission and distribution networks, the system operator, suppliers, generators, consumers, and market mechanisms, will need to evolve to realize the full potential of smart grids. At the heart of this evolution is the integration of information and communication technology, and energy infrastructures for the increasingly decentralized development, monitoring, and management of a resilient grid [5]. It is important to plan for the integration of magnetic field based measurement into existing power systems.

## 7.2.1 Chances

Although the electrical power industry is undergoing rapid change, the generation and transmission technologies themselves are changing slowly because of the slow progress in these technical areas and resistance from utilities and end users. However, the field instrumentation on the grid is quickly reaching its lifecycle limit, which adversely affects overall grid reliability and efficiency. This present chances for integrating magnetic field based measurement into existing power system measurement, instrumentation, and control.

The first chance to integrate new types of sensors is the renewal and replacement of old measurement and instrumentation devices. For example, in most distribution networks, in accordance with requirement of smart grids, more measurement points should be deployed in the distribution system to facilitate monitoring, fault location, and identification. Traditional current measurement sensors are generally difficult to install. Magnetic field based measurement devices, because of their non-contact characteristics, are very easy to install in an existing distribution system without intervention of any services. The magnetic field based measurement adjusts to the existing installation architecture like no other system. There are no tangled cables, no additional housings, and no additional requirements. This allows a perfect overview of the installation at all times and for options for expansion and modification in the long term.

In addition, nowadays utilities tend to build sensor networks composed of a group of tiny, typically battery-powered devices and a wireless infrastructure that monitors and records conditions in any number of environments. The sensor network connects to the Internet, an enterprise WAN or LAN, or a specialized industrial network so that collected data can be transmitted to back-end systems for analysis and use in applications. By using sensor networks, it is possible to build a real-time data network that is more robust, more flexible, with larger area of coverage, and more powerful data processing capability. Furthermore, it is possible to integrate the currently available security infrastructure (such as SCADA or WAMS) by proper design. Such a network would play an important role in power system distributed monitoring and control for the following reasons [6]:

- The sensor network can integrate the distributed monitoring and control system into the networked management with relatively low cost.
- The sensor network forms a robust measurement network, hence it can provide more robust information for the power system.
- Certain failure information is easier to detect by the coordination of many distributed sensor networks in a large area.
- It is possible to reduce the cost of the sensor network and avoid the problem of data flooding through optimization and selection of sensor nodes.

- Sensor networks provide efficient data access, storage, process, and management, making the best use of information.

Nowadays, with advances in sensor and communication technology, it is possible to combine the sensors and communication network into a single structure that provides grid data services to applications in a highly flexible and scalable manner. Various services can be integrated into the sensor network via attached servers or through integration into network management systems. These include standard network management and security functions as well as grid-specific capabilities such as sensor meta-data management, IEC 61850 CIM (Common Information Model) interface services, and grid topology/connectivity.

The sensor network architectural view treats sensors and the communication network as an integrated structure. Grid sensor architecture must consider the underlying physical system structure, the relationship to the communications network structure, and the relationship or relationships to applications that make use of sensor data. Magnetic field based measurement devices, which can be conveniently digitized and integrated with modern digital communication, hence can be easily interfaced with other smart sensors and actuators [7], and will play an important role in such sensor network construction.

Another drive is from big power data. Nowadays utilities are building their own big data system in which a huge amount of measurement data, including production, operation, control, trading, and consumption, are continuously collected, communicated, and processed to extract useful information needed to meet the fast-growing demands of the high-accuracy and real-time performance of modern power and energy systems. The reason for using big data technology in power systems is because of te deployment of sensors. Recent developments in monitoring systems and sensor networks have dramatically increase the variety, volume, and velocity of measurement data in electricity transmission and distribution. The development of big data will require more sensors to be deployed in return [8].

## 7.2.2 Challenges

So far, it can be safely said that magnetic field based measurement and instrumentation are just local solutions and cannot be generalized from one solution to another. To extend the applications of magnetic field based measurement in future electric power systems, many issues must still be solved.

The big barrier is the lack of standardization of magnetic field based measurement systems. Since magnetic field based measurement generally obtains information in a non-contact manner, the configuration of the measurement system is highly dependent on the physical system to be measured because the information is delivered in the space. it is therefore hard to standardize magnetic field based measurement technology.

Magnetic field based measurement is broad in its scope, so the potential standards landscape is also very large and complex. The fundamental issue is organization and prioritization to achieve an interoperable and secure measurement system. The interface, as well as the streaming communication of the sensors, must therefore be standardized, while the devices should be modular, easy to install, and reliable.

New tools should be developed to integrate magnetic based measurement devices to achieve a flexible and scalable architecture, while improving the efficiency, robustness, and reliability of the state estimation system. New algorithms should be developed to extend the mixed emerging measurement infrastructure to advanced monitoring and control functionalities such as harmonics, transient etc. The algorithms for modelling, simulation, and network analysis of smart grids need to enable operators to develop effective measurement strategies, optimal sensor placement plans, and cryptographic infrastructures for grid security [9].

The grid upgrade may also face resistance from regulators (as well as end users) because some of the benefits are difficult to measure. Regulators are responsible for ensuring that utilities make wise investments that restrain the price of electricity, but improvement cannot easily be quantified.

## 7.3   Future Development

The utilities industry across the world is trying to address numerous challenges, including generation diversification, optimal deployment of expensive assets, demand response, energy conservation, and reduction of the industry's overall carbon footprint [10]. Smart grids are expected to revolutionize existing electrical grids by allowing two-way communications to improve the efficiency, reliability, economics, and sustainability of the generation, transmission, and distribution of electrical power. A higher level of intelligence to enable distributed data acquisition and decentralized decision-making will be the focus of future smart grids. Advanced sensors and automation devices enabling distributed intelligence to be applied to achieve faster self-healing methodologies and fault location/identification will be extensively deployed.

This process requires not only knowledge in the field of electric energy but also of many other fields, such as information technology, communication, control and automation, and nanotechnology. This universal effort will be effective if and only if the management and control of energy includes reliable information from accurate measurement methods and voltage and current sensors.

Smart grids rely on accurate real-time data to ensure that the equipment that controls power delivery, and which interfaces to the grid, is kept running at peak efficiency. The data captured by sensors on the grid can spot potential trouble spots forming and alert operators to the problem or activate functions that can perform remedial action. This is leading to the installation of devices that perform accurate measurements of delivered power and its characteristics.

Although a number of current-sensing options exist, sensors based on the magnetic effect provide a combination of features that are highly suited to those applications, including voltage transient survival, current inrush handling, space constraints, and modularity. The technology can support contactless sensing, ensuring intrinsic isolation and protection against large voltage transients and inrush currents.

Research and development efforts should be focused in the following areas in future:

- enhanced performance of measurement systems
- standardization of sensors, measurement approach, interface and communication
- innovative applications.

### 7.3.1 Performances

The quality of a measurement system is determined by the perfect interaction and strength of all of its individual parts. For a magnetic field based measurement system, it is necessary to optimize every component and feature, compact size, technology, measurement results, user-friendliness or flexibility, perfectly for practical application and function.

Future magnetic field based measurement system may include the following desirable features:

- independent use or collaboration with other measurement devices to measure important physical parameters, i.e. voltage, current, power, power factors, frequency, harmonics, and spatial operation parameters
- sensors with high bandwidth (from DC up to 10 MHz) and/or a wide dynamic range (e.g. ±30 Gs)
- high sensitivity and fast responsiveness
- modular and integrated design
- high accuracy and precision
- self-calibration, adaptability, and resilience
- communication capabilities (preferably wireless communication), standardized interface and interoperable
- low cost, compact size, easy to install and low power consumption
- high electromagnetic compatibility
- self-diagnostic (measure device health and lifetime)
- synchronized time-stamp functions
- sophisticated algorithms and data visualization.

Some of these desirable features are actually requirements, e.g. low cost and easy installation, as these are fundamental requirements for extensive deployment in existing or newly built systems.

Efficient detection of magnetic fields is central to many areas of research and technology. Research and development in magnetic sensors is very active, e.g. some new ultrasensitive magnetic field detection sensors, which uses an atomic system to achieve high sensitivity results from quantum coherence [11], have been developed. The best method for an application can be selected by balancing the requirements of the field to be mapped to the typical field measurement range, reproducibility and accuracy, mapped volume and field geometry, and time bandwidth.

### 7.3.2 Standardization

Standardization can help to maximize compatibility, interoperability, safety, repeatability, and quality. There have been significant developments in the field of measurement, particularly in smart sensors, intelligent instruments, micro-sensors, digital signal processing, digital recorders, digital field buses, and new methods of signal transmission.

Standards will play a key role in advanced magnetic field based measurement capabilities. New standards and measurement methods will be needed to ensure higher performance under future distributed and non-centralized grid conditions, and help to ensure that data are consistent and usable by multivendor equipment and applications.

Interoperability reduces overall costs and improves data integration. It is important to develop standards for measurement devices and practices for application of magnetic field based measurement. For measurement devices, it is important to develop standards to make the measurement device modular and scalable, to remove the market barrier:

- **Interface standards**: development of standards to support plug and play devices and interoperability, cyber security, self-identification, self-describing, self-configuration, self-calibration, and self-diagnostic test of sensors or measurement devices.
- **Testing methods**: methods and procedures to test and evaluate new sensor technologies, including connectivity and communication-related issues.

For these applications, it is also important to develop related standards, guides, and recommended practices.

### 7.3.3 Applications

Obviously the various applications described in this book cannot include the entire potential of magnetic field based measurement technology.

For power transmission systems, applications can be extended to the shielding wire (current measurement and fault location), insulators (leakage current), and structure (lightning stroke), etc. Large lightning current can be measured indirectly based on a non-contact method, which should be further studied.

For substation systems, applications can be extended to transformers (leakage, bias magnetization, etc.), post, bushing, and other external insulation equipment (leakage current), traveling wave detectors, fault indicators, etc.

For power distribution systems, various modular current sensors should be developed to facilitate various monitoring and control purposes. Application can be extended to underground cables, where fields produced by distribution cables vary significantly with the installation pattern. Parameters such as the relative position of various conductors and circuits, phase placement, depth of burial, etc. affect the magnetic field produced around the cable considerably. Current measurement in power distribution units has never been so compact and perfectly integrated. It is finally possible to monitor the individual circuits of an installation. Measuring current close to electrical loads creates a completely new level of transparency. Magnetic field based measurement provides versatile solutions for current monitoring in smart-grid applications, which sets a new standard in both transparency and user friendliness.

For power generation systems, various applications can be extended to other types of generators and motors. In a future renewable generation highly penetrated energy internet system, where peer-to-peer energy transactions may be implemented, traceable on-site energy measurement systems for ensuring fair energy trade might be a good potential application. Measurement and verification can provide a common tool for standardization to support performance-based contracting, financing, and emissions trading.

Another novel development is Poynting vector measurement [12], which could potentially be used for power flow loss analysis, fault detection, etc. Magnetic field based measurement can be combined with electric field measurement to realize an integrated Poynting vector measurement device.

## Bibliography

1 L. Cheng, N. Qi, F. Zhang, H. Kong, and X. Huang, "Energy internet: Concept and practice exploration," in *IEEE Conference on Energy Internet and Energy System Integration (EI2)*, 2017, pp. 1–5.

2 Z. Dong, T. Xu, Y. Li, P. Feng, X. Gao, and X. Zhang, "Review and application of situation awareness key technologies for smart grid," in *IEEE Conference on Energy Internet and Energy System Integration (EI2)*, 2017, pp. 1–6.

3 C. Root, H. Presume, D. Proudfoot, L. Willis, and R. Masiello, "Using battery energy storage to reduce renewable resource curtailment," in *IEEE Power Energy Society Innovative Smart Grid Technologies Conference (ISGT)*, 2017, pp. 1–5.

4 J. Klimstra and M. Hotakainen, *Smart Power Generation*. Helsinki: Avain Publishers, 2011.

5 S. H. Laskar, S. Khan, M. Moursheda, S. Robert, A. Ranalli, T. Messervey, D. Reforgiato, R. Contreau, A. Becue, K. Quinn, Y. Rezgui, and Z. Lennarh, "Smart grid futures: Perspectives on the integration of energy and ICT services," in *Energy Procedia*, vol. 75, 2015, pp. 1–6.

6 Q. Huang, S. Jing, C. Zhang, and Y. Chen, "Design and implementation of a power system sensor network for wide-area measurement," in *2008 International Symposium on Computer Science and Computational Technology*, vol. 2, Dec 2008, pp. 796–799.

7 IEEE, "IEEE draft standard for a smart transducer interface for sensors and actuators – transducer to microprocessor communication protocols and transducer electronic data sheet (TEDS) formats," *IEEE P1451.2/D20, February 2011*, pp. 1–28, March 2011.

8 X. He, Q. Ai, R. C. Qiu, W. Huang, L. Piao, and H. Liu, "A big data architecture design for smart grids based on random matrix theory," *IEEE Transactions on Smart Grid*, vol. 8, no. 2, pp. 674–686, 2017.

9 W. Lei and X. Li, "Research on PMU/SCADA mixed measurements state estimation algorithm with multi-constraints," in *IEEE Symposium on Electrical Electronics Engineering (EEESYM)*, 2012, pp. 32–35.

10 H. Farhangi, "The path of the smart grid," *IEEE Power and Energy Magazine*, vol. 8, no. 1, pp. 18–28, January 2010.

11 M. Bal, C. Deng, J.-L. Orgiazzi, F. Ong, and A. Lupascu1, "Ultrasensitive magnetic field detection using a single artificial atom," *IEEE P1451.2/D20, February 2011*, vol. 3, pp. 1–8, 2012.

12 J. A. B. ao Faria, "Poynting vector flow analysis for contactless energy transfer in magnetic systems," *IEEE Transactions on Power Electronics*, vol. 27, no. 10, pp. 4292–4300, Oct 2012.

# Index

*Magnetic Field Measurement with Applications to Modern Power Grids*, First Edition.
Qi Huang, Arsalan Habib Khawaja, Yafeng Chen and Jian Li.
© 2020 John Wiley & Sons Ltd. Published 2020 by John Wiley & Sons Ltd.